M. P. Gough.

Space Plasma Group ENGG.

University of Sussex.

MAGNETOSPHERIC PLASMA PHYSICS

DEVELOPMENTS IN EARTH AND PLANETARY SCIENCES

Editor:

T. RIKITAKE (Nihon University)

Editorial Board:

MAGNETOSPHERIC PLASMA PHYSICS

Edited by Atsuhiro Nishida

Written by A.Galeev, T.Sato, A.Nishida,
G.Haerendel and G.Paschmann,
M.Ashour–Abdalla and C.F.Kennel

Developments in Earth and Planetary Sciences

04

Center for Academic Publications Japan/Tokyo
D. Reidel Publishing Company/Dordrecht · Boston · London

Library of Congress Cataloging in Publication Data

DATA APPEARS ON SEPARATE CARD

ISBN 90–277–1345–6

Published by Center for Academic Publications Japan, Tokyo, in co-publication with D. Reidel Publishing Company, P. O. Box 17, 3300 AA Dordrecht.

Sold and distributed in Japan, China, Korea, Taiwan, Indonesia, Cambodia, Laos, Malaysia, Philippines, Thailand, Vietnam, Burma, Pakistan, India, Bangla Desh, Sri Lanka by Center for Academic Publications Japan, 4–16, Yayoi 2-chome, Bunkyo-ku, Tokyo 113, Japan.

Sold and distributed in the U.S.A. and Canada by Kluwer Boston Inc., 190 Old Derby Street, Hingham, MA 02043, U.S.A.

Sold and distributed in all other countries by Kluwer Academic Publishers Group, P. O. Box 322, 3300 AH Dordrecht, Holland.

Printed in Japan

PREFACE

Studies related to the earth and planets along with their surroundings are of great concern for modern scientists. Global geodynamics as represented by plate tectonics has now become one of the most powerful tools by which we can study the causes of earthquakes, volcanic eruptions, mountain formation and the like. Various missions sent out to space, manned or unmanned, brought out geoscientific features of the moon, Mars, Venus and other planets. Earthquake prediction that was the business of astrologers and fortune-tellers some twenty years ago, has now grown up to be an important science. A number of destructive earthquakes were successfully forecast in the People's Republic of China.

In the light of the above-mentioned and other accomplishments in geosciences, we feel that it is a good thing to publish a series of monographs which review selected topics of earth and planetary sciences. We are of course well aware of the fact that similar monographs have been and will be published from overseas publishers. The series, which we plan to publish, will therefore stress Japanese work. But we hope that the series will also include review articles by distinguished overseas authors.

The series, which is named the "Developments in Earth and Planetary Sciences" will be published by the Center for Academic Publications Japan and the D. Reidel Publishing Company. It is my great pleasure to work as the Editor of the series. I should like to have comments on what subjects we shall choose in future publications. I shall be greatly obliged if anyone would suggest suitable subjects and potential authors for the series to me.

Tsuneji Rikitake
Editor

PREFACE TO VOLUME 4

It is often said that 99% of the universe is in the plasma state. Certainly the atmospheres of astronomical bodies, and the gases that fill the space between them, are ionized and electrically conducting. The Earth's atmosphere, too, becomes a plasma beginning about 100 km above the Earth's surface. About twenty years ago, the Earth's plasma atmosphere was given the name "magnetosphere", because its dynamics is controlled by the interaction of the solar wind with the Earth's magnetic field. Since then, spacecraft have discovered magnetospheres at Venus, Mars, Jupiter, and Saturn, whose operating principles, while different from the Earth's, may still be understood in the language of plasma physics. And astrophysicists have given the name "magnetosphere" to the plasma environments of pulsars, X-ray sources, and certain radio sources.

Although they are minute in energy content relative to the astrophysical magnetosphere, the magnetospheres of the planets are essential links in the chain of studies that connect plasmas in the laboratory to cosmic plasmas. They, and the solar wind, are the largest plasmas that can be studied *in situ*. By comparing remote and direct observations of planetary magnetospheres, we begin to perceive how faintly the complex plasma behavior in astrophysical magnetospheres is illuminated by remote observations.

Our measurements of planetary plasmas, and particularly the Earth's, are now so complete that they rival those made in the laboratory. In some cases, they exceed laboratory measurements in resolution and detail. And in all cases, they provide plasma parameters for study that differ from those found in most laboratory experiments. Studies of solar system plasmas therefore enrich our knowledge of basic plasma physics.

One of the six principal courses of the Autumn College of Plasma Physics, International Center for Theoretical Physics, Trieste, Italy, October 16–November 23, 1979, was devoted to space plasma physics. This volume contains review articles drawn from the five lecture series comprising the space plasma physics course given on that occasion. Although the audience was drawn to the Autumn course because of its emphasis on controlled thermonuclear fusion, many could not resist the lure and enchantment of space plasma observations, which are understandable in terms of the same basic plasma principles that underline laboratory research.

Our book deals largely with the Earth's magnetosphere, because our understanding of it is comparatively complete. However, we shall refer also to our growing knowledge of other planetary magnetospheres. The first, introductory chapter deals with the origin of the plasma that fills planetary magnetospheres. The means by which solar wind energy and momentum is communicated to the magnetosphere is discussed in the second chapter. Our understanding of this problem has recently been advanced by the ISEE 1, 2, and 3 spacecraft. The energy thus supplied is stored in the tails of planetary magnetospheres in the form of magnetic energy. The third chapter discusses a basic instability of magnetospheric tails that converts stored magnetic energy into flow kinetic energy and into accelerated particles. The dissipation of the tail magnetic energy of the Earth leads to auroral substorms, the major sink of energy in the Earth's magnetosphere. The coupling of hydromagnetic flows to the Earth's ionosphere involves magnetic field-aligned currents. Space physicists have recently begun to understand how these currents generate large electrostatic potentials that accelerate the particles responsible for the structure of the *aurora*. Chapter 4 presents results of advanced numerical simulation of auroral acceleration processes. Even where there is no auroral acceleration, the particles heated in substorm events are scattered into the Earth's atmosphere by microscopic plasma turbulence. This produces a diffuse auroral glow in the Earth's polar atmosphere. Chapter 5 reviews the observations and theory of the electrostatic electron cyclotron harmonic waves responsible for the diffuse aurora.

We are grateful for the warm hospitality extended to us by the International Center for Theoretical Physics, its director, A. Salam, and its staff. The Autumn College was funded by the IAEA and by UNESCO. We thank the directors of the 1979 Autumn College, B. Kadomtsev, and M. Rosenbluth, for incorporating space plasma physics into its curriculum, and particularly B. MacNamara, without whose efforts the Autumn College would have been much the poorer.

March 1981

A. Nishida

CONTENTS

Chapter 3

MAGNETOSPHERIC TAIL DYNAMICS A. A. GALEEV 143

Chapter 4

AURORAL PHYSICS T. SATO 197

Chapter 5

ELECTROSTATIC WAVES AND THE STRONG DIFFUSION OF
MAGNETOSPHERIC ELECTRONS

C. F. KENNEL and M. ASHOUR-ABDALLA　245

Chapter 1

ORIGIN OF MAGNETOSPHERIC PLASMA

A. NISHIDA

*Institute of Space and Astronautical Science, Komaba, Meguro-ku,
Tokyo, Japan*

1. Introduction

When a planet has an intrinsic magnetic field, the space surrounding the
planet is strongly influenced by the planetary magnetic field. The magnetic
field acts as a barrier to charged particles that travel toward the planet,
because gyrating motions of the particles deflect direction of the flight. The
situation is illustrated schematically in Fig. 1. Due to this effect, plasma from
external origins tends to be excluded from the vicinity of magnetized planets.

The principal origin of plasma external to planets is the sun. Continuous
emission of plasma from the sun is now well established and it has been
designated as the *solar wind*. As the planetary magnetic field prevents the
access of the solar wind toward the planet, a cavity is carved out around a
magnetized planet in the domain of the streaming solar plasma. The cavity,
inside which field lines of the planetary magnetic field are confined, is called
the *magnetosphere* of the planet.

In fact, the preceding description of the magnetosphere has to be taken as
a first-order approximation, since the interaction between the solar wind
and planetary magnetic fields involves greater complexity and an accurate
definition of the magnetosphere is not so simple. Small fraction of the solar
wind plasma does penetrate deep inside the planetary magnetic field and
precipitates down to $\sim 100\,\mathrm{km}$ altitude. Planetary magnetic field is not
entirely contained inside the cavity and small fraction of field lines is believed
to be connected with magnetic field lines carried by the solar wind. These
complexities represent central issues of the magnetospheric physics and will

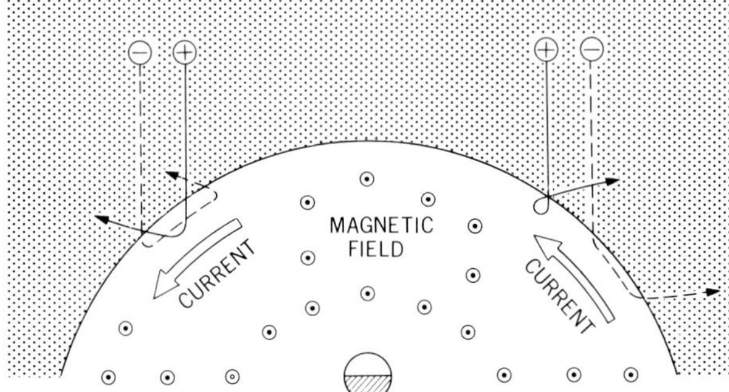

FIG. 1. Schematic illustration of trajectories of positive ions ($+$) and electrons ($-$) incident on the planetary magnetic field. The electric field arising from separation of incident ions and electrons almost equalizes their penetration depths.

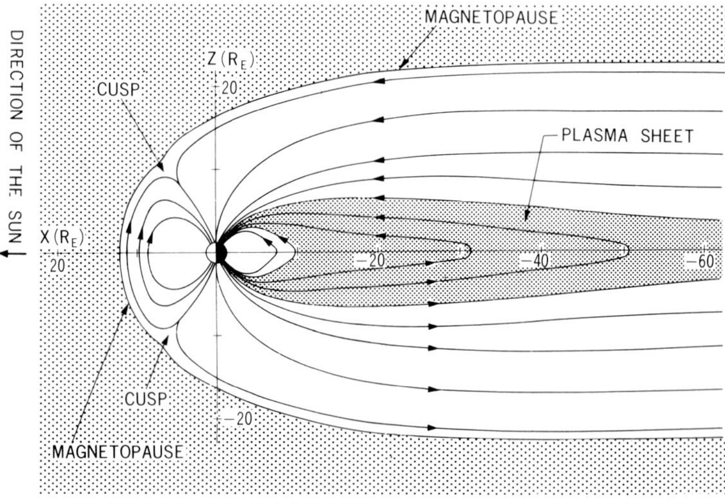

FIG. 2. Schematic illustration of the noon-midnight section of the magnetosphere.

be taken up for detailed discussions in later chapters of this book. For the present, however, we shall continue with the first-order description which has virtue of simplicity and is sufficiently valid for the purpose of this chapter.

The cross section of the magnetosphere by the noon-midnight meridian is illustrated in Fig. 2. The solar wind flows almost parallel to the sun-planet line since aberration due to orbital motion of the planet is small (about 4° in the case of the Earth). The attitude of the magnetosphere with respect to the sun-planet line depends on the attitude of the dipole moment of the planetary magnetic field and hence it changes with the season and also with Universal Time when the dipole moment is not aligned with the rotational axis. For simplicity Fig. 2 is depicted for an idealized situation when the dipole moment makes 90° with the sun-planet line.

The bounding surface of the magnetosphere, called *magnetopause*, is approximately a hemi-ellipsoid on the sunward side. The minor axis of this ellipsoid is directed toward the sun in the idealized situation illustrated in Fig. 2. The length R_m of the minor axis, namely the distance from the center of the planet to the subsolar point on the magnetopause, is a function of the dipole moment M of the planetary magnetic field and the momentum flux of the solar wind, and it is given by

$$R_m = (\alpha^2 M^2 / 8\pi\beta nmv^2)^{1/6} \tag{1}$$

where n, m, and v represent number density, mean ionic mass and bulk velocity of the upstream solar wind prior to the interaction with the magnetosphere. α is the enhancement factor of the planetary magnetic field immediately inside the magnetopause at the subsolar point. The enhancement occurs because the solar wind plasma creates electric current flow parallel to the magnetopause as it is reflected; as seen in Fig. 1 positive ions and electrons are displaced in opposite directions during their reflection.

To explain the meaning of the factor β, we note that Eq. (1) is an approximate expression of the pressure balance condition at the magnetopause:

$$P_{sw} + \frac{B_{sw}^2}{8\pi} = P_{MG} + \frac{1}{8\pi}\left(\frac{\alpha M}{R_m^3}\right)^2 \tag{2}$$

where P and B represent pressure and magnetic induction, and suffixes SW and MG designate quantities on the solar wind side and magnetosphere side, respectively. The pressure P_{sw} exerted on the magnetopause is dependent upon both the dynamic pressure nmv^2 and the thermal pressure of the upstream solar wind. In the solar wind, however, dynamic pressure is about two orders of magnitude greater than thermal pressure, and hence P_{sw} is determined essentially by the upstream dynamic pressure only. We have equated therefore P_{sw} to βnmv^2. Numerical value of β has been estimated to be about 0.8 on the basis of gas dynamic modeling of the interaction of a supersonic stream with a blunt body (SPREITER *et al.*, 1966). (Note that $\beta = v/c$).

TABLE I. Dipole moment, equatorial surface field, polarity, angle between the dipole moment and the rotational axis, and R_m, the distance of the subsolar magnetopause in the unit of the planetary radius (Siscoe, 1979). 1 γ is 1 nT or 10^{-5} Gauss.

	Dipole moment M_p (G-cm^3)	Equatorial surface field	Polarity	Angle (Ω, M) (deg.)	R_m $\overline{R_\mathrm{p}}$
Mercury	5×10^{22}	0.350γ	N	~10	1.6
Earth	8×10^{25}	0.31G	N	11.5	11
Jupiter	1.5×10^{30}	4.1G	R	10	50
Saturn	4×10^{28}	0.2G	R	1	20

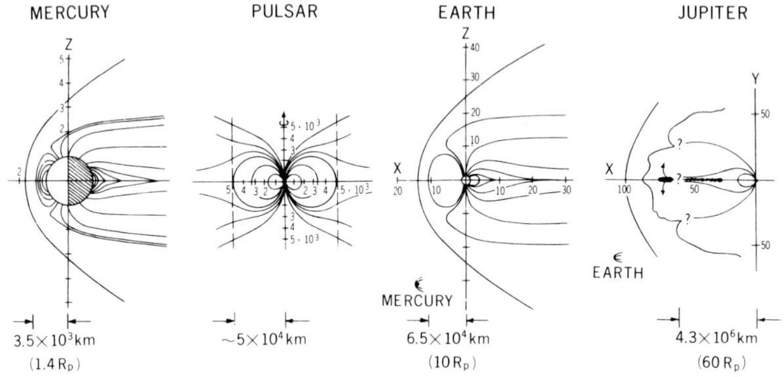

FIG. 3.   Noon-midnight sections of magnetospheres of Mercury, Pulsar, Earth, and Jupiter. Sizes of planetary magnetospheres are based on spacecraft observations (NASA).

To yield Eq. (1) from (2) it has been assumed further that the magnetic pressure is negligibly small as compared to the thermal pressure on the solar wind side ($P_\mathrm{sw} \gg B_\mathrm{sw}^2/8\pi$), while the opposite is the case on the magnetosphere side ($P_\mathrm{MG} \ll (\alpha M/R^3_\mathrm{m})^2/8\pi$). The numerical value of the factor α has been estimated to be approximately 2.

In our solar system, presence of the intrinsic magnetic field has been established for planets Mercury, Earth, Jupiter, and Saturn. Magnetic dipole moment, surface field at the equator, polarity, angle between the dipole moment and the rotational axis are tabulated in the first four columns of Table I. The fifth column gives R_m divided by radius R_p of the planet. For estimating R_m solar wind speed of 400 km/sec and density of $5r^{-2}/\mathrm{cm}^3$ have been used, where r represents sun-planet distance in the unit of the Astronomical Unit. The mean ionic mass has been set equal to proton mass.

Figure 3 illustrates noon-midnight cross sections of magnetospheres of

Mercury, Pulsar, Earth, and Jupiter. To emphasize the difference in size, the Mercury's magnetosphere is illustrated below the Earth's magnetosphere with the same scaling, and similarly Earth's magnetosphere is compared with the Jovian magnetosphere. Pulsar's magnetosphere is also shown for reference. Since the solar wind flows toward the planets with supersonic and super-Alfvenic speeds, a *bow shock* of nearly paraboroidal shape is formed in front of the magnetosphere. Section of the bow shock is also illustrated in Fig. 3.

On the nightside, planetary magnetospheres have an extended tail called magnetospheric tail (or, magnetotail). In the case of the Earth the magnetotail clearly extends at least to the lunar distance ($\sim 60\ R_E$) which is about six times the subsolar distance $R_m \sim 10\ R_E$ of the Earth's magnetopause (R_E denotes Earth's radius). Further extension to about 3,000 R_E has been indicated by observations of the solar wind cavity far downstream from the Earth (INTRILIGATOR *et al.*, 1979). Although magnetic field lines in the magnetotail originate from field lines of the planetary magnetic field, their extended configuration has little resemblence to dipole field lines. The tail magnetic field is generated by the electric current that flows across the magnetotail and closes by flowing on the surface of the magnetotail (as illustrated in Fig. 4 as a cross-tail current and tail magnetopause current). The

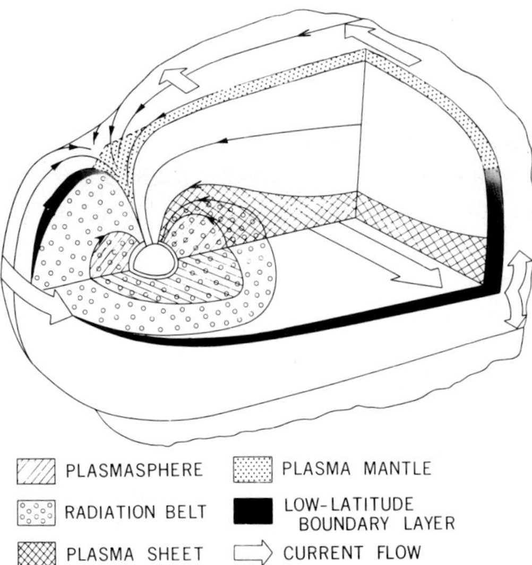

FIG. 4. Schematic view of the internal structure of the magnetosphere (NASA, modified by the author.)

midplane of the magnetotail which separates northern and southern halves of the magnetotail is sometimes referred to as *neutral sheet,* but this term should not be taken literally because there usually is small but non-zero component of the magnetic field across the midplane.

Magnetic field lines that cover the magnetopause converge from both dayside and tail and enter the interior of the magnetosphere at the *cusp* of one hemisphere (Fig. 2). Similarly, field lines emerge from the interior and spread over the magnetopause at the cusp of the other hemisphere.

The magnetotail has been detected also for the Mercury's magnetosphere. In this magnetosphere where the planetary magnetic field is very weak, field lines show an extended configuration even at a distance as close as ~ 0.3 R_{H} (where R_{H} is Mercury's radius) from the center of the planet (NESS *et al.*, 1975).

The origin of the magnetotail, or equivalently, the generator of the cross-tail electric current, is believed to be the transfer of momentum from the solar wind to the magnetosphere. Namely, the solar wind exerts drag on magnetospheric plasma as it flows by and thus pulls field lines, which are frozen in the plasma, away from the planet. The nature of this "drag" is an important issue of the magnetospheric physics, but we pass the subject in the present first-order description of the magnetosphere structure leaving it to later chapters.

In the case of the Jovian magnetosphere, stress exerted by the internal plasma also acts to extend the field lines away from the planet. Since the angular velocity Ω of the Jupiter's rotation is nearly twice as fast as that of the Earth and the size of the Jovian magnetosphere is nearly one hundred times larger than that of the Earth, Jovian plasma experiences much stronger centrifugal force than the Earth's plasma. The centrifugal force significantly modifies the magnetic field structure when the kinetic energy density of the corotational motion becomes comparable to the magnetic field energy density, namely when

$$\frac{1}{2} nm(\Omega R)^2 \sim \frac{1}{8\pi} \left(\frac{M}{R^3}\right)^2 . \tag{3}$$

The above equality can be satisfied at $R = 50 R_{\mathrm{J}}$ (where R_{J} is the Jovian radius) if number density n is $10^{-1}/\mathrm{cm}^3$. Observations by Pioneer and Voyager spacecraft have revealed that magnetic field lines in the Jovian magnetosphere have an extended configuration beyond distances of $\sim 20 R_{\mathrm{J}}$ all around the planet. The structure is often referred to as Jupiter's *magnetodisc.* Since the field lines are pulled out by the centrifugal force the magnetic field in the magnetodisc cannot be properly represented by the dipole field and Eq. (1) can fail to give a correct value of R_m. Indeed, when Pioneer 10 and 11 entered the Jovian magnetosphere, the magnetopause was located at $\sim 100\ R_J$ from the center of the planet.

Now let us turn our attention to plasmas that populate planetary magnetospheres. Figure 4 gives an inside view of the Earth's magnetosphere. Solid lines are magnetic field lines and thick arrows indicate flows of electric currents. An important feature of the magnetospheric plasma is that it comprises several populations which have distinct characteristics both in spatial distribution and in energy spectrum. Hence it has become a common practice to define several regions inside the Earth's magnetosphere according to the plasma population that dominates the region. These regions are called plasmasphere, radiation belt, plasma sheet, tail lobe, and plasma mantle.

Plasmasphere is the domain of the low energy plasma whose energies are less than about 20 eV. It is located at the central core of the magnetosphere, and the bounding surface of the plasmasphere is called *plasmapause*. The plasmapause is covered by magnetic field lines. Distance of the plasmapause from the center of the Earth varies from time to time but the equatorial distance of about 5 R_E is most frequently observed. At the plasmapause the density of low-energy plasma drops sharply from the order of $10^2/cm^3$ to the order of $1/cm^3$ or less.

Radiation belt, or trapped particles, is the domain of high energy particles whose energy is greater than about 10 keV and reaches MeV or more. In the earthward part it overlaps the plasmasphere but the radiation belt occupies a much greater space than the plasmasphere. The boundary surface of the radiation belt, sometimes referred to as the trapping boundary, is also covered by magnetic field lines and it makes contact with the magnetopause in the subsolar meridian.

Plasma sheet is the domain of medium energy plasma that occupies the central part of the magnetotail. Its boundary surface on the earthward side lies in the vicinity of the nightside plasmapause. Energy of plasmas in the plasma sheet lies mostly between 0.1 keV to 10 keV, which is intermediate between energies of the plasmaspheric plasma and the trapped particles. *Auroral electrons*, namely ~ 1 keV electrons that precipitate into the atmosphere and excite auroral emissions, originate from the plasma sheet. Number density in the plasma sheet is usually $\sim 0.1/cm^3$ or more, and it drops by more than an order of magnitude at high-latitude boundaries. High latitude magnetotail outside the plasma sheet is called *lobe*.

Plasma mantle is a thin layer of plasma having energies of the solar wind but flowing in high-latitude region of the magnetotail. The layer is bound on the sunward side in the cusp region. The plasma mantle apparently contains plasma that has penetrated the magnetopause into the *entry layer* in the cusp region where the magnetic field on the magnetopause is not strong enough to repel the incident solar wind plasma. The mantle plasma flows along magnetotail field lines.

Plasma having energies of the solar wind has also been found in the dayside low-latitude region immediately inside the magnetopause. This region has been designated as the *low latitude boundary layer* (HAERENDEL *et al.*, 1978).

The formation of these characteristic domains reflects spatial distribution of plasma sources and trajectories of charged particles that are placed under electric and magnetic field existing in the magnetosphere. In the case of the Earth's magnetosphere, plasma sources are ionizations of the planetary atmosphere and entries of the solar wind. Ionizations of satellite atmospheres have also to be counted as a source in magnetospheres such as Jupiter's. Plasmas originating from the atmospheric and the solar-wind sources are expected to have an intrinsic difference in temperature, because planetary atmospheres have temperatures of 10^{-2} to 10^{-1} eV while the solar wind protons have energies of about 1 keV. Although high speed photoelectrons are produced at ionization they can hardly heat the atmospheric plasma to solar-wind energies because photoelectron energies are only about 10 eV. This suggests that the low-energy population constituting the plasmasphere is of atmospheric origin and the medium- and high-energy population in the plasma sheet and radiation belt are of solar-wind origin.

The correspondence between plasma sources and observed populations suggested above is not entirely exclusive, however, because various acceleration processes have been known to operate inside planetary magnetospheres. Ions like O^+ and He^+, which are produced copiously from atmospheric ionization but are scarce in the solar wind, have been detected in the Earth's magnetosphere with energies of $1 \sim 10$ keV (SHELLEY, 1979). This means that the atmospheric plasma can be accelerated to medium or even high energies in the Earth's magnetosphere. Nevertheless, the fact remains that the majority of plasmas constituting the Earth's plasmasphere have energies of less than a few tens of eV: in most cases, plasma produced from atmospheric sources flows upward and fills the magnetosphere without experiencing such a drastic change in energy.

Trajectories of charged particles across magnetic field are governed by drift motions of the guiding center. The drift velocity consists of electric-field

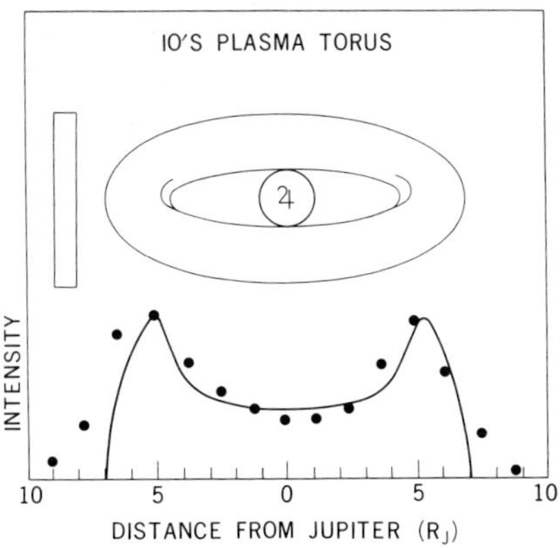

FIG. 5. Schematic illustration of Io's plasma torus, and measured intensity of 685 Å emission (BROADFOOT *et al.*, 1979).

drift, gradient-B drift and curvature drift. The electric field is generated by planetary rotation and by interaction between solar wind and the magnetosphere. Since gradient-B and curvature drift velocities depend upon kinetic energy of particles, trajectories are energy dependent. At low energies (at energies less than about 1 keV in the earth's magnetosphere) the electric field drift is dominant, and this drift is often referred to as *convection* of magnetospheric plasma. Domains of various plasma populations are defined by trajectories that are connected to source regions, or by trajectories that are not connected to sinks. The earth's plasmasphere is made of low-energy plasmas whose trajectories are closed around the earth.

In the Jovian magnetosphere low-energy plasma is provided mainly from the satellite Io. Volcanic gases become ionized both by impact of ambient plasma and energetic particles and by illumination of the solar UV radiation. When ionized, a particle is heated by receiving energy from the electric field associated with the planetary rotation, at the same time as it is led to corotate with the planet. Because a particle starts to corotate once it becomes ionized, it is separated from Io and forms a torus along the orbit of Io. Figure 5 illustrates position and size of the Io's plasma torus which were inferred from the observation of UV emissions reproduced in the lower part of the figure (BROADFOOT *et al.*, 1979). Plasmas of the Io origin are characterized by heavy concentration of sulphur and oxygen ions: cold plasma observed within $\sim \pm 1$ R_J of Io's orbit seems to consist almost 100% of these heavy ions (BAGENAL *et al.*, 1980).

In what follows, basic concepts on the origin of magnetospheric plasma are reviewed. We shall successively look at supply of atmospheric plasma and photoelectrons by upward streaming, supply in the Jovian magnetosphere from the satellite Io, and supply of solar-wind plasma through penetration into the magnetotail.

2. Atmospheric Source of Magnetospheric Plasma

The height where the photoionization rate maximizes in the atmosphere is determined by competition of two parameters. One of the parameters is number density n of the neutral atmosphere which decreases with height, and the other is flux ϕ of the solar ultraviolet radiation which increases with increasing height as more of the absorbing atmosphere is left behind. For production of ions from a neutral species designated by suffix a by solar radiation with wavelength λ, the maximum rate occurs at an altitude where the following relation is satisfied:

$$n_a = (\sigma_\lambda H_a \sec \chi)^{-1} \tag{4}$$

where σ is absorption cross section, χ is zenith angle of the incident solar radiation, and H is the scale height (see, for example, BANKS and KOCKERTS, 1973).

In planetary atmospheres the height of the maximum photoionization rate is very low as compared to the scale of the magnetosphere. In the Earth's atmosphere production rates of O^+, O_2^+, and N_2^+ ions maximize in the height range of $120 \sim 160$ km and decrease sharply with increasing height. Above an altitude of about 130 km ion densities are governed, in the daytime, by balance between the photoionization rate and the loss rate by the following reactions:

$$O^+(^4S) + O_2 \rightarrow O_2^+ + O, \quad O_2^+ + e \rightarrow O + O$$

$$\tag{5}$$

$$N_2^+ + O \rightarrow NO^+ + N, \quad NO^+ + e \rightarrow N + O$$

as experimentally confirmed by TORR *et al.* (1979). The resulting ion composition is that above an altitude of about 160 km the atomic O^+ ions are more abundant than the molecular O_2^+ and N_2^+ ions.

Hydrogen ions, namely protons, which are the principal constituents of the plasmasphere are produced from O^+ by a charge exchange reaction:

$$O^+(^4S) + H \leftrightarrows O(^3P) + H^+ \tag{6}$$

This reaction proceeds rapidly because difference in energy between the left-hand and the right-hand side is much smaller than typical energies of atoms and ions at the height range concerned. For daytime $[O^+]$ of $\gtrsim 10^3/\text{cm}^3$ that prevail up to $\sim 1{,}400$ km or higher altitudes (BREIG and HOFFMAN, 1975), the reaction (6) is far more effective for producing protons than photo-ionizations of hydrogen atoms (BATES and PATTERSON, 1961). The ratio between H^+ and O^+ densities is given by

$$\frac{[H^+]}{[O^+]} = \frac{9}{8}\frac{[H]}{[O]}\left(\frac{T_n}{T_i}\right)^{1/2} \tag{7}$$

where T_n and T_i denote neutral and ion temperatures, respectively.

Ion density distributions determined by photochemical processes are not tailored to satisfy the condition of the static equilibrium. Rather, ion pressure gradients associated with such distributions act to drive the plasma upward to higher levels of the magnetosphere. Plasma distributions in the magnetosphere are thus governed not only by photochemical but also by dynamical processes. In the case of the Earth's magnetosphere, steady-state distributions above the height of ~ 400 km are determined basically by the following set of equations (MARUBASHI, 1970): Equation of motion of H^+ (denoted by suffix 1)

$$m_1 n_1 v_1 \frac{\partial v_1}{\partial s} + \frac{\partial}{\partial s}(n_1 k T_1) + m_1 n_1 g = R_1 + n_1 eE \tag{8}$$

where m, n, and T denote mass, number density, and temperature, s denotes distance along a magnetic line of force measured upward from a reference level, v, g, and eE are components of velocity, gravitational force, and electric force in the direction of the magnetic field, and R is the same component of the frictional force which is given by

$$R_1 = -\alpha_{12} n_1 n_2 v_1 - \alpha_{1n} n_1 n(O) v_1 \ . \tag{9}$$

The first term of R_1 represents the friction due to binary Coulomb collisions between H^+ and O^+ and the second term between H^+ and O whose density is expressed as $n(O)$. Velocities of O^+ and O are neglected as they are much less than the velocity of much lighter H^+ ion. We have considered only motions parallel to the magnetic field because flow velocities in the perpendicular directions are much smaller than the parallel velocity.

Equation of motion of O^+ (represented by suffix 2):

$$\frac{\partial}{\partial s}(n_2 k T_2) + m_2 n_2 g = R_2 + n_2 eE \tag{10}$$

where

$$R_2 = \alpha_{12} n_1 n_2 v_1 . \tag{11}$$

Inertia term has been neglected in Eq. (10) because of the smallness of the O^+ velocity.

Equation of motion of electrons:

$$\frac{\partial}{\partial s}(n_e k T_e) + m_e n_e g = -n_e e E \tag{12}$$

where condition of the charge neutrality

$$n_e = n_1 + n_2 \tag{13}$$

is implied.

Equation of continuity of the H^+ flow:

$$\frac{1}{A}\frac{\partial}{\partial s}(n_1 v_1 A) = q_1 - l_1 \tag{14}$$

where A denotes cross section of a magnetic tube of force. q and l are production and loss rates of H^+ which are given by

$$q_1 = Kn(H)n_2 \tag{15}$$

and

$$l_1 = \frac{8}{9} Kn(O)n_1 \tag{16}$$

where K is the rate coefficient. (Note that Eq. (7) which represents the chemical equilibrium obtains when $q_1 - l_1$ is equated to zero.)

In order to close the foregoing set of equations, equations of energy are also needed. Let us first assume, however, that temperature is uniform and the same for all species ($T = T_1 = T_2 = T_e$). The electric field E that appears in equations of motion arises from the tendency of electrons to spread more easily than ions, and its strength is limited by the condition (13) of the charge neutrality. By eliminating E and n from Eqs. (8) through (14), one obtains a differential equation for v_1;

$$v_1 \frac{\partial v_1}{\partial s}\left\{1 - \frac{n_1 + n_2}{n_1 + 2n_2}\frac{c_1^2}{v_1^2}\right\} = -\left\{\frac{2(n_1 + n_2)}{n_1 + 2n_2}\frac{\alpha_{12}}{m_1}n_2 + \frac{\alpha_{1n}}{m_1}n(O)\right\}v_1$$

$$-\frac{n_1 + (2 - m_2/m_1)}{n_1 + 2n_2}g - \frac{n_1 + n_2}{n_1 + 2n_2}\frac{\partial c_1^2}{\partial s}$$

$$-c_1^2 \frac{n_1+n_2}{n_1+2n_2} \frac{1}{I} \frac{\partial I}{\partial s} + c_1^2 \frac{n_1+n_2}{n_1+2n_2} \frac{1}{A} \frac{\partial A}{\partial s} \quad (17)$$

where $c_1^2 = (2kT/m_1)^{1/2}$ represents mean thermal speed of H^+ ions and $I(s)$ represents H^+ flux through a tube of force;

$$I(s) \equiv n_1(s)v_1(s)A(s) . \qquad (18)$$

Equation (17) has a singularity when v_1 takes a value

$$v_1 = \left(\frac{n_1+n_2}{n_1+2n_2} c_1^2 \right)^{1/2} . \qquad (19)$$

The critical point is located at a distance where the right-hand side of Eq. (17) vanishes. In the Earth's magnetosphere this occurs at an altitude of about 2,000 km.

The nature of the solution to (17) depends both on the state of the ionizing region and on the plasma pressure in the high altitude magnetosphere towards which plasma flows. Figure 6 shows H^+ velocity profiles as a function of altitude for five different conditions of pressure at $s \to \infty$. For all the cases the same $[O^+]$ $(2 \times 10^5$ ions/cm$^3)$ is used at the lower boundary. The

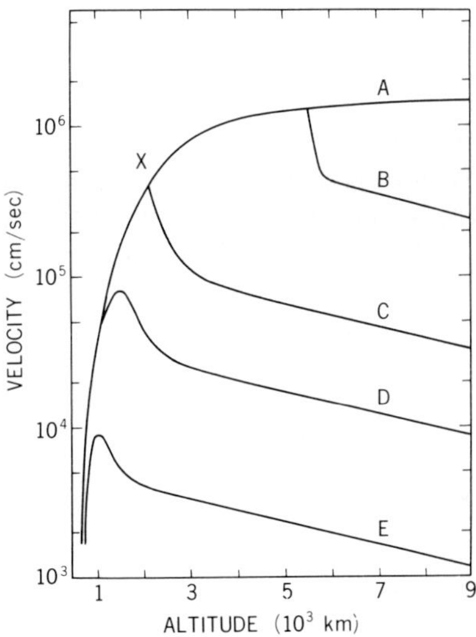

FIG. 6. Velocity profiles of upflowing plasma. Branches A through E correspond to different values of pressure at infinity. X is the critical point of the flow velocity (MARUBASHI, 1970).

branch A corresponds to a situation where the pressure p_∞ at $s \to \infty$ is zero. This solution passes the critical point X and is supersonic beyond that point since $v_1 > c_1$. The branch C also passes the critical point but it corresponds to a finite pressure p_c at $s \to \infty$. When the pressure p_∞ is intermediate between 0 and p_c the solution takes a form of branch B where a transition from the supersonic to a subsonic flow takes place at a shock front. When p_∞ exceeds p_c the solution becomes like branches D and E which stay subsonic everywhere, and when p_∞ is increased further there comes a point when the equations can be satisfied without invoking motion. This is called *diffusive equilibrium*. For even greater p_∞ the flow becomes directed downward (BANKS and HOLZER, 1969).

Solution of the branch A applies outside the plasmasphere where the plasma pressure at high altitudes is very low because the following processes operate to remove plasma. One is the escape along high-latitude field lines which extend to the magnetotail. The other is transport to the magnetopause due to convective motion of the magnetospheric plasma. Outflows represented by the branch A are called *polar wind*. Solutions of branches B, C, and D apply when the plasma density builds up in tubes of force where rapid removal of plasma has ceased to operate. Solutions like E represent upward motions of plasma in the daytime caused by daily increase in plasma density in the ionizing region. In this diurnal variation the magnetosphere acts as a reservoir of plasma and a downward flow is produced in the nighttime.

Figure 7 shows calculated altitude profiles of the upward H^+ flux and ion densities for the polar wind solution. The flux is expressed for a tube of force having a $1\,cm^2$ cross section at the geomagnetic latitude of $70°$ on the surface of the Earth. Solid and dashed lines correspond to two different choices of O^+ density at the lower boundary. The H^+ flux increases in the $\lesssim 1,500\,km$ height range where H^+ is produced from O^+ by the reaction (6), and then it stays almost constant. The H^+ density has a peak around the $1,000\,km$ height where O^+ and O densities are still high enough to suppress the outflow velocity of H^+. Note that transition of the dominant ion species from O^+ to H^+ occurs at less than $1\ R_E$ from the ground. Density profiles of H^+ and O^+ in the presence of upward H^+ streaming will be discussed again in the next section.

The calculated daily up and down fluxes of H^+ inside the plasmasphere are shown in Fig. 8b. In this figure position of a magnetic tube of force is expressed by local time (abscissa) and the geocentric distance of its equatorial crossing point L (ordinate). Fluxes of the H^+ outflow at the $3,000\,km$ height

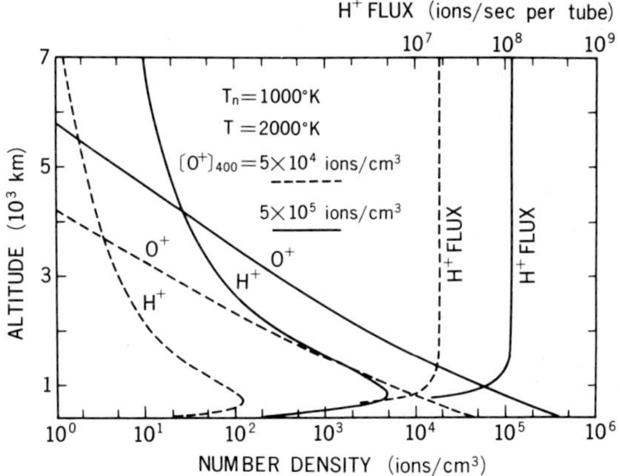

FIG. 7.   H$^+$ and O$^+$ densities and upward H$^+$ flux that correspond to a polar wind solution (MARUBASHI, 1970).

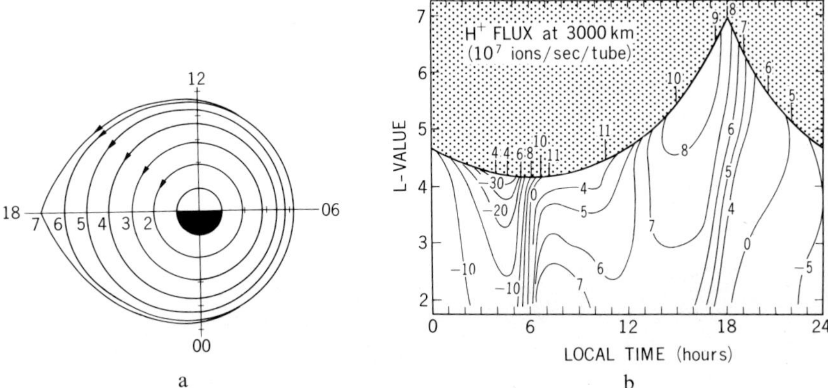

FIG. 8.   a. Assumed trajectories of equatorial crossing points of magnetic lines of force.   b. Diurnal variation of upward H$^+$ flux and its dependence on the equatorial distance L of field lines. Field lines are supposed to follow the trajectory of (a) (MARUBASHI, 1979).

are given in the unit of 10^7 ions/sec per unit tubes of force which have a cross section of 1 cm^2 at the 1,000 km level. Equatorial distances of tubes of force are assumed to vary with local time as illustrated in Fig. 8a. The deviation of these trajectories from exactly concentric circles that correspond to a full corotation is due to influence of the convective motion induced by the solar wind. It is seen that flux of the outflow has two daily maxima. One occurs in early morning when plasmas which are freshly created by morning sunshine surge upward. The other occurs in the afternoon because volume of each tube of force is increased in the noon-dusk sector as tubes follow the trajectory of Fig. 8a and consequently plasma pressure in high altitudes is reduced in that sector. The former maximum is more pronounced for field lines having lower L as they pass the ionizing region in lower latitudes where ion production rates are higher, while the latter maximum is more pronounced at higher L where the increase in the volume of tubes of force due to non-circular rotation has a greater amplitude.

In the nighttime the H^+ flux is directed downward. The highest downward flux is expected to occur just before the sunrise not only because plasma density in the $100 \sim 300$ km height range has been reduced by recombination throughout the nighttime but also because plasma pressure at high altitudes is enhanced by contraction of tubes of force enforced by motions of Fig. 8a. As compared to the daily average of the plasma content in a tube of force, the fraction that is changed by diurnal up and down flows is about 10% at $L=3$ and is greater or smaller than this at lower or higher L (MARUBASHI, 1979).

An example of simultaneous observations of the ion composition and H^+ flow velocity is reproduced in Fig. 9. The observations were made on

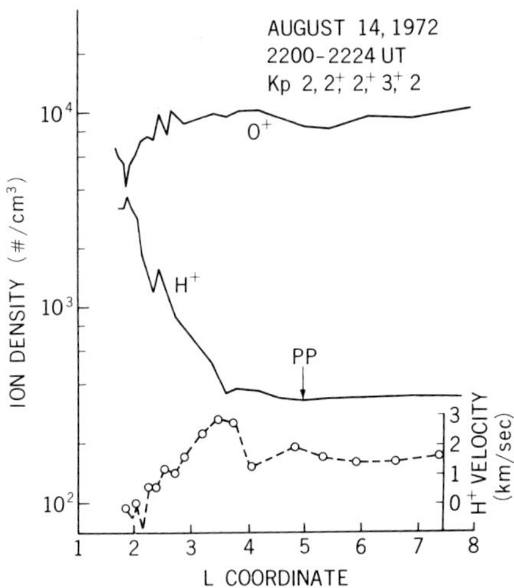

FIG. 9. Comparison of O^+ density, H^+ density, and H^+ outflow velocity observed at an altitude of 1,400 km as the satellite cruised from L of ~ 2 to ~ 8 (GREBOWSKY *et al.*, 1978).

board a polar orbiting satellite in the afternoon (around 1700 LMT) at an altitude of 1,400 km. The latitude is expressed by the L coordinate which for the present purpose can be said to indicate the equatorial crossing distance of the magnetic line of force that pass the observing point. It is seen that upward H^+ flow with velocities of a few km/sec existed at latitudes higher than about $50°$ (where $L \doteqdot 2.5$), and that H^+ density was significantly reduced but O^+ density slightly increased in the region of the H^+ outflow. The observed relation between the upward H^+ flow velocity and the ion density is exactly what is expected from theoretical model of the polar wind (GREBOWSKY *et al.*, 1978). The symbol pp in the figure indicates projection, along a magnetic field line, of the plasmapause identified by high-altitude plasma observations.

3. Temperature of the Atmospheric Plasma

In the last section we discussed solutions of a simplified set of equations in which temperature was assumed to be constant and the same for all species. The temperature has to be determined in fact by solving the energy equation self-consistently. In a steady state these equations are:

for H^+ ions,

$$\frac{\partial}{\partial s}\left(\frac{3}{2}n_1 k T_1 v_1\right) + n_1 k T_1 \frac{\partial v_1}{\partial s} - \frac{\partial}{\partial s}\left(\kappa_1 \frac{\partial T_1}{\partial s}\right)$$

$$= \beta_{1e}(T_e - T_1) + \Sigma\{\beta_{1j}(T_j - T_1) + \gamma_{1j}v_1^2\} \tag{20}$$

for O^+ ions,

$$-\frac{\partial}{\partial s}\left(\kappa_2 \frac{\partial T_2}{\partial s}\right) = \beta_{2e}(T_e - T_2) + \beta_{21}(T_1 - T_2) + \beta_{2n}(T_n - T_2) + \gamma_{21}v_1^2 \tag{21}$$

for electrons,

$$-\frac{\partial}{\partial s}\left(\kappa_e \frac{\partial T_e}{\partial s}\right) = \beta_{e1}(T_1 - T_e) + \beta_{e2}(T_2 - T_e) + \gamma_{e1}(v_e - v_1)^2 \tag{22}$$

where β represents coefficients of heat transfer by Coulomb collisions and γ represents coefficients for the heat generated by Coulomb collisions between particles having different flow speeds. The electron flow velocity v_e is related to ion velocities by the condition that the flow maintains the charge neutrality:

$$n_e v_e = n_1 v_1 + n_2 v_2 . \tag{23}$$

Summation in Eq. (20) is over O^+ ions and neutral species. Density and temperature (T_n) of neutral species are taken from a separate model.

Assumptions that have been made in Eqs. (20), (21), and (22) are, (i) flow velocities of O^+ ions and neutral species are much less than the H^+ flow velocity, and (ii) electron cooling to neutral gases is negligible. The diffusion-thermal effect has been neglected, and equations of motion that are solved with Eqs. (20), (21), and (22) are still in a simplified form that does not include the thermal force term in R. In addition to O, neutral species O_2 and N_2 are considered from now on since the thermal structure in the 200 to 3,000 km height range will be examined. Coefficients β and γ are given in RAITT *et al.* (1975). A vertical magnetic field is assumed and hence vertical flows are considered.

At the lower boundary (200 km), we assign O^+ and H^+ temperatures obtained by equating local heating and cooling rates because transport processes are not important in low altitudes. Electron temperature at the lower boundary is chosen to match the observation. At the upper boundary (3,000 km) a downward electron heat flux is assigned and for ions $\partial T_1/\partial s = \partial T_2/\partial s = 0$ is assumed. O^+ density at 300 km is set to $2 \times 10^5/\text{cm}^3$.

Density and temperature profiles obtained from coupled continuity, momentum and energy equation are shown in Figs. 10 and 11. Four curves (a) through (d) in each figure correspond to different values of the H^+ outflow velocity at 3,000 km height which reflect different values of plasma pressure at much greater distances. The outflow velocity is 0.06, 0.34, 5, and

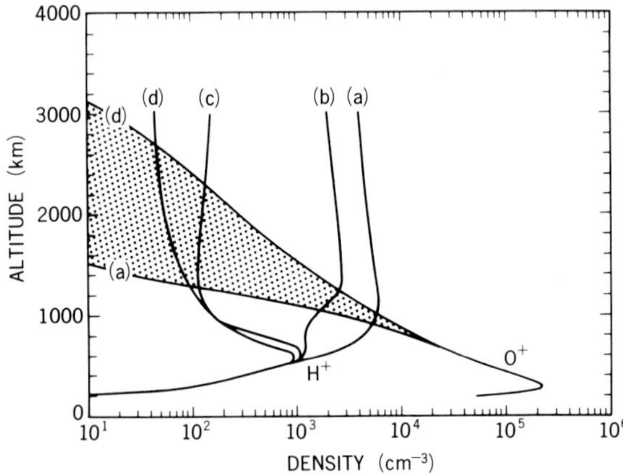

FIG. 10.   Calculated density profiles of O^+ and H^+ ions corresponding to different values of the upward outflow velocity of the H^+ ion (RAITT *et al.*, 1975).

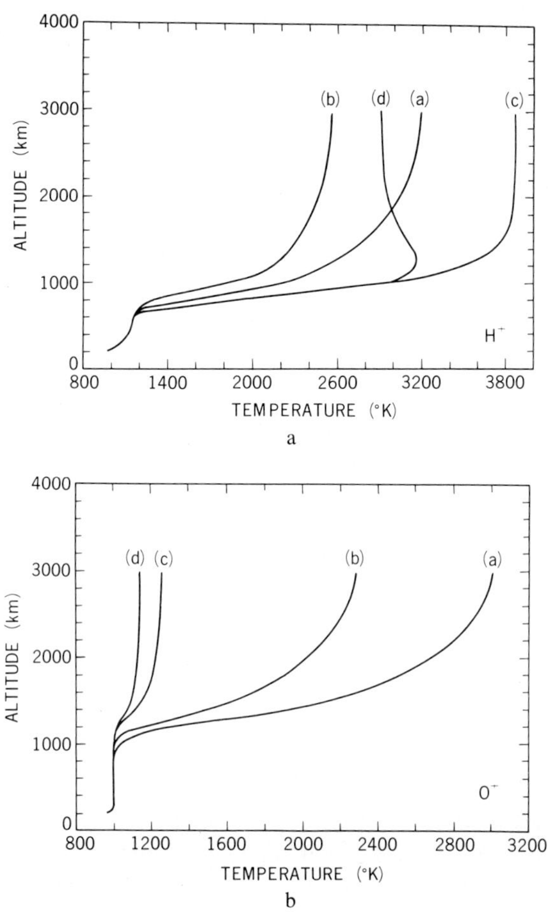

FIG. 11.   a. Calculated H^+ temperature profiles.   b. Calculated O^+ temperature profiles (RAITT *et al.*, 1975).

20 km/sec for cases a, b, c, and d, respectively. In terms of the thermal speed of H^+ at 3,000 km the outflow is subsonic in cases (a) and (b) and supersonic in cases (c) and (d). Below ~ 500 km height the H^+ density profile is independent of the outflow velocity because in this height range time scale of the oxygen-hydrogen charge exchange reaction is much shorter than the time scale of diffusion of H^+. Above ~ 500 km the influence of the outflow on the density profile is evident and a sharp outward gradient in $-\nabla n$ exists when the outflow is supersonic. The H^+ density becomes monotonically smaller as

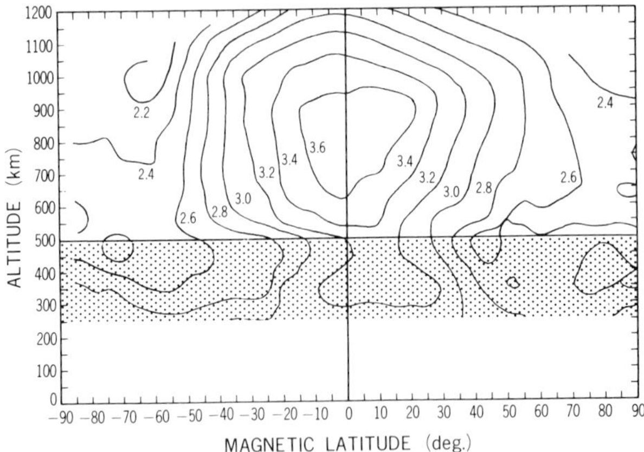

FIG. 12. An altitude/latitude contour map of the H^+ density at 1500 LMT in summer months under geomagnetically quiet conditions (RAITT and DORLING, 1976).

the outward flow velocity is increased because flux of the H^+ outflow is limited by friction exerted by O^+. On the other hand, the fall-off rate of the O^+ density with increasing height is reduced as the H^+ outflow velocity is increased. This is because the electric field produced by electrons can lift more oxygen ions when H^+ density is reduced due to the higher H^+ outflow velocity.

Statistical summary of the observed H^+ density is presented in Fig. 12 in the form of an altitude/latitude profile at 1500 LMT. Data used were taken during the period of March–September 1973 and with K_p in the range of 0–4. Contours are labelled by $\log_{10}[n(H^+)]$. (The shaded region indicates the altitude range where the H^+ data are not reliable in this study.) It is seen that the contours in the 600–1,000 km range are nearly vertical in low latitudes ($\lesssim 50°$) but nearly horizontal in high latitudes. The horizontal contours in the latter region can be compared with the sharp outward gradient of $-\nabla n$ that has been noted in the supersonic outflow solutions (c) and (d) of Fig. 10. Low latitude profiles correspond to subsonic solutions (a) and (b) where the density stays relatively unchanged with height, and upright contours reflect latitude dependence of the ionospheric condition (RAITT and DORLING, 1976).

The calculated temperature of electrons (not shown in the figures) increases monotonically from 1,000°K at 200 km to 4,650°K at 3,000 km. This profile is independent of the adopted value of the outflow velocity, and it is essentially consistent with the observed middle-latitude temperature profile

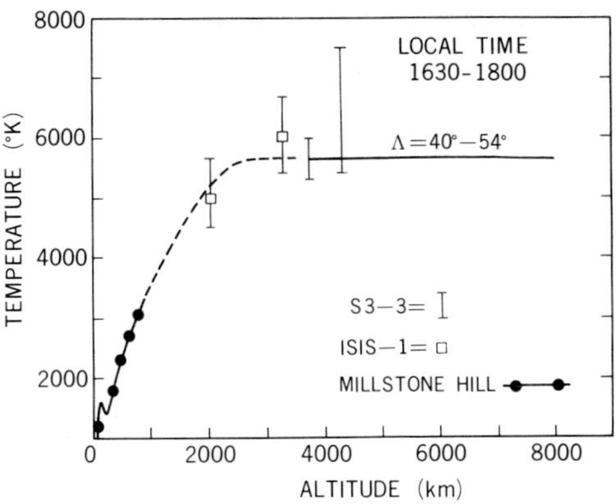

FIG. 13.   Temperature profile of electrons on field lines that cross the Earth at latitudes of 40° to 54° (RICH *et al.*, 1979).

of Fig. 13 which is obtained by combining temperature observations by radar backscatter and satellite-borne probe. The satellite data are averages of non-storm time observations when the satellite was on magnetic field lines which cross the ground in the indicated range of geomagnetic latitudes Λ. The observed decrease in the temperature gradient beyond $\sim 3,000$ km altitude has been suggested to be related to the $T_e^{5/2}$ dependence of the electron heat conductivity (RICH *et al.*, 1979). (Applicability of the heat conductivity, however, can be questioned because of large mean free path at such high altitudes.) The heat flux conducted downward through the electron gas is transferred to ions through Coulomb collisions and acts as the energy source of the flow-out process.

Calculated temperature of H^+ ions (Fig. 11a) is independent of the outflow velocity up to ~ 600 km height where heat loss to O^+ is in balance with heat gain from electrons. Above ~ 600 km the H^+ temperature rises rapidly as thermal conduction from higher altitudes becomes effective and the heat generated by the collision of streaming H^+ ions with almost stationary O^+ ions comes into play. These two heating effects vary with the H^+ flow velocity in an opposite way; while the collisional heating rate is increased with increasing H^+ velocity, the heat conductivity in H^+ ions, being proportional to the ratio n_1/n_e between H^+ density and electron density, decreases with increasing H^+ velocity in $\lesssim 1,200$ km range where H^+ is the minor ion. This is why the H^+ temperature drops from a to b and then increases from b to c as the outflow velocity is increased. Once the flow velocity has reached the supersonic range, further increases in the outflow velocity cause substantial decreases in coefficients β_{12} and γ_{12} representing H^+-O^+ temperature coupling and Joule heating by the H^+ flow through O^+, respectively. The decrease in β_{12} means less heat lost to O^+, but the decrease in the heat production rate due to decrease in γ_{12} is more important and causes decrease in H^+ temperature with a supersonic increase in the H^+ speed. At high altitudes above $\sim 1,400\,km$ energy balance is primarily determined by thermal conduction and H^+-e coupling under conditions of the subsonic flow, and by balance between convection $(-n_1 k T_1(\partial v_1/\partial s))$ and conduction in supersonic flows.

Calculated temperature profiles of O^+ ions (Fig. 11b) are largely determined by collisions with the neutral gas below $1,000\,km$, which keeps the O^+ temperature near the neutral gas temperature. At high altitudes O^+-H^+ coupling causes the O^+ temperature to rise toward the H^+ temperature when H^+ outflow speed is low because O^+ is a minor ion and the factor n_2/n_e in κ_2 results in a very small O^+ thermal conductivity. As H^+ flow speed increases, O^+ becomes a major ion up to higher altitudes and hence the O^+ thermal conductivity increases to such an extent that O^+ thermal conduction dominates the energy balance. This results in O^+ temperatures that are nearly constant with altitude. The value of the O^+ temperature at all altitudes shows a monotonic decrease with an increase in H^+ flow velocity (RAITT *et al.*, 1975).

Recently, attempts were made to improve the foregoing theoretical model of the plasma structure by using basic conservation equations which are more detailed. One of the points that have been made is to include the diffusion-thermal effect in the H^+ energy equation. While in Eq. (20) the H^+ heat flux q_1 was equated to heat conduction $-\kappa_1 \partial T/\partial s$ alone, account should have been made of the diffusion-thermal term and q_1 should have been written as

$$q_1 = -\kappa_1 \frac{\partial T_1}{\partial s} + 1.13\, n_1 k T_1 (v_1 - v_2). \tag{24}$$

The origin of the diffusion-thermal effect lies in the proportionality to (relative velocity)$^{-3}$ of the frequency of Coulomb collisions. Due to this dependence particles moving faster in the direction of the streaming suffer less resistance by collisions than those moving slower, and resultant skewing of the distribution function gives rise to an additional heat flux (BRAGINSKII, 1965). Inclusion of the diffusion-thermal effect results in reduction of the temperature of H^+ above $\sim 900\,km$ by $\sim 500°K$ when the outflow velocity is 5 or $10\,km/sec$ (SCHUNK *et al.*, 1978).

A more ambitious attempt is the "13-moment transport equations" advocated by SCHUNK (1975). In the conventional theory where only five velocity moments of the Boltzman equation are taken, there are only five independent parameters, namely density, bulk velocity vector and temperature. In the "13-moment" formulation, on the other hand, stress tensor and heat flux are also adopted as independent parameters. To express the even higher moments, the distribution function f is assumed to be the following function of the independent variables:

$$f = n\left(\frac{m}{2\pi kT}\right)^{3/2} \exp\left(-\frac{mc^2}{2kT}\right)\left[1 + \frac{m}{2kT_{\mathrm{p}}}\tau : cc - \left(1 - \frac{mc^2}{5kT}\right)\frac{m}{kT_{\mathrm{p}}}q\cdot c\right] \quad (25)$$

where c represent random velocity, q the heat flux and τ the stress tensor. An advantage of this formulation is that the temperature anisotropy can be derived from the calculated stress tensor. The anisotropy thus estimated in the 2,000 to 10,000 km height range is 10 to 20%, which is much less than the anisotropy that can cause plasma instabilities (SCHUNK and WATKINS, 1979).

As noted previously, heat is conducted downward through the electron gas to the height range below $\sim 3,000$ km from higher levels of the magnetosphere. The downward heat flux that is required to explain the thermal structure on a quiet day is shown in Fig. 14 as a function of local time. The figure is based on radar backscatter observations at Millston Hill which is at the geomagnetic latitude of 54°. It is noteworthy that the heat input is higher in the daytime than in the night-time. The input rises sharply at sunrise and drops sharply at sunset. This suggests that the high-altitude heat source is closely related to the sunlit conditions at lower altitudes.

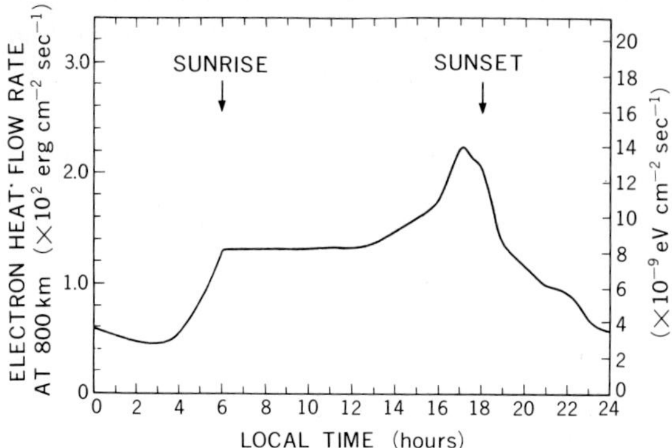

FIG. 14.   Diurnal variation of the estimated heat flow in the electron gas from the magnetosphere at 800 km (ROBLE, 1975).

Hence, as a possible candidate of the heat source, photoelectrons that have been produced at lower altitudes, streamed upward and trapped at higher altitudes have been suggested. Indeed the daily average downward heat input rate of 7.5×10^9 eV/cm$^2 \cdot$ sec is comparable to the average upward-directed flow of energy carried by photoelectrons (ROBLE, 1975). Physics of the photoelectrons will be discussed in the next section.

Another possible source of energy residing in high altitudes is energetic particles in the radiation belt. Although transfer of their energy to low energy plasma through Coulomb collisions is too inefficient to be of any practical importance, protons with energies of $10 \sim 100$ keV can excite ion cyclotron waves and yield energy to the plasma when these waves decay through Landau damping. The process has been considered to be responsible for the excitation of *Stable Auroral Red (SAR) Arcs*, which represent a belt of emissions of the atomic oxygen 6,300 Å line with a width of several hundred kilometers in the meridional direction. The occurrence frequency is relatively high in the latitude range of 45° to 60°. This hypothesis on the origin of the SAR arcs is based on the observation that SAR arcs appear during magnetic storms when energetic particles are injected into the radiation belt (as reviewed by REES and ROBLE, 1975), and in support of the hypothesis enhanced pitch angle scattering of the $100 \sim 200$ keV protons has been observed at the location of SAR arc (LUNDBLAD and SØRAAS, 1978). However, it is not clear if the same process can produce the downward heat flux of the required amount, over a much wider range of latitudes, even on quiet days, and it seems difficult to expect the daytime enhancement in the heat flux from this process.

It has been reported recently that tenuous protons observed outside the plasmasphere have energies of $10 \sim 25$ eV (BEZRUKIKH and GRINGAUZ, 1976; HORWITZ and CHAPPELL, 1979). Protons of this energy have been detected also in the plasmasphere (LENNARTSSON *et al.*, 1979). The heating mechanism of these intermediate energy protons represents an interesting new issue. Acceleration by field-aligned electric field is a possibility, but its occurrence in low latitudes is yet to be confirmed.

4.   Transport and Heating Effects of Photoelectrons

The photoionization of atmospheric gases by solar EUV radiation converts a fraction of the solar energy into photoelectron kinetic energy. These photoelectrons have initial energies of around 10 eV which is much higher than the mean energies of the ambient neutral, ionic and electron gases (which are of the order of 0.1 eV). The excess energy is lost gradually to ambient thermal electrons through Coulomb collisions, but individual photoelectrons are transported over large distances before their energy is degraded to the thermal energy. It is therefore necessary to discuss the transport of photoelectrons separately from transport of the ambient plasma.

The distribution function $f(r, v, t)$ of photoelectrons follows the Boltzman equation;

$$\frac{\partial f}{\partial t} + v \cdot \nabla_r f + \frac{dv}{dt} \cdot \nabla_v f = Q(r, v, t) + \frac{\partial f}{\partial t}\bigg|_c \tag{26}$$

where Q is the local production rate of the primary electrons due to solar radiation and $\dfrac{\partial f}{\partial t}\bigg|_c$ represents the change in f due to elastic and inelastic collisions with other constituents. We assume that photoelectrons are transported predominantly along magnetic field lines and that the atmosphere is horizontally stratified along these field lines. This approximation allows us to assume azimuthal symmetry about a magnetic field line and also to write the spatial gradient in Eq. (26) as a derivative along a direction z which is parallel to the magnetic field.

We rewrite Eq. (26) by introducing the flux $\phi(t, z, E, \mu)$ which is related to f through

$$\phi(t, z, E, \mu) = v(E) f(t, z, E, \mu) \tag{27}$$

where

$$E = \frac{1}{2} m_e v^2 .$$

The equation for ϕ is,

$$\frac{1}{v}\frac{\partial \phi}{\partial t} + \mu \frac{\partial \phi}{\partial z} + \left(\frac{1}{v}\right)\frac{dE}{dt}\left(\frac{\phi}{2E} - \frac{\partial \phi}{\partial E}\right)$$

$$= Q(t, z, E, \mu) + \frac{1}{v}\frac{\partial \phi}{\partial t}\bigg|_c \tag{28}$$

where μ represents the cosine of the pitch angle, the angle between v and B.

It has become a common practice to express the degradation of the photoelectron energy due to interactions with thermal electrons by the $\dot{v}\nabla_v f$ term on the left hand side of Eq. (26).

$$\frac{\mathrm{d}E}{\mathrm{d}t}=-\frac{\omega_{\mathrm{p}}^2 e^2}{v}\begin{cases}\ln\dfrac{m_{\mathrm{e}}v^3}{\gamma\omega_{\mathrm{p}}e^2}, & kT\ll E<\dfrac{m_{\mathrm{e}}e^4}{2\hbar^2}\\[2ex]\ln\dfrac{m_{\mathrm{e}}v^2}{\hbar\omega_{\mathrm{p}}}, & \dfrac{m_{\mathrm{e}}e^4}{2\hbar^2}<E\end{cases}\tag{29}$$

where γ is Euler's constant and ω_{p} is the plasma frequency

$$\omega_{\mathrm{p}}=\left(\frac{4\pi ne^2}{m_{\mathrm{e}}}\right)^{1/2}.\tag{30}$$

Eq. (29) expresses the combined effects of two-body Coulomb collisions and the collective effect of Cerenkov radiation of plasma waves. $E_0=m_{\mathrm{e}}e^4/2\hbar^2$ is the critical energy beyond which the de Broglie wavelength is greater than the classical distance of closest approach and a quantum mechanical treatment is required, and it is about 14 eV. Energy loss to ions has been shown to be negligible.

We transform Eq. (28) further by introducing the electron scattering depth τ which is defined by

$$\mathrm{d}\tau=\sum_l n_l(z)\sigma_l(E)\,\mathrm{d}z\tag{31}$$

where $n_l(z)$ is the number density and $\sigma_l(E)$ is the total scattering cross section for species l. When it is assumed that the steady state distribution function gives an adequate description of the photoelectron flux, Eq. (28) becomes

$$\mu\frac{\mathrm{d}\phi}{\mathrm{d}\tau}(\tau, E, \mu)=\frac{P(E)}{\sum\limits_l n_l(z)\sigma_l(E)}\left(\frac{\partial\phi}{\partial E}-\frac{\phi}{2E}\right)-\phi(\tau, E, \mu)$$

$$+\int\chi(\mu', \mu, E', E)\phi(\tau, E', \mu')\,\mathrm{d}E'\,\mathrm{d}\mu'\tag{32}$$

$$+S(\tau, E, \mu)$$

where

$$P(E)=\left(\frac{m_{\mathrm{e}}}{2E}\right)^{1/2}\frac{\mathrm{d}E}{\mathrm{d}t}\tag{33}$$

and

$$S=Q\left/\sum_l n_l\sigma_l\right..\tag{34}$$

μ' and μ are cosines of the pitch angles associated with the incident and outgoing directions, and E' and E are incident and outgoing electron energies. The second term on the right-hand side of Eq. (32) represents particles leaving a volume element specified by E and μ, and the third term describes how particles are scattered into that volume element. The scattering process includes elastic scattering of photoelectrons by neutrals (which are assumed here to be isotropic) and energy loss of photoelectrons due to excitation and ionization of neutral particles. Inelastic collisions between photoelectrons and neutrals are the main way in which photoelectrons below 300 km lose energy (ORAN and STRICKLAND, 1978).

Figure 15 shows the calculated photoelectron spectrum as a function of pitch angles at two altitudes (247 km and 545 km) which are above the height (~ 160 km) of the maximum production of photoelectrons. The calculation was performed with the assumption that the solar zenith angle is $48°$ and that photoelectrons are produced with an isotropic pitch angle distribution. It is seen that the pitch angle distribution turns from near isotropy at 247 km to strong upward-streaming anisotropy at 545 km. Figure 16 shows upward and downward fluxes at selected energies as a function of height. According to this figure the total flux of upgoing photoelectrons integrated over energy is as high as $2 \times 10^9/\text{cm}^2\text{sec}$, which is almost an order of magnitude greater than the upward flux of the ambient plasma discussed in the last section. The energy flux carried by photoelectrons is about $2 \times 10^{10}\,\text{eV}/\text{cm}^2\text{sec}$. If this energy can be used to heat ambient electrons at the height range of 1,000 to

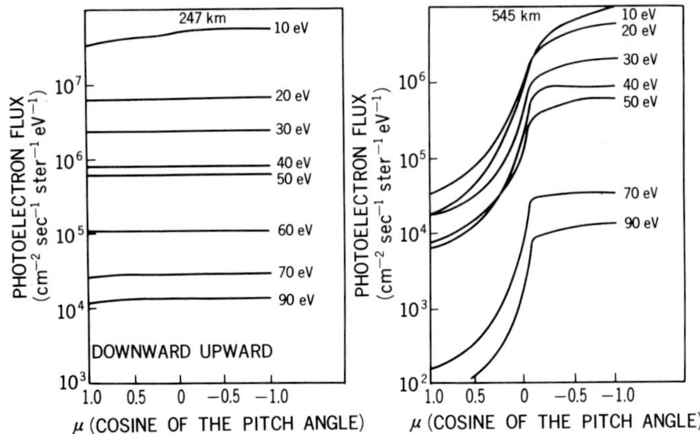

FIG. 15. Calculated photoelectron spectrum as a function of pitch angle at two altitudes (ORAN and STRICKLAND, 1978).

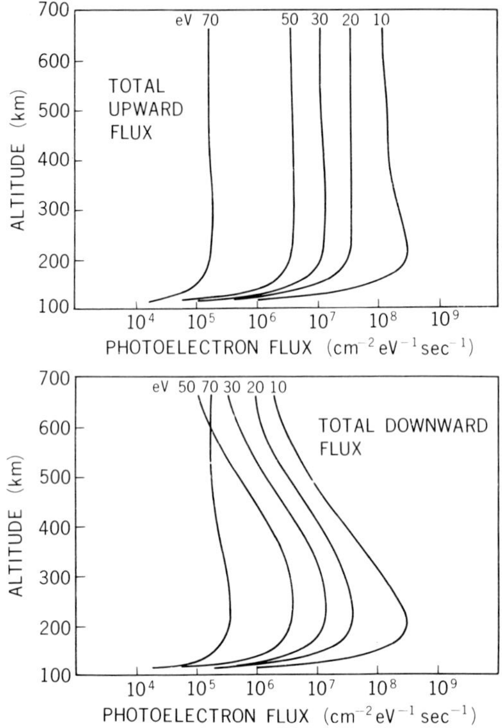

Fig. 16.   Upward and downward flux of photoelectrons at selected energies (Oran and Strickland, 1978).

3,000 km, observed profile of the electron temperature and the inferred downward flux of heat through ambient electrons is not hard to understand.

In view of the high intensity of the upward photoelectron flux, it does not seem to be justifiable to discuss transport of the ambient plasma in the low-latitude dayside region independently of the transport of photoelectrons; that is, the electric field produced by photoelectrons should be taken into account when the transport of ambient plasma is considered. This electric field acts to suppress the upward flow of ambient electrons (or cause downward flow of ambient electrons) and enhance the upward flow of ions. The associated deceleration of the upstreaming photoelectrons would not be significant, however, because the energy needed to bring a proton to the radial distance of 5 R_E, for example, against the Earth's gravitational force is only 0.1 eV. At altitudes above $\sim 1{,}000$ km energy density of photoelectrons can be comparable to that of ambient plasma.

The thermal gas heating rate by photoelectrons can be estimated by

$$Q(z) = \int \frac{dE}{dt}\, \Phi(E,\, z)\, dE \tag{35}$$

where Φ is the total photoelectron flux that is obtained by integrating ϕ from $\mu = -1$ to 1. The heating rate is shown in Fig. 17 for the three zenith angles. The maximum heating rate occurs at a height of about 200 km.

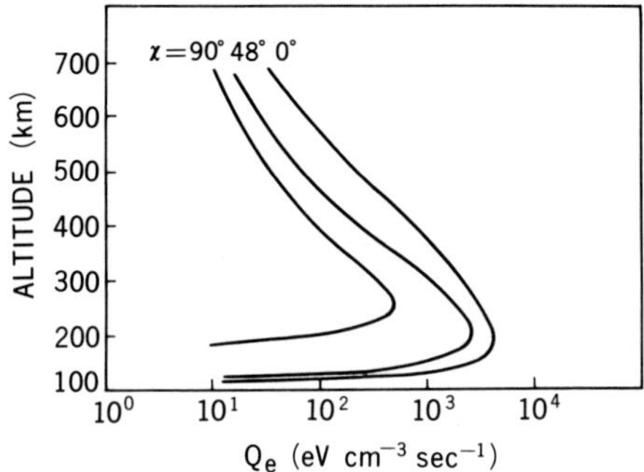

FIG. 17.   Thermal electron heating rate by photoelectrons as a function of altitude (ORAN and STRICKLAND, 1978).

As the escaping photoelectrons travel toward higher altitudes, their pitch angle decreases because of the conservation of the first adiabatic invariant;

$$\frac{1-\mu^2}{B} = \text{const.} \tag{36}$$

Scatterings that take place at high altitudes act to transform the field-aligned pitch angle distribution to a more isotropic distribution, and this means that some of the escaping photoelectrons becomes trapped in a magnetic bottle. According to model calculations, about a half of the photoelectrons that would have reached the conjugate ionosphere if the pitch angle scattering were absent becomes trapped in the middle of the field line (LEJENNE and WORMSER, 1976). For example, steady-state photoelectron flux on a field line that crosses the equatorial plane at a radial distance of 1.42 R_E is increased by about 50% when the pitch angle scattering is taken into account (MANTAS et al., 1978). Thus the scattering acts to enhance the content of the energy

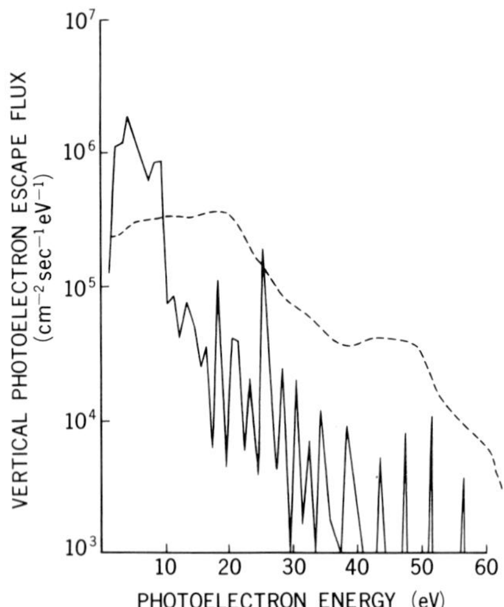

Fig. 18.   Comparison of photoelectron escape fluxes from Jupiter (solid curve) and the Earth (dashed curve, scaled) (Swartz *et al.*, 1975).

reservoir in high altitudes.

The thermalization time τ of an energetic electron of initial energy E_i (eV) in a completely ionized plasma of uniform density N_e (cm^{-3}) and temperature of a few thousand degrees Kelvin is approximately $\tau \simeq 4,300$ $(E_i^{3/2}/N_e)$ sec. For 20 eV electron placed in $N_e = 5 \times 10^3/\text{cm}^3$ plasma, τ is 80 sec (Mantas *et al.*, 1978).

Photoelectron escape fluxes have been estimated also for the Jovian atmosphere. The escape flux energy spectrum for Jupiter is shown in Fig. 18 and compared with the scaled down (by a factor of 5.2^{-2}) spectrum of the Earth's photoelectrons. The difference in the spectral shape reflects difference in the major neutral species in the escape flux region (Swartz *et al.*, 1975). In the Jovian atmosphere molecular hydrogen is the major species while in the Earth's atmosphere atomic oxygen is dominant.

In Jupiter's strong gravitational field, kinetic energy of a few electron volts is needed for an atmospheric proton streaming along a magnetic line of force to cross the maximum of the gravitational potential and reach higher levels of the magnetosphere. The potential maximum, beyond which the centrifugal force exceeds the gravity, is located at a distance of about $2\ R_J$ from the rotational axis except on very high-latitude field lines. Since atmospheric protons have kinetic energies of the order of 0.1 eV only, they cannot by themselves cross the barrier. However, they can be lifted beyond the barrier by virtue of the electric field which is generated by escaping photoelectrons who have enough energy to overcome the barrier (IOANNIDIS and BRICE, 1971). Prior to the discovery of the plasma supply from Io, this mechanism was considered as the most important origin of plasma in the Jovian magnetosphere.

5. Plasma from Jupiter's Satellite Io

The magnetosphere of Jupiter is unique among the family of planetary magnetospheres in that the source of plasma is located within the magnetosphere. Ionization of volcanic gases emitted from the satellite Io can very efficiently provide the magnetosphere with its plasma because the gravitational force of Jupiter does not act as a barrier.

As soon as a neutral atom or molecule is ionized it is influenced by electric and magnetic fields. The dominant electric field in the Jovian magnetosphere is one that is associated with the corotational motion of plasma. This is given in the inertia frame by

$$E = -\frac{1}{c}(\boldsymbol{\Omega} \times \boldsymbol{R}) \times \boldsymbol{B} \tag{37}$$

where $\boldsymbol{R}$ is the position vector from the center of Jupiter and $\boldsymbol{\Omega}$ is the angular velocity vector of Jupiter's rotation. This electric field is directed outward. In terms of the "frozen-in field line" theorem one can say that the above electric field is associated with rotation of magnetic field lines with Jupiter. For a newly formed ion or electron from the Io's atmosphere, the electric field experienced is

$$E' = -\frac{1}{c}(\Delta\boldsymbol{\Omega} \times \boldsymbol{R}) \times \boldsymbol{B} \tag{38}$$

where $\Delta\Omega = \Omega - \Omega'$ is difference between Ω and the angular velocity Ω' of Io's orbital motion. Motion of such a particle is a cycloid as illustrated schematically in Fig. 19 in a frame of reference fixed to Io. Here initial energy of the particle is assumed to be zero in Io's frame of reference because it is much less than the energy to be acquired from the electric field. Immediately after their birth ions and electrons are moved in the direction of the electric force and accelerated thereby, but the motion is turned into gyration as velocity vectors are deflected due to the effect of the magnetic field. After each gyration they return to the point where the electric potential is the same as at starting point, but in the meantime they have moved in the direction of

A. NISHIDA

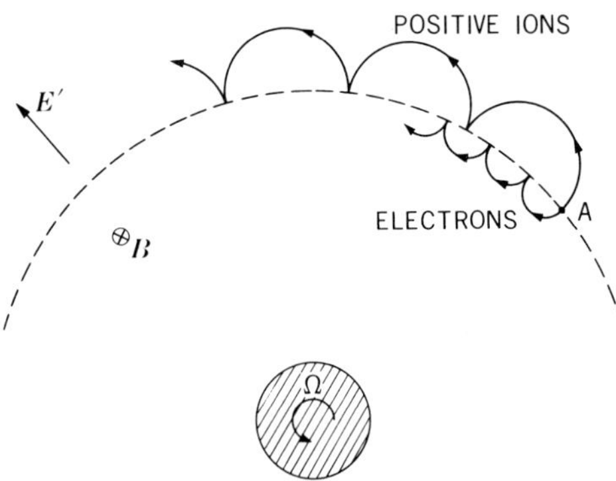

Fig. 19. Trajectories of a positive ion and an electron that are formed at point A in the Jovian magnetosphere. The corotational electric field is directed away from the planet.

the corotation. When the motion of each particle is separated into the guiding center motion in Io's frame of reference and the gyrating motion, both motions have the same speed. This speed is given by $|\varDelta\boldsymbol{\Omega}\times\boldsymbol{R}|$, which is about 59 km/sec. The plasma produced from Io's atmosphere is removed from the point of production with this speed and form a torus along the orbit of Io. Plasma torus along Io's orbit has been detected by Voyager (BROADFOOT *et al.*, 1979).

Quantitatively, however, this simple picture need be modified since the corotational electric field $\boldsymbol{E}$ (or $\boldsymbol{E}'$, depending on the frame of reference) is reduced due to current flowing in the Io's neighbourhood. Three types of currents can flow in the direction of $\boldsymbol{E}'$. The first is associated with the cycloidal motion of the newly created plasma. Since an ion or an electron is radially displaced following its production by $(2/\pi)R_g$, where R_g denotes Larmor radius, in the direction of the electric force, current density of the first type is related to the production rate $\dot{n}_i$ of the i-th species by

$$\boldsymbol{j}^1 = \sum_i q_i \dot{n}_i R_{gi} \frac{\boldsymbol{E}'}{E'}$$

$$= \sum_i m_i \dot{n}_i c^2 \frac{\boldsymbol{E}'}{B^2}. \tag{39}$$

The second is due to the Pedersen mobility of charged particles in Io's ionosphere where collision frequency is not negligibly small as compared to gyrofrequency.

$$j^{\mathrm{II}} = \sigma_{\mathrm{p}} E' \ .$$ (40)

We are not concerned here with Hall current which yields no divergence in a uniform magnetic field. The current of the third type is the polarization current:

$$j^{\mathrm{III}} = \frac{1}{4\pi} \frac{c^2}{V_{\mathrm{A}}^2} \frac{\mathrm{d}E'}{\mathrm{d}t}$$ (41)

where

$$V_{\mathrm{A}}^2 = \frac{B^2}{\sum_i 4\pi n_i m_i} \ .$$

The electric field E' is to be determined self-consistently from

$$\frac{\partial}{\partial t}(\nabla \cdot E') = -\nabla \cdot \left[\frac{c^2}{V_{\mathrm{A}}^2} \left\{ \frac{\sum_i m_i \dot{n}_i}{\sum_i m_i n_i} E' + 4\pi \frac{V_{\mathrm{A}}^2}{c^2} \sigma_{\mathrm{p}} E' + \frac{\mathrm{d}E'}{\mathrm{d}t} + V_{\mathrm{A}}(\nabla \cdot E'_{\perp}) \frac{B}{B} \right\} \right]$$ (42)

where the following relation is used to express field-aligned current $j_{\parallel}$ in terms of E':

$$j_{\parallel} = \frac{c^2}{4\pi V_{\mathrm{A}}} \nabla \cdot E'_{\perp} \frac{B}{B} \ .$$ (43)

When explicit time dependence is neglected, Eq. (42) becomes

$$\nabla \cdot \left(\frac{E'_{\perp}}{\tau_{\mathrm{p}}} + V_{\mathrm{A}} \frac{\mathrm{d}}{\mathrm{d}s} E'_{\perp} \right) = 0$$ (44)

where

$$\tau_{\mathrm{p}}^{-1} = \frac{\sum m_i \dot{n}_i}{\sum m_i n_i} + 4\pi \frac{V_{\mathrm{A}}^2}{c^2} \sigma_{\mathrm{p}}$$

and

$$\frac{\mathrm{d}}{\mathrm{d}s} = \frac{1}{V_{\mathrm{A}}} \{ v_{\perp} \cdot \nabla_{\perp} + V_{\mathrm{A}} \nabla_{\parallel} \} \ .$$

(Uniformity of V_{A} is assumed for simplicity.) Electric field in Io's neighbourhood is to be obtained from the solution of Eq. (44) (GOERTZ, 1980).

Measurements of the electric field itself have not been conducted in the Jovian magnetosphere, but E' can be inferred from thermal speed of ions. Plasma observations by Voyager have revealed that thermal speed of ions in

34 A. NISHIDA

the vicinity of Io's orbit is 15 km/s for sulphur ions and 20 km/s for oxygen
ions (BAGENAL *et al.*, 1980). These values are substantially less than
59 km/sec that is expected when the electric field is not modified by current
flows, and this indicates that the corotational electric field is reduced to
about 30% of the original strength.

According to the foregoing basic model the radial extent of the plasma
torus originating from Io is given by the range of particle drift shells that pass
Io's atmosphere. If the Jovian magnetic field were perfectly dipolar and its
axis were perfectly aligned with the rotational axis, the radial extent would
have been a few times of Io's radius. Since these conditions are not met in
reality the plasma torus can be more widely spread and the radial extent of
about 1 R_J can be expected. The observed extent, however, was even greater
than this. Figure 20 is the contour of the electron density deduced from
radiowave observations by Voyager I (WARWICK *et al.*, 1979). That the
electron density peaks at the orbital distance of Io clearly suggests that Io's
atmosphere is the principal source of plasma in the Jovian magnetosphere,
but the radial extent of the plasma is much wider than 1 R_J. It appears that at
least two processes are operative radially to spread the plasma. One is
diffusion caused by resonances with time-varying electric and magnetic field
and the other is outward transport related to the centrifugal force.

Radial diffusion of charged particles has been studied extensively, in the
Earth's magnetosphere, for the energetic radiation-belt particles but not for

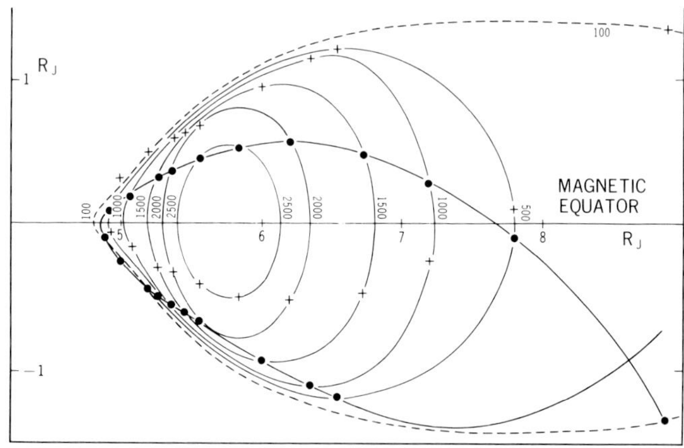

FIG. 20. Contour of electron density deduced from the upper hybrid resonance
frequency observed by Voyager I. Dots are density points along the spacecraft
trajectory and crosses are their mirror images with respect to the equator (WARWICK
et al., 1979).

the ambient plasma. One of the reasons is that the ambient plasma in the
Earth's magnetosphere is provided from the planetary atmosphere and hence
the source is not localized and the effect of the diffusion is hard to identify.
Moreover, time variations of the source condition due to the diurnal
variation of the atmosphere (discussed in Section 2) and the dominance of
the solar-wind induced convective motion in the outer magnetosphere make
slow process of the radial diffusion relatively unimportant. In the Jupiter's
magnetosphere dominance of the Io's atmosphere as the source and that of
the corotation as the motion seem to set a condition for the effect of the
diffusion to be significant.

The radial diffusion is caused by a resonant interaction between a time
dependent electric or magnetic field and azimuthal motion of a particle. For
illustration, let us consider the simplest case where the electric field is
spatially uniform and varies in time with the period of the azimuthal drift
motion. In Fig. 21 two representative trajectories of a positive ion under the
downward (relative to the sheet of the paper) magnetic field are depicted.
Difference between trajectories a and b lies in the phase relation between the
particle position and the electric field. In the case (a) the phase relation is
such that the electric field always has a positive component in the direction of
the particle velocity. In this case the electric-field drift velocity $\boldsymbol{u} = c(\boldsymbol{E} \times \boldsymbol{B}/B^2)$
is directed toward the planet everywhere, so that the particle is displaced
toward lower L shells. In the case (b) where the electric field always has a
negative component in the direction of the particle velocity, an outward
displacement takes place. If the phase is randomly distributed, the result is a
spreading of the drift shell. The diffusion coefficient is given by

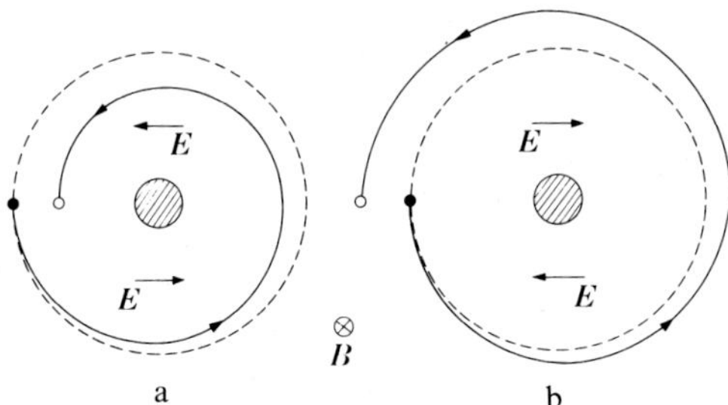

FIG. 21.   Trajectory of resonant particles placed under constant magnetic field and
time-varying electric field. Cases a and b correspond to different phase relations
between particle position and electric field direction.

$$D(L, \omega) = \frac{\sum\limits_{n=1}^{\infty} P_n(n\omega, L)}{8B_0^2} \tag{45}$$

for a particle whose azimuthal motion has an angular frequency ω. P_n is the power spectrum of the electric field at the specified frequency and on field line shell with equatorial crossing distance LR_J. Effect of diffusion due to time-varying magnetic perturbation field can be similarly assessed (FÄLTHAMMAR, 1965).

The angular frequency of the drift motion that enters the diffusion coefficient as a parameter is dependent on the particle energy. This is because the drift velocity arising from the non-uniformity of the magnetic field depends on kinetic energy of particle's thermal motion. In the Jovian magnetosphere, however, the corotation with the planet dominates the particle motion unless the kinetic energy is greater than about 10 MeV, so that the diffusion coefficient of low-energy ions is essentially the same as that of energetic particles that have energies as high as 1 MeV. This coefficient has been estimated to be $D=dL^3$, $d=5\times10^{-11}R_J^2/\mathrm{sec}$, which means that at Io's orbit an ions takes almost a year to diffuse across a radial distance of 1 R_J.

If outflow due to centrifugal force can be neglected, plasma density satisfies the diffusion equation

$$\frac{\partial}{\partial L}\left[\frac{D}{L^2}\frac{\partial}{\partial L}(L^2 N)\right] + S - R = 0 \tag{46}$$

where N is number of particles per unit field line shell, and S and R are production and loss rates which are equated to zero except in the source region and at the boundary. Since the plasma originating from Io's atmosphere is confined to the equatorial region, N can be related to number density n per unit volume roughly by

$$N = 4\pi\theta R_J^3 L^2 n \qquad L < L_0 \tag{47}$$
$$= 4\pi\theta R_J^3 L L_0 n \qquad L > L_0$$

where L_0 is the L value of the Io's orbit and θ is the half angle of the thickness of the plasma disc at the satellite orbit as measured from the center of the planet. From Eqs. (46) and (47) we obtain

$$n = \frac{\Phi ln(L_b/L_0)}{4\pi\theta R_j^3 dlnL_b} \frac{lnL}{L^4} \qquad L < L_0$$

$$= \frac{\Phi lnL_0}{4\pi\theta R_j^3 dlnL_b} \frac{ln(L_b/L)}{L_0 L^3} \qquad L > L_0 \tag{48}$$

where Φ is the total ion production rate and n is assumed to be zero both at the inner boundary ($L=1$) and the outer boundary ($L=L_b$) (SISCOE, 1978).

The solution expressed by Eq. (48) has a maximum at $L=1.28$ and beyond that point monotonically decreases with an increasing L: in contrast to the observed profile no density peak is produced at Io's orbit. In order to reproduce the density peak at Io's orbit and also to express a steeper density decline inside than outside of the peak as observed, in a pure diffusion model one is forced to assume that the process is not operating continuously but the build-up of the plasma has started shortly ($1 \sim 100$ days) before the Voyager observation. It is also necessary to assume that the diffusion coefficient D is about 50 times greater outside Io's orbit than inside (RICHARDSON *et al.*, 1980).

The above asymmetry in the effective value of D means that the plasma is transported preferentially outward and suggests that the centrifugal drive cannot be neglected. There are two conceivable ways through which flow under the centrifugal force can influence the density distribution. One is by moving particles outward each time they are scattered. The particle current Γ is expressed in general by

$$\Gamma = -D\nabla n + \mu nF \tag{49}$$

where μ represents mobility under the action of the force F. If scattering is due to binary Coulomb interactions the mobility term is given by

$$\mu nF = \frac{v}{\Omega_b^2} nr\Omega^2 \tag{50}$$

where v and Ω_b represent collision frequency and gyrofrequency, respectively. When the mobility term is included, Eq. (46) is rewritten as

$$\frac{\partial}{\partial L}\left[\frac{D}{L^2 R_J}\frac{\partial}{\partial L}(L^2 N)\right] - \frac{\partial}{\partial L}\left[\frac{vR_J\Omega^2}{\Omega_b^2}LN\right] = 0 \tag{51}$$

in regions where $S=R=0$. Writing $T=T_0(L_0/L)^\alpha$ (hence $v=\xi nT_0^{-3/2}(L/L_0)^{(3/2)\alpha}$ where ξ is a factor that depends on mass and charge)

38   A. NISHIDA

and $\Omega_b = \Omega_0(L_0/L)^3$, we obtain at $L < L_0$ (where $N \propto L^2 n$)

$$\frac{\partial}{\partial L}(L^4 n) = \frac{\xi T_0^{-3/2} R_J^2 \Omega^2}{d\Omega_0^2} \left(\frac{L}{L_0}\right)^{6+(3/2)\alpha} L^2 n^2 . \tag{52}$$

Equation (52) can be integrated to give

$$L^4 n = \frac{L_0^4 n_0}{1 + \eta \dfrac{n_0}{\left(1 + \dfrac{3}{2}\alpha\right)L_0}\left[1 - \left(\dfrac{L}{L_0}\right)^{1+(3/2)\alpha}\right]} . \tag{53}$$

The parameter $\eta = \xi T_0^{-(3/2)} \Omega^2 / d\Omega_0^2$ is estimated to be about 3×10^{-8} cm^3 on the basis of mean ionic mass of 5×10^{-23} g, ion mean thermal energy of $4 \times 10^{5\circ}$K and diffusion coefficient $d = 10^{-11} R_J^2/\text{sec}$. In order that the mobility effect to be significant as compared to the diffusion, $\eta n_0 \gtrsim 1$ is required. However, according to observations $n_0 \sim 2 \times 10^3/\text{cm}^3$ even at the maximum, so that for the centrifugal force to cause appreciable outward transport of the scattered particles the scattering frequency has to be about 10^4 times higher than the frequency of binary Coulomb collisions.

Another way by which centrifugal force can drive plasma outwards is through dynamical instability which it induces on the plasma distribution. Since density of plasma decreases in the direction of the centrifugal force Rayleigh-Taylor instability and ballooning instability can arise. Dispersion equation for an incompressible perturbations leading to these instabilities is given by

$$\omega^2 = k_\parallel^2 V_A^2 + r\Omega^2 \frac{\partial}{\partial r}[\ln\, n(r)] \tag{54}$$

where $k_\parallel$ is the wavenumber in the direction of the magnetic line of force. If both ends of field lines are fixed to the ionosphere during the perturbation, $k_\parallel = \pi/l$ where l is length of the field line, and the sufficient condition for instability can be written as

$$2WL^2 \left(\frac{mn(L)L^2 R_J^2 \Omega^2}{B_J^2/4\pi}\right)^{1/2} \left(-L\frac{\partial}{\partial L}n\right)^{1/2} = \pi \tag{55}$$

where W is the half thickness of the plasma disc in the unit of R_J and this will be taken to be 2. B_J is the equatorial field strength on the surface of Jupiter (HASEGAWA, 1980; contribution from the centrifugal force arising from the bouncing motion of ions has been neglected as it would not be significant at

Io's orbit where the magnetic field is nearly dipolar). Gradient of n immediately outside Io's orbit is proportional to L^{-5} (FROIDEVAUX, 1980). Equation (55) tells us that the ballooning instability can be set up at Io's orbit ($L=5.9$) if number density of sulphur or oxygen ions is of the order of $10^5/\mathrm{cm}^3$. Nevertheless, we should note that the ballooning instability by itself does not really cause dumping of plasma from the Jovian magnetosphere because field lines on which plasma is loaded have their ionospheric ends kept fixed.

Whether or not the ionospheric ends of field lines have to be thought of as fixed depends upon the magnitude and time scale of the perturbation. The velocity of the ionospheric end of field lines induced by the centrifugal force can be estimated as follows: The density of the azimuthal current associated with drift motions caused by the centrifugal force is given, when integrated over thickness, by

$$I_\mathrm{d} = \frac{cmnLW\Omega^2 R_\mathrm{J}^2}{B} \tag{56}$$

while the ionospheric Pedersen current that flows along the same L shell is given by

$$I_i = \Sigma_\mathrm{p} E_i = \frac{1}{c} \Sigma_\mathrm{p} v_i B_i \tag{57}$$

where suffix i denotes parameters at ionospheric heights. If fraction α of the current I_d is connected to I_i (the rest being closed in the magnetosphere in the equatorial region), we obtain by assuming a dipolar geometry

$$v_i \sim \frac{\alpha c^2 mnL^6 W\Omega^2 R_\mathrm{J}^2}{\Sigma_\mathrm{p} B^2} . \tag{58}$$

The corresponding outward drift velocity in the magnetosphere is

$$v \sim L^2 v_i . \tag{59}$$

By substituting B and n at Io's orbit and assuming $\Sigma_\mathrm{p} \sim 1$ mho ($\sim 10^{12}$ statmho), we obtain $v_i \sim 4 \times 10^5 \ \alpha\,\mathrm{cm/sec}$ and $v \sim 1.6 \times 10^7 \ \alpha\,\mathrm{cm/sec}$. These values are exceedingly large. Note, for comparison, that with $D = 5 \times 10^{-11} L^3 R_\mathrm{J}^2/\mathrm{sec}$ the diffusion over distance of 1 R_J ($= 7 \times 10^9\,\mathrm{cm}$) takes about 10^8 sec at $L=5.9$; v is comparable to this diffusion rate even when α is as small as 10^{-5}. Hence in the outward flow process that competes with the diffusion the ionospheric ends of field lines need not be considered to be fixed. It is probable that the interchange motion (for which $k_\parallel = 0$) that is excited by Rayleigh-Taylor instability acts to transport plasma effectively outward from the Io's orbit.

From observed density n_0 of $\sim 10^3/\text{cm}^3$ and the ion production rate of $\sim 10^{29}/\text{sec}$ deduced from optical observations (BROADFOOT *et al.*, 1979) the outward flow speed can be estimated to be of the order of 10^4 cm/sec if the outflow is proceeding continuously. This indicates that $\alpha \sim 10^{-4}$, namely that the azimuthal density perturbation, which causes ionospheric channeling of azimuthal current in the magnetosphere, is stabilized at $\Delta n/n \sim 10^{-4}$. Or, it may be that a much faster outflow is induced transiently each time the density exceeds some threshold value.

On account of conservation of adiabatic invariants plasma is cooled as it is transported outward. Hence temperature is expected to decrease with increasing L unless heating mechanism is operating in the outer magnetosphere. The observation of very hot (~ 30 keV) plasma at $L \gtrsim 30$ (KRIMIGIS *et al.*, 1980) may be taken to testify to the presence of a very efficient heating process, but recent analysis seems to indicate that these hot plasma and the cold plasma represent separate populations (BELCHER *et al.*, 1980). The hot population may originate from penetration of solar wind particles into the magnetosphere and acceleration of Jovian ionospheric plasma by field-aligned electric field.

Plasma transported inward from Io's orbit is expected to be heated on account of the conservation of adiabatic invariants (MACHIDA and NISHIDA, 1978). Observations by Voyager have demonstrated, however, that the temperature sharply drops inside Io's orbit quite contrary to that expectation

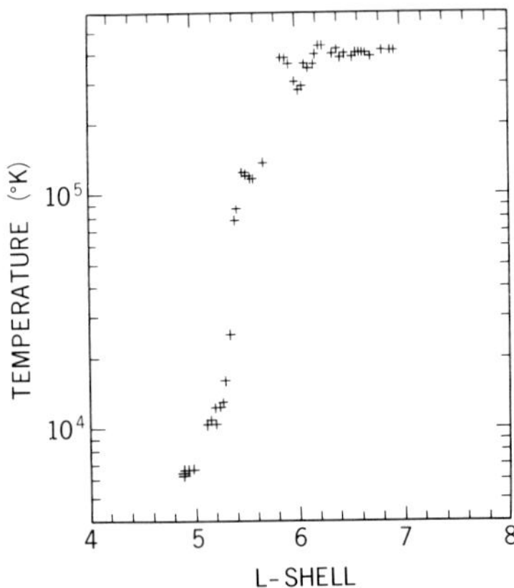

FIG. 22.   Profile of ion temperature in the Jovian magnetosphere (BAGENAL *et al.*, 1980).

(Fig. 22). It has been suggested that ions are excited by collisions with electrons and cooled by generation of ultraviolet and optical emissions (BAGENAL *et al.*, 1980).

6. Entry of Solar-Wind Plasma into the Plasma Sheet

Planetary magnetospheres have extended magnetic structures such as magnetotail in the Earth's magnetosphere and magnetodisc in the Jupiter's magnetosphere. These structures are supported internally by pressure of hot plasma that constitutes plasma sheet and plasma disc in respective magnetospheres. Intuitively such plasma populations are considered to originate from the solar wind because their mean energy, which is of the order of 1 keV in the plasma sheet and 10 keV in the plasma disc, is much closer to energy of the solar wind than to temperature of planetary or satellite ionospheres and also because these populations are found in outer regions of the magnetosphere with part of their bounding surfaces coinciding with the magnetopause.

The actual mode of entry of the solar wind plasma from the magnetosheath into the plasma sheet, however, still remains largely as an unresolved issue. According to observations at radial distances of about 20 to 60 R_E, ions in the plasma sheet are streaming more often earthward than tailward. Thus it is at great distances beyond the lunar orbit that the solar wind plasma gains access into the plasma sheet. Earthward streaming is particularly prominent when the plasma sheet is expanding and is observed in northern and southern boundary regions of the plasma sheet with a thickness of the order of 1 R_E (LUI *et al.*, 1977).

It has been suggested that the solar wind plasma is injected into the plasma sheet by way of the plasma mantle (PILIPP and MORFILL, 1978). The mantle plasma is thought to flow from high to low latitudes along the streamlines drawn by dashed lines in Fig. 23. Magnetic field lines in the high latitude magnetotail extend to great distances where field lines become either irregular or connected with interplanetary field lines, but low-latitude field lines are closed. The plasma that has drifted equatorward and entered this region of closed field lines becomes trapped, because particles are reflected at the far end of closed field lines. Plasma sheet can be formed on closed field lines in this way.

Streamlines of Fig. 23 represent a combination of field-aligned anti-earthward flow and drift motion toward lower latitudes. Equatorward drift motions have frequently been detected at radial distances of ~ 60 R_E, and

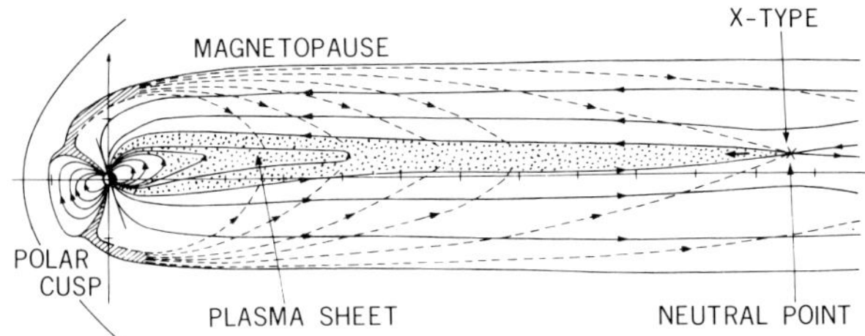

FIG. 23.   Streamlines of inward motion of plasma from the mantle (dashed lines). The streamline depends on velocity of the field-aligned streaming motion and is not the same for all particles originating from the same point in the cusp (PILIPP and MORFILL, 1978; slightly modified).

they have been interpreted to be the drift under the dawn-to-dusk electric field produced by influence of the solar wind (McCOY *et al.*, 1975). The velocity of the electric field drift is independent of energy of the particle, so that the distance which a particle travels in the anti-earthward direction before reaching the last closed field line is greater for a faster particle. Figure 24 shows contours of constant bulk velocity estimated on the basis of two model calculations. In both models the tail midplane is represented by the X-Y plane and the tail magnetopause is set at a distance of 18 R_E from this plane. Model 1 (top) assumes that all particles are supplied at the $x=0$ plane within a thickness of $z=18$ R_E to 16 R_E. This model is intended to simulate entry of magnetosheath plasma only at the polar cusp. Model 2 (bottom) assumes that the magnetosheath plasma traverses the magnetopause everywhere. Both model calculations confirm that, at any given x coordinate, plasma that is found in lower latitudes has lower bulk velocities as compared to plasma that is streaming in higher latitudes. Thus the process acts as an energy spectrum analyzer and the particles which enter closed field lines represent lower-energy side of the spectrum of the particles that have entered the magnetosphere. The implication is that the plasma injected into the plasma sheet is less energetic, on the average, than the solar wind in the magnetosheath.

According to observations, however, plasma in the plasma sheet is much more energetic than the mantle plasma and is even hotter than plasma in the magnetosheath (HARDY *et al.*, 1979). In order that an injection through the mantle be a significant process, therefore, it is necessary that plasma can acquire almost an order of magnitude higher energy inside the plasma sheet. The dawn-to-dusk electric field can contribute to the acceleration process in a variety of ways. Since closed field lines in the plasma sheet are drifting

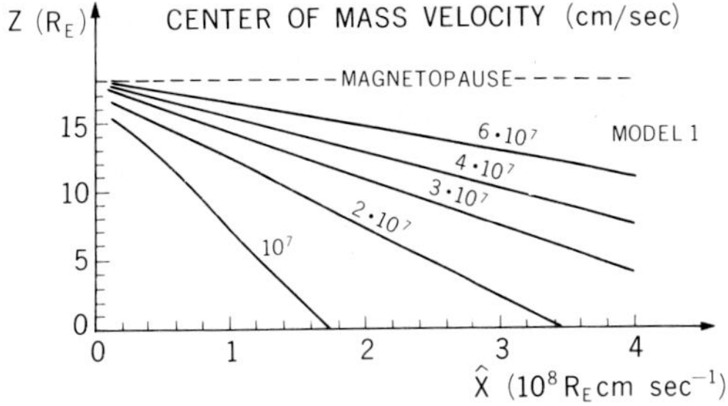

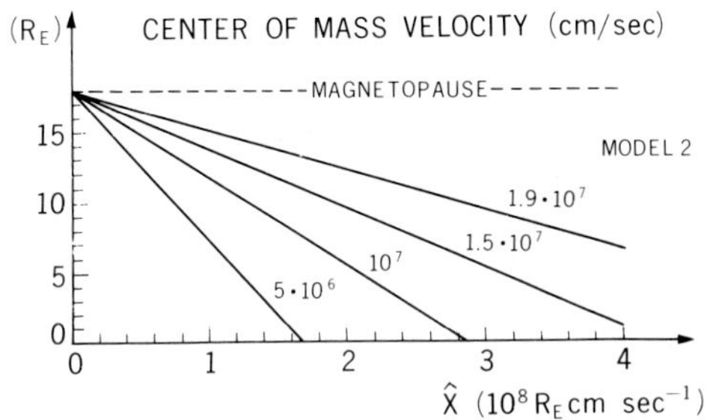

FIG. 24. Contours of the center of mass velocity of plasmas that entered the magnetotail and drifted toward the X-Y plane with drift velocity of 20 km/sec (PILIPP and MORFILL, 1978). $\hat{X}$ is product of X coordinate and drift speed toward the equatorial plane.

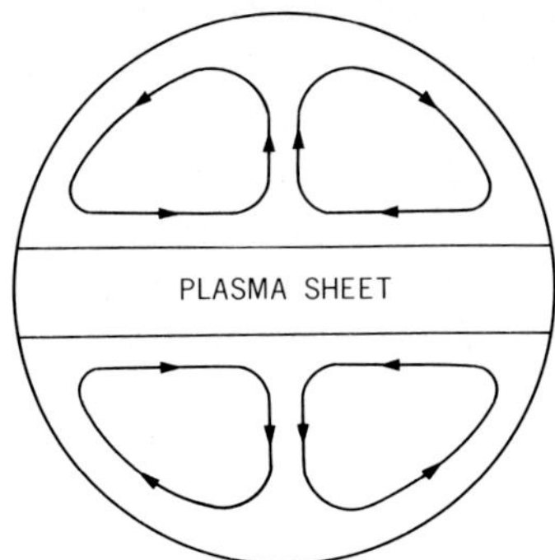

FIG. 25. Streamlines of the plasma convection generated in the tail in consequence of the dayside reconnection with northward interplanetary field lines (MAEZAWA, 1976).

earthward under this electric field, the reflection of the injected plasma, which leads to the trapping, is performed by an earthward moving wall. It is conceivable that ions can gain energies of several keV in this way, and furthermore plasma loaded on closed field lines is adiabatically heated as field lines shrink in the course of their earthward motion (COWLEY, 1980). The last closed field lines in northern and southern hemispheres meet at an X-type neutral line where reconnection proceeds in the presence of the dawn-to-dusk electric field. Acceleration and heating would occur also at this neutral line and the slow shock associated therewith. Earthward streaming particles observed in boundary regions of the plasma sheet may represent freshly trapped particles that have just been subjected to these processes.

In the foregoing model of the plasma injection into the plasma sheet, electrons are supposed to follow streamlines of ions in order to satisfy the condition of the charge neutrality. Since electrons are usually 2 or 3 times less energetic than protons in the magnetosheath, electrons probably have lower mean energy than protons in the plasma mantle too and it is likely that streamlines are essentially determined by the ion dynamics. Electrons from the mantle also need be heated in order that they attain energies of plasma sheet electrons.

Another conceivable way of the plasma supply into the plasma sheet is entry from the low-latitude boundary layer. Plasma having spectral characteristics similar to the solar wind in the magnetosheath has been found not only in the high-latitude plasma mantle but also in low-latitude boundary regions of the magnetotail (AKASOFU *et al.*, 1973). Drift parallel to the equatorial plane from this boundary layer can feed plasma laterally into the plasma sheet. The drift motion in this case can be of the ∇B-drift and curvature-drift type where positive ions drift from dusk to dawn while electrons drift from dawn to dusk. If magnetic field has strength of 20 γ and varies with scale length of $l\,R_E$, drift speed of a 3 keV particle is $30\,l^{-1}$ km/sec with which the particle can traverse diameter ($\sim 40\,R_E$) of the tail in about $10^4\,l$ sec. Since flow in the plasma sheet as observed at radial distances of $20 \sim 30\,R_E$ is directed more often earthward than anti-earthward, it has to be thought, according to this model, that beyond that distance there is a region where efficiency of the lateral entry is higher than it is earthward of $20 \sim 30\,R_E$. It may be suggested that at greater distances the neutral sheet becomes more sharply defined so that l is reduced and the drift speed is raised. Another possible process of entry of plasma from the low-latitude boundary layer into the plasma sheet is due to dayside reconnection of the Earth's magnetic field lines with interplanetary field lines when the latter have a northward polarity. It has been suggested on the basis of ground magnetic observations that plasma convection as depicted in Fig. 25 is generated in the magnetotail in consequence of such reconnection (MAEZAWA, 1976) and it has been noted that a similar flow pattern can be generated under more general circumstances (CROOKER, 1979). This convection is of the nature of the electric field drift.

The mechanism of the plasma-sheet formation represents an important missing link in our understanding of magnetospheric physics. It is hoped that the GTL (Geomagnetic Tail Laboratory) spacecraft that is being planned to be sent beyond the lunar orbit will provide information for advancing the proposed ideas on this important issue.

I gratefully acknowledge helpful comments given by Drs. W. J. Raitt and O. Ashihara to Sections 2, 3, and 4 and by Dr. T. Terasawa to Sections 5 and 6 of this Chapter.

REFERENCES

AKASOFU, S.-I., E. W. HONES, Jr., S. BAME, J. R. ASBRIDGE, and A. T. Y. LUI, Magnetotail and boundary layer plasma at a geocentric distance of ~ 18 R_E: Vela 5 and 6 observations, *J. Geophys. Res.*, **78**, 7257, 1973.

BAGENAL, F., J. D. SULLIVAN, and G. L. SISCOE, Spatial distribution of plasma in the Io torus, *Geophys. Res. Lett.*, **7**, 41, 1980.

BANKS, P. M. and T. E. HOLZER, High-latitude plasma transport: the polar wind, *J. Geophys. Res.*, **75**, 6317, 1969.

BANKS, P. M. and G. KOCKERTS, *Aeronomy*, Academic Press, New York, 1973.

BATES, D. R. and T. N. L. PATTERSON, Hydrogen atoms and ions in the thermosphere and exosphere, *Planet. Space Sci.*, **5**, 257, 1961.

BELCHER, J. W., C. K. GOERTZ, and H. S. BRIDGE, The low energy plasma in the Jovian magnetosphere, *Geophys. Res. Lett.* **7**, 17, 1980.

BEZRUKIKH, V. V. and K. I. GRINGAUZ, The hot zone in the outer plasmasphere of the Earth, *J. Atmos. Terr. Phys.*, **38**, 1085, 1976.

BRAGINSKII, S. I., Transport processes in a plasma, in *Reviews of Plasma Physics*, edited by M. A. Leontovich, Vol. 1, Consultants Bureau, New York, 1965.

BREIG, E. L. and J. H. HOFFMAN, Variation in ion composition at middle and low latitudes from Isis 2 satellite, *J. Geophys. Res.*, **80**, 2207, 1975.

BROADFOOT, A. L., M. J. S. BELTON, P. Z. TAKACS, B. R. SANDEL, D. E. SHEMANSKY, J. B. HOLBERG, J. M. AJELLO, S. K. ATREYA, T. M. DONAHUE, H. W. MOOS, J. L. BERTAUX, J. E. BLAMONT, D. F. STROBEL, J. C. MCCONNELL, A. DALGARNO, R. GOODY, and M. B. MCELROY, Extreme ultraviolet observations from Voyager 1 encounter with Jupiter, *Science*, **204**, 979, 1979.

COWLEY, S. W., Plasma populations in a simple open model magnetosphere, *Space Sci. Rev.*, **26**, 217, 1980.

CROOKER, N. U., Dayside merging and cusp geometry, *J. Geophys. Res.*, **84**, 951, 1979.

DeCOSTER, R. J. and L. A. FRANK, Observations pertaining to the dynamics of the plasma sheet, *J. Geophys. Res.*, **84**, 5099, 1979.

FÄLTHAMMAR, C-G., Effect of time-dependent electric fields on geomagnetically trapped radiation, *J. Geophys. Res.*, **70**, 2503, 1965.

FROIDEVAUX, L., Radial diffusion in Io's torus: Some implications from Voyager 1, *Geophys. Res. Lett.*, **7**, 33, 1980.

GOERTZ, C. K., Io's interaction with the plasma torus, *J. Geophys. Res.*, **85**, 2949, 1980.

GREBOWSKY, J. M., J. H. HOFFMAN, and N. C. MAYNARD, Ionospheric and magnetospheric "plasmapauses," *Planet. Space Sci.*, **26**, 651, 1978.

HAERENDEL, G., G. PASCHMANN, N. SCKOPKE, H. ROSENBAUER, and P. C. HEDGECOCK, The frontside boundary layer of the magnetosphere and the problem of reconnection, *J. Geophys. Res.*, **83**, 3195, 1978.

HARDY, D. A., H. K. HILLS, and J. W. FREEMAN, Occurrence of the lobe plasma at lunar distance, *J. Geophys. Res.*, **84**, 72, 1979.

HASEGAWA, A., Ballooning instability and plasma density limitation in Jovian magnetosphere, 1980.

HORWITZ, J. L. and C. R. CHAPPELL, Observations of warm plasma in the dayside trough at geosynchronous orbit, *J. Geophys. Res.*, **84**, 7075, 1979.

INTRILIGATOR, D. S., H. R. COLLARD, J. D. MIHALOV, O. L. VAISBERG, and J. H. WOLFE, Evidence for earth magnetospheric tail associated phenomena at 3,100 R_E, *Geophys. Res. Lett.*, **6**, 585, 1979.

IOANNIDIS, G. and N. BRICE, Plasma densities in the Jovian magnetosphere: Plasma slingshot or Maxwell demon?, *Icarus*, **14**, 360, 1971.

KRIMIGIS, S. M., T. P. ARMSTRONG, W. I. AXFORD, C. O. BOSTROM, C. Y. FAN, G. GLOECKLER, L. J. LANZEROTTI, E. P. KEATH, R. D. ZWICKL, J. F. CARBARY, and D. C. HAMILTON, Low-energy charged particle environment at Jupiter: a first look, *Science*, **204**, 998, 1980.

LEJEUNE, G. and F. WORMSER, Diffusion of photoelectrons along a field line inside the plasmasphere, *J. Geophys. Res.*, **81**, 2900, 1976.

LENNARTSSON, W., E. G. SHELLEY, R. D. SHARP, R. G. JOHNSON, and H. BALSIGER, Some initial ISEE-1 results on the ring current composition and dynamics during the magnetic storm of December 11, 1977, *Geophys. Res. Lett.*, **6**, 483, 1979.

LUI, A. T. Y., E. W. HONES, Jr., F. YASUHARA, S.-I. AKASOFU, and S. J. BAME, Magnetotail plasma flow during plasma sheet expansions: Vela 5 and 6 and Imp 6 observations, *J. Geophys. Res.*, **82**, 1235, 1977.

LUNDBLAD, J. Å. and F. SØRAAS, Proton observations supporting the ion cyclotron wave heating theory of SAR arc formation, *Planet. Space Sci.*, **26**, 245, 1978.

MACHIDA, S. and A. NISHIDA, Thermal structure of the Jovian plasmasphere, *Planet. Space Sci.*, **26**, 745, 1978.

MAEZAWA, K., Magnetospheric convection induced by the positive and negative Z components of the interplanetary magnetic field: Quantitative analysis using polar cap magnetic records, *J. Geophys. Res.*, **81**, 2289, 1976.

MANTAS, G. P., H. C. CARLSON, and V. B. WICKWAR, Photoelectron flux buildup in the plasmasphere, *J. Geophys. Res.*, **83**, 1, 1978.

MARUBASHI, K., Escape of the polar-ionospheric plasma into the magnetospheric tail, *Rep. Ionos. Space Res. Japan*, **24**, 322, 1970.

MARUBASHI, K., Effects of convection electric field on the thermal plasma flow between the ionosphere and the protonosphere, *Planet. Space Sci.*, **27**, 603, 1979.

McCOY, J. E., R. P. LIN, R. E. McGUIRE, L. M. CHASE, and K. A. ANDERSON, Magnetotail electric fields observed from lunar orbit, *J. Geophys. Res.*, **80**, 3217, 1975.

NESS, N. F., K. W. BEHANNON, R. P. LEPPING, and Y. C. WHANG, The magnetic field of Mercury, 1, *J. Geophys. Res.*, **80**, 2708, 1975.

ORAN, E. S. and D. J. STRICKLAND, Photoelectron flux in the Earth's ionosphere, *Planet. Space Sci.*, **26**, 1161, 1978.

PILIPP, W. G. and G. MORFILL, The formation of the plasma sheet resulting from plasma mantle dynamics, *J. Geophys. Res.*, **83**, 5670, 1978.

RAITT, W. J. and E. B. DORLING, The global morphology of light ions measured by the ESRO-4 satellite, *J. Atmos. Terrt. Phys.*, **38**, 1077, 1976.

RAITT, W. J., R. W. SCHUNK, and P. M. BANKS, A comparison of the temperature and density structure in high and low speed thermal ion flows, *Planet. Space Sci.*, **23**, 1103, 1975.

REES, M. H. and R. G. ROBLE, Observations and theory of the formation of stable auroral red arcs, *Rev. Geophys. Space Phys.*, **13**, 201, 1975.

RICH, F. J., R. C. SAGALYN, and P. J. L. WILDMAN, Electron temperature profiles measured up to 8,000 km by S3-3 in the late afternoon sector, *J. Geophys. Res.*, **84**, 1328, 1979.

RICHARDSON, J. D., G. L. SISCOE, F. BAGENAL, and J. D. SULLIVAN, Time dependent plasma injection by Io, *Geophys. Res. Lett.*, **7**, 37, 1980.

ROBLE, R. G., The calculated and observed diurnal variation of the ionosphere over Millstone Hill on 23–24 March 1970, *Planet. Space Sci.*, **23**, 1017, 1975.

SCHUNK, R. W., Transport equations for aeronomy, *Planet. Space Sci.*, **23**, 437, 1975.

SCHUNK, R. W. and D. S. WATKINS, Comparison of solutions to the thirteen-moment and standard transport equations for low speed thermal proton flows, *Planet. Space Sci.*, **27**, 433, 1979.

SCHUNK, R. W., W. J. RAITT, and A. F. NAGY, Effect of diffusion-thermal processes on the high-latitude topside ionosphere, *Planet. Space Sci.*, **26**, 189, 1978.

SHELLEY, E. G., Heavy ions in the magnetosphere, *Space Sci. Rev.*, **23**, 465, 1979.

SISCOE, G. L., Jovian plasmasphere, *J. Geophys. Res.*, **83**, 2118, 1978.

SPREITER, J. R., A. L. SUMMERS, and A. Y. ALKSNE, Hydromagnetic flow around the magnetosphere, *Planet. Space Sci.*, **14**, 223, 1966.

SWARTZ, W. E., R. W. REED, and T. R. MCDONOUGH, Photoelectron escape from the ionosphere of Jupiter, *J. Geophys. Res.*, **80**, 495, 1975.

TORR, D. G., M. R. TORR, H. C. BRINTON, L. H. BRACE, N. W. SPENCER, A. E. HEDIN, W. B. HANSON, J. H. HOFFMAN, A. O. NIER, J. C. G. WALKER, and D. W. RUSCH, An experimental and theoretical study of the mean diurnal variation of O^+, NO^+, O_2^+, and N_2^+ ions in the mid-latitude F_1 layer of the ionosphere, *J. Geophys. Res.*, **84**, 3360, 1979.

WARWICK, J. W., J. B. PEARCE, A. C. RIDDLE, J. K. ALEXANDER, M. D. DESCH, M. L. KAISER, J. R. THIEMAN, T. D. CARR, S. GULKIS, A. BOISHOT, C. C. HARVEY, and B. M. PEDERSEN, Voyager 1 planetary radio astronomy observations near Jupiter, *Science*, **204**, 995, 1979.

Chapter 2

INTERACTION OF THE SOLAR WIND WITH THE DAYSIDE MAGNETOSPHERE

G. HAERENDEL and G. PASCHMANN

*Max-Planck-Institut für Physik und Astrophysik, Institut für
Extraterrestrische Physik, Garching, West Germany*

The most obvious traces of the frontside interaction of the solar wind
with the earth's magnetic field are: the bow shock, the flow diversion around
the magnetospheric obstacle (magnetosheath), the sharp interface with the
earth's magnetic field (magnetopause), the polar cusps, the internal
boundary layer, the system of field-aligned currents between the latter and
the ionosphere, and the characteristic convection and current patterns set
up in the polar ionosphere.

Physical processes that play a major role in establishing these
phenomena are collisionless dissipation mechanisms in the shock,
collisionless reconnection of external and internal magnetic field lines,
anomalous Ohmic dissipation in the magnetopause current layer and the
accompanying diffusion, eddy convection in the boundary layer, Kelvin-
Helmholtz instability of the flow inside this layer, and some kind of
anomalous resistance between the magnetopause region and the ionosphere,
causing field-aligned potential drops and a certain degree of decoupling of
these two regions.

At the time of writing, much of this is in the state of exploration. We
have not yet developed the theoretical tools for a self-consistent description
of the dominant processes. The following presentation of the subject does
not intend to provide the reader with a collection of limited theoretical
results, but to survey the subject in a rather phenomenological way
supported by selected data sets and simple cartoons. We will, however, keep
an eye on the quantitative success of some of the theoretical concepts. The

most intriguing and least understood processes, for which we have just begun to obtain direct measurements, are probably reconnection and the mass transport to and inside the boundary layer. They will, therefore, receive a disproportionate coverage in this treatment. Although we will come to some conclusions constraining the choice of relevant processes, there will be no final answers. Much work has yet to be done.

1. Solar Wind Flow around the Magnetopause

As the highly conductive solar wind interacts with the earth's magnetic field, it creates a well defined sharp interface, the magnetopause, outside of which we find the solar wind plasma, and inside of which the magnetic field can be traced back to the earth. This does not imply that these field lines are totally confined to the magnetospheric cavity (so-called "closed" field lines). A small fraction of the magnetic flux on the frontside can be linked to the solar wind ("open" field lines) due to ongoing reconnection; and all of the polar cap field is commonly thought to extend through the tail to the solar wind far behind the earth. So the magnetopause does no necessarily separate "closed" internal field lines from those which are linked to the solar wind, but is rather a flow boundary. This statement needs specification as well. The magnetopause need not be impermeable to the solar wind flow (tangential discontinuity), but may have a small normal component and represent a rotational discontinuity. This would be the consequence of reconnection, which according to VASYLIUNAS (1975) can be generally defined as flow across a surface separating regions of different magnetic topology. On the frontside, the topology changes from "entirely confined within the solar wind" to "open field lines" in the above defined sense. In view of the "openness" of the magnetosphere it is somewhat surprising that experimentally one rarely has difficulties to define a magnetopause when combining the magnetic with the flow signatures. Exceptions exist in the polar cusp regions.

The magnetopause is a free boundary established by the solar wind flow around the earth's magnetic field. Its theoretical derivation is not a simple task, since reconnection, tangential stresses and mass transfer resulting therefrom should be included, as well as possibly some form of diffusion. In addition, the solar wind is being shocked before it reaches the magnetopause, and the flow behind the shock (magnetosheath) is highly turbulent. Position and shape of the shock depend in turn on position and shape of the yet undetermined magnetopause. Furthermore, the plasma pressure inside the magnetosphere is not at all negligible in comparison with the magnetic pressure, $B^2/8\pi$. In its full extent, the bow shock - magnetosheath flow - magnetopause - reconnection problem should be solved self-consistently. It is a non-stationary problem, even if the solar wind should remain constant for a while. The reason is twofold. Very likely the reconnection process is of transient nature, and the flow past the polar cusps cannot be regarded as laminar, but leads to vortex formation and separation (see Subsection 1.2 and Section 3). Another more trivial reason is that, due to the earth's rotation, the dipole axis is wobbling around the pole axis.

The theoretical efforts to describe shape and position of magnetopause and bow shock have neglected most of the complications mentioned. The main processes have been decoupled from each other as much as possible. In treating the magnetopause, the bow shock is normally completely ignored and a free molecular flow model is adopted for the impinging solar wind. The magnetopause is regarded as impermeable, and the internal gas pressure is set equal to zero. Once its shape has been established, either by the simplified theory, or phenomenologically from observations, one can treat the problem of the flow around the magnetosphere. Here, too, considerable simplifications have been introduced. The flow is taken as laminar, the pressure as isotropic. The dynamical role of the magnetic field, in particular any shear stresses arising from it, are neglected. A gas dynamic analogue is treated which contains only implicitly the magnetic field in the scalar pressure. After this problem has been solved, one proceeds to work out the distribution of the magnetic field by considering it as frozen-in and neglecting its dynamic role. A further refinement of the flow problem is the consideration of the density depletion near the sub-solar stagnation point which arises from the stresses exerted by the magnetic field when draped around the magnetopause.

In the following subsections we shall summarize the results obtained by the various simplifying steps with some quick glances at the quality of the agreement with observations. Finally, we consider the microstructure of the magnetopause.

1.1   Shape of the magnetopause

The free boundary problem of the magnetosphere has been solved by assuming (a) a very simple pressure law on the outside, (b) zero gas pressure on the inside and (c) zero magnetic field on the outside (e.g. SLUTZ, 1962; MIDGLEY and DAVIS, 1963). Only recently, bow shock and magnetopause have been determined self-consistently by numerical solution of the initial value problem (WU *et al.*, 1980). The older approach consisted essentially in searching for a surface which results in zero magnetic field outside, whereby the field sources are the earth's dipole and the magnetopause currents. According to Ampère's law they are related to the normal vector, n_{MP}, of the magnetopause by:

$$j = \frac{c}{4\pi}\, n_{\mathrm{MP}} \times B .\tag{1}$$

Starting with a trial field and surface one can calculate by Biot-Savart's integral the contribution, B_{sc}, of the surface currents at the magnetopause to the total field, B:

$$B = B_{\mathrm{D}} + B_{\mathrm{sc}} .\tag{2}$$

This field is further subject to the condition of pressure balance:

$$\frac{B_{\mathrm{MP}}^2}{8\pi} = p_{\mathrm{MP}} .\tag{3}$$

B_{MP} and p_{MP} are magnetic field and pressure on the respective sides of the magnetopause. The most popular approach to describing the pressure is to take the Newtonian flow formula:

$$p = p_{\mathrm{st}} \cdot \cos^2 \psi \tag{4}$$

where p_{st} is the pressure at the stagnation point given by:

$$p_{\mathrm{st}} = K\rho_\infty v_\infty^2 \tag{5}$$

and ψ is the angle between boundary normal, n_{MP}, and velocity vector at infinity, v_∞.

A trial surface and magnetic field will in general not be consistent with the pressure balance. But Eqs. (1) and (2) lend themselves to an iterative improvement of B_{sc} by varying the shape of the surface until pressure balance is achieved (BEARD and CHOE, 1974).

The remaining question concerns the quality of the Newtonian pressure formula which stems from a free molecular flow model of the solar wind plasma without allowance for a finite temperature nor a deflection by the bow shock. If the ions, which carry almost all the momentum, were elastically reflected, K in Eq. (5) should have the value $K=2$; for inelastic reflection $1 < K < 2$. SPREITER *et al.* (1966) compared the gas dynamic result for the stagnation pressure in front of a blunt object in a supersonic stream with the upstream dynamic pressure, $\rho_\infty v_\infty^2$. For $\gamma = 5/3$ it is:

$$K = \frac{p_{st}}{\rho_\infty v_\infty^2} = \frac{256}{225}\left(\frac{3}{5}\right)^{1/2}\left(\frac{M_\infty^2}{M_\infty^2 - 1/5}\right)^{3/2}. \tag{6}$$

With upstream Mach numbers $M_\infty^2 \gg 1$ this value approaches 0.881, slightly less than that predicted by the Newtonian model for completely inelastic interaction. A further problem is the validity of the Eq. (4) for the pressure variation along the surface. Comparison for a trial surface showed that the fit is very good up to $\psi \approx 50°$. With higher angles the pressure predicted by the Newtonian formula may be off by a factor of 2. Since the magnetic pressure varies essentially as the inverse sixth power of the geocentric distance, r, the resulting corrections for the surface are small. In conclusion, one can get very reasonable models for the shape of the magnetopause by this simple approach.

The stand-off distance of the stagnation point can be calculated easily from Eqs. (3) and (5), if one realizes that $B_{MP} = 2\,B_D$, since the contribution of the surface currents must be approximately such that the unperturbed dipole field is cancelled outside the magnetopause (this amounts to a neglect of the magnetic tensions of the internal field). Since $B_D = \mathcal{M}/r^3$, we get

$$r_{st} = \left(\frac{\mathcal{M}^2}{K2\pi\rho_\infty v_\infty^2}\right)^{1/6} \tag{7}$$

where $\mathcal{M}$ is the earth's dipole moment.

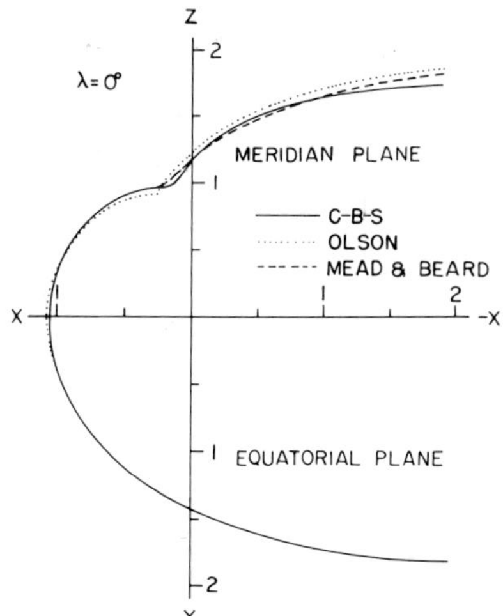

FIG. 1. Meridian plane and equatorial plane cross-sections of the magnetosphere for perpendicular incidence (from BEARD and CHOE, 1974).

This relation allows a dimensionless scaling of magnetopause models for varying values of M_∞, p_∞, and v_∞. Figure 1 shows the trace of the magnetopause in meridian and equatorial plane from three different model calculations (BEARD and CHOE, 1974). Here it is assumed that $\mathcal{M}$ is perpendicular to the sun-earth line. In reality, this angle can vary by as much as $34°$ in either direction. The shape varies correspondingly.

1.2   Flow around the magnetopause

Theoretical treatments of the flow around the magnetosphere are almost exclusively based on the assumption that the solar wind can be adequately described by the continuum equations for a perfect gas. Furthermore, except for recent numerical MHD models (WU et al., 1980), the shape of the magnetopause was taken as given, e.g. by the approach described in the preceding subsection. The fundamental MHD equations, in conservation form for mass, momentum, and entropy, together with the relevant Maxwell's equations, are:

$$\frac{\partial \rho}{\partial t} = -\mathbf{\nabla} \cdot (\rho \mathbf{v}) \tag{8a}$$

$$\frac{\partial (\rho \mathbf{v})}{\partial t} = -\mathbf{\nabla} \cdot \left\{ \rho \mathbf{v v} + \mathbf{I} \left(p + \frac{B^2}{8\pi} \right) - \frac{\mathbf{B B}}{4\pi} \right\} \tag{8b}$$

$$\frac{\partial (\rho s)}{\partial t} = -\mathbf{\nabla} \cdot (\rho s \mathbf{v}) \tag{8c}$$

$$\frac{\partial \mathbf{B}}{\partial t} = \mathbf{\nabla} \times (\mathbf{v} \times \mathbf{B}) \tag{8d}$$

(Throughout this chapter electrostatic units are used.)

An infinitely conducting, ideal gas is assumed, with γ being the ratio of specific heats. All symbols have the conventional meaning. $\mathbf{I}$ is the unit tensor. The entropy s is given by:

$$s = s_0 + \ln \left\{ \frac{p/p_0}{(\rho/\rho_0)^\gamma} \right\}. \tag{8f}$$

The bow shock and magnetopause are regarded as ideal discontinuities and are treated by solving the following jump relations and matching the continuous fluid solutions on either side. In a reference system attached to the discontinuities, the jump relations are:

$$[\rho v_{\mathrm{n}}] = 0 \tag{9a}$$

$$\left[\rho v v_{\mathrm{n}}+\left(p+\frac{B^2}{8\pi}\right)\boldsymbol{n}-\frac{1}{4\pi}\boldsymbol{B}_{\mathrm{t}}B_{\mathrm{n}}\right]=0 \tag{9b}$$

$$\left[\left(\frac{\rho}{2}v^2+h+\frac{B^2}{8\pi}\right)v_{\mathrm{n}}-\frac{\boldsymbol{v}\cdot\boldsymbol{B}}{4\pi}\right]=0 \tag{9c}$$

$$[B_{\mathrm{n}}\boldsymbol{v}_{\mathrm{t}}-\boldsymbol{B}_{\mathrm{t}}v_{\mathrm{n}}]=0 \tag{9d}$$

$$[B_{\mathrm{n}}]=0 \tag{9e}$$

Square brackets denote the difference of the respective quantity on either side of the discontinuity. The indexes "n" and "t" refer to the components normal and tangential to the discontinuity. Since the entropy is not necessarily conserved, the conservation of energy (Eq. (9c)) is introduced instead. $h=[\gamma/(\gamma-1)]p$ is the specific enthalpy.

The use of fluid equations for the collisionless gas, in particular the consideration of discontinuites, is justified by the presence of a magnetic field which decreases the basic interaction lengths to something of the order of the gyroradii, depending on the situation. Plasma waves excited by anisotropy driven instabilities will have the tendency to restore a near isotropic pressure tensor whenever this has been grossly perturbed, which occurs, for example, when plasma moves across a discontinuity. Therefore, for modelling the flow around the magnetopause an assumption of $\gamma=5/3$ appears to be the best possible one within the limitations of the one-fluid approach.

SPREITER *et al.* (1966) have simplified the task of calculating the flow around the magnetosphere considerably by dropping all terms containing the magnetic field, $\boldsymbol{B}$, from the fluid equations and solving Maxwell's equations separately. This approach is justified by the observation that the flow velocities, except near the stagnation point, are largely super-Alfvénic, i.e. the Alfvénic Mach-number:

$$M_{\mathrm{A}}=\frac{v}{v_{\mathrm{A}}}>1 \tag{10}$$

with $v_{\mathrm{A}}=B(4\pi\rho)^{-1/2}$. A typical value for v_{A} with $B=10\gamma$ and $n=10\,\mathrm{cm}^{-3}$ is 70 km/sec, whereas even behind the bow shock the plasma flow velocity is typically $>100\,\mathrm{km/sec}$. The magnetic stresses are related to the inertial term in the equation of momentum conservation by $M_{\mathrm{A}}^{-2}\ll1$. The stationary flow problem has thus been treated by solving the following gas dynamic set of equations:

$$\boldsymbol{\nabla}\cdot(\rho\boldsymbol{v})=0 \tag{11a}$$

$$\rho(\boldsymbol{v}\cdot\boldsymbol{V})\boldsymbol{v}+\boldsymbol{V}p=0 \tag{11b}$$

$$(\boldsymbol{v}\cdot\boldsymbol{V})s=0 \tag{11c}$$

with s given by Eq. (8f).

Maxwell's equations are solved separately:

$$\boldsymbol{V}\times(\boldsymbol{v}\times\boldsymbol{B})=0 \tag{12a}$$

$$\boldsymbol{V}\cdot\boldsymbol{B}=0 \tag{12b}$$

Once $\boldsymbol{v}$ has been found from Eqs. (11), the frozen-in magnetic field which is passively transported by the fluid can be calculated from Eqs. (12). It may so happen, in particular near the stagnation point, that the thus calculated $\boldsymbol{B}$ yields magnetic stresses which even exceed the gas pressure and the inertial terms ($M_A^2 < 1$). In this case, the gas dynamic approximation breaks down.

Another simplification in solving the gas dynamic Eqs. (11) is to adopt cylindrical symmetry for the magnetopause. The method of calculations is described in detail by SPREITER et al. (1966). It uses an iterative procedure of finding the shape of the shock. In this model, the shock is a normal hydrodynamic one. Its position and shape probably do not depend very much on the simplifications, as long as $M_A^2 \gg 1$ and the value of γ is chosen appropriately.

A few results are presented in Figs. 2 through 4. The first figure shows the streamlines for a Mach-number $M_A = 8$ and $\gamma = 5/3$. In addition, characteristic or Mach lines are shown (by dashed lines) which can be thought of as standing compression or expansion waves of infinitesimal

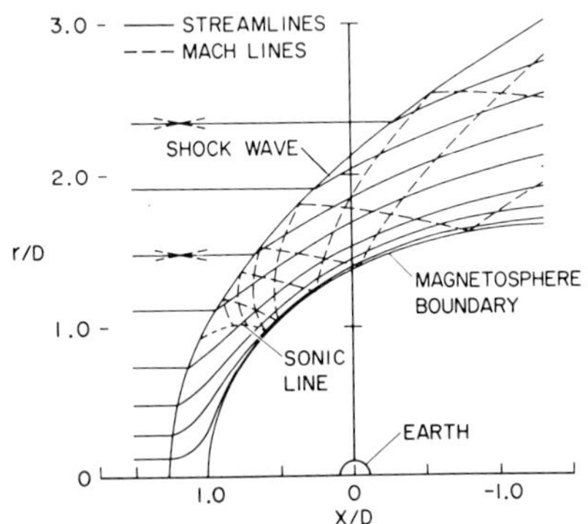

FIG. 2. Streamlines and wave patterns for supersonic flow past the magnetosphere; $M_\infty = 8$, $\gamma = 5/3$ (from SPREITER et al., 1966).

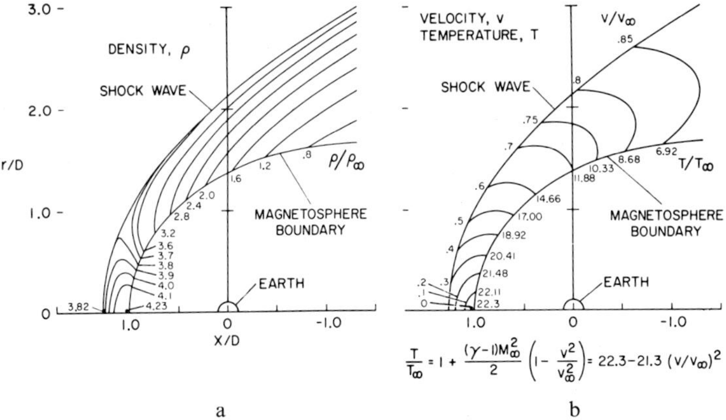

FIG. 3.   Density, velocity and temperature fields for supersonic flow past the magnetosphere, $M_\infty = 8$, $\gamma = 5/3$ (from SPREITER *et al.*, 1966).

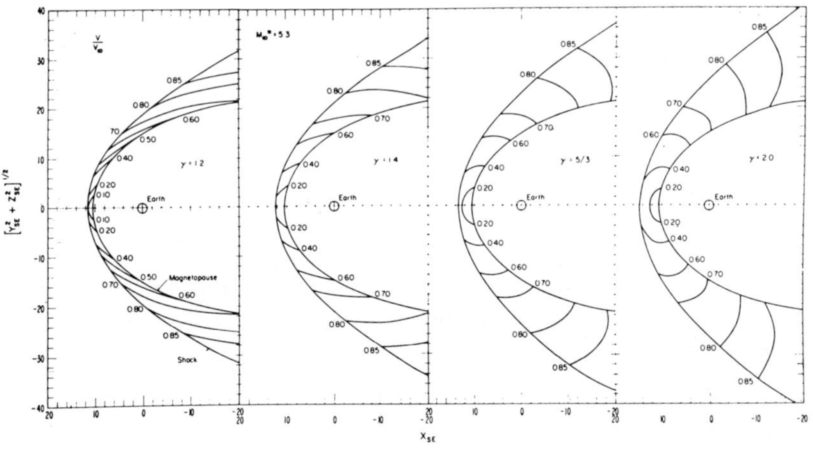

FIG. 4.   Velocity distributions for $M_\infty = 5.3$, and $\gamma = 1.2$; 1.4; 5/3; 2.0 (from DRYER, 1970).

amplitude crossing the streamlines at such an angle that the flow velocity component perpendicular to them equals the local velocity of sound. Hence, Mach lines exist only where the flow is supersonic. Figures 3a, b contain the variation of density, flow velocity and temperature through the magnetosheath in terms of their upstream values for the same choice of M_∞ and γ. Temperature and velocity are related by:

$$\frac{T}{T_\infty} = 1 + \frac{(\gamma-1)M_\infty^2}{2} \cdot \left(1 - \frac{v^2}{v_\infty^2}\right). \tag{13}$$

The density stays close to its maximum value $(\gamma+1)/(\gamma-1)=4$ throughout the subsonic region. Near the stagnation point, the compression is somewhat increased. The dependence of the solutions on the choice of γ is demonstrated in Fig. 4 (from DRYER, 1970). One can notice a great sensitivity to the value of γ, which may be exploited to assess the quality of the gasdynamic model by future data analyses.

The stand-off distance, r_{s}, of the bow shock is given in terms of the distance, r_{st}, of the stagnation point (Eq. (7)) by (SPREITER *et al.*, 1966):

$$\frac{r_{\mathrm{s}}}{r_{\mathrm{st}}} = 1 + 1.1 \frac{\rho_{\infty}}{\rho_{\mathrm{b}}} \tag{14}$$

where the ratio of densities on either side of the nose of the shock is:

$$\frac{\rho_{\infty}}{\rho_{\mathrm{b}}} = \frac{(\gamma-1)M_{\infty}^2 + 2}{(\gamma+1)M_{\infty}^2}. \tag{15}$$

The lower the Mach-number, M_{∞}, the smaller is the increase in density and the larger the stand-off distance of the shock.

The preceding equations provide a simple means by which a dimensionless comparison of actually observed flow parameters in the magnetosheath with the gas dynamic models can be carried out. This has yet to be done. One of the results would be an indication of the best choice of γ (compare Fig. 4). A collection of data defining the average locations of bow shock and magnetopause but uncorrected for the varying solar wind conditions has been presented by FAIRFIELD (1971) (Fig. 5). They confirm qualitatively the validity of the described models, but do not allow more detailed conclusions.

A very interesting question is posed by the flow past the cusp regions. In the simple closed models of the magnetosphere, due to div $\boldsymbol{B}=0$, there are two singular points where the internal magnetic field vanishes. For a vacuum magnetosphere, the normal external pressure must be zero, as well. Therefore, the normal vector of the magnetopause should be at right angles to the flow velocity (Eq. (4)). Immediately beyond, the angle must assume a value much smaller than 90°. This is the reason for the formation of a cusp-like indentation of the magnetopause (see Fig. 1). In reality, of course, the external pressure cannot drop to zero, nor can the interior of the magnetopause be considered as a vacuum (see Section 3). Field lines may even be "open", i.e. link the magnetosheath with the interior magnetosphere. The question remains what the best representation of the flow past these cusp regions would be.

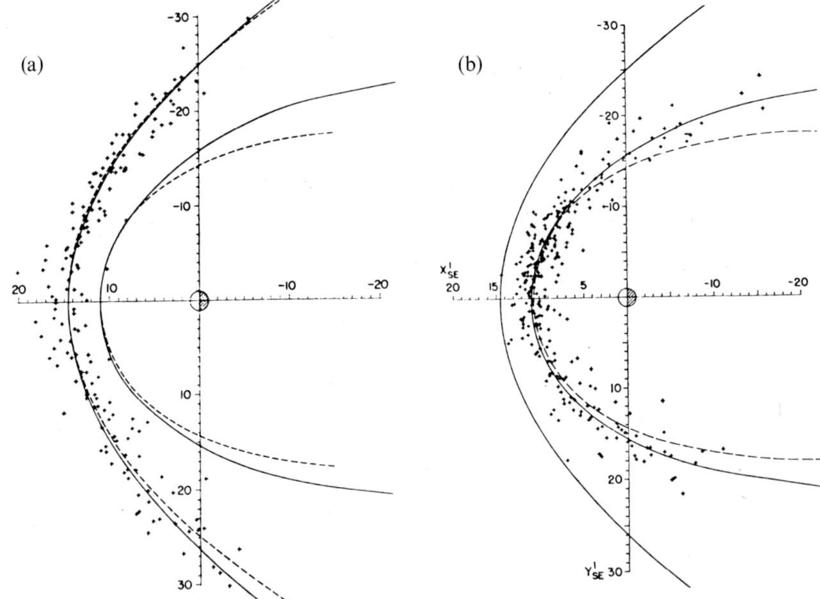

FIG. 5. (a) Position of the bow shock in the solar ecliptic plane as determined by measurements on five IMP spacecraft, 1963–1968. Crosses represent the average location on individual passes, and the solid line hyperbola represents the best-fit curve to the points. $|Z_{SE}| < 7$ R_E (after FAIRFIELD, 1971). (b) Position of the magnetopause in the solar ecliptic plane. Crosses represent the average location on individual passes, and the solid line ellipse represents the best-fit curve to the points. $|Z_{SE}| < 7$ R_E (after FAIRFIELD, 1971).

WALTERS (1966) proposed the existence of attached shocks which enables the already supersonic flow to adjust to the grossly concave boundary. This model was based on an analogy with gasdynamic flows past a similarly shaped solid obstacle as shown in Fig. 6. Spreiter and co-workers criticized the proposal and argued that the magnetopause is a free surface which must adjust to pressure jumps as would be caused by an attached shock. However, Maxwell's equations do not permit solutions with a balancing jump of the internal magnetic pressure (see SPREITER *et al.*, 1968). They suggested instead the existence of another flow boundary separating plasma trapped in the exterior cusp region from the free flow in the magnetosheath as shown in Fig. 7. We shall present observations which pertain to this question in Subsection 3.1, but mention already here that the effective viscosity provided by the magnetic field does not permit a strict

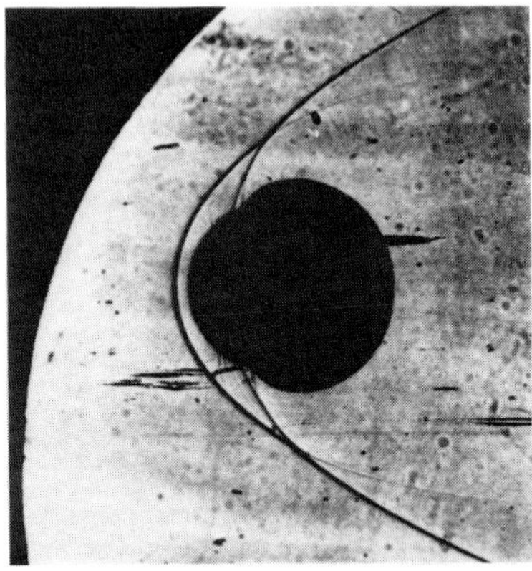

Fig. 6. Focussed shadowgraph of model of magnetosphere in flight at Mach-number 5.03 through argon ($\gamma = 5/3$). Meridian plane $P_s/P_{st} = 1$ (from SPREITER *et al.*, 1968).

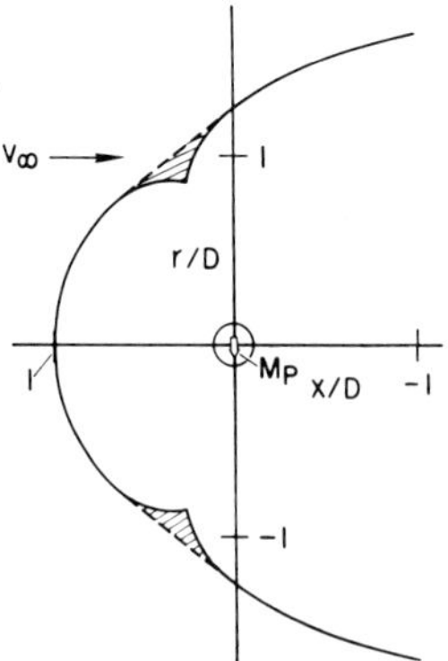

Fig. 7. Separation of free streaming from trapped magnetosheath plasma (from SPREITER *et al.*, 1968).

separation of a free magnetosheath flow from a plasma trapped in the exterior cusp region. Formation and separation of vortices will undoubtedly occur (HAERENDEL, 1978).

The draping of magnetic field lines around the magnetosphere which can be claculated with Eqs. (12) using the flow field from a gasdynamic solution is a three-dimensional problem and is not easy to visualize. Figures can be found in SPREITER *et al.* (1968). Near the stagnation point the shear stresses can become so strong as to affect the shape of the magnetosphere. No attempt has yet been made to cope with the full MHD problem in a self-consistent way.

1.3 *Plasma depletion near the stagnation point*

The gasdynamic model of the flow around the magnetopause becomes inaccurate near the stagnation point, since the Alfvénic Mach-number is no longer large compared with unity and the magnetic stresses cannot be neglected. A consequence of these stresses is a density depletion set up near the stagnation point, for two reasons. The deflection of the flow by the bow shock leads to an overall divergence of the flow with respect to a given flux tube and, secondly, the compressional stress exerted by the magnetic field tends to squeeze the plasma out along field lines. The process has been investigated by ZWAN and WOLF (1976), from whose paper the sketch of the depletion process is shown in Fig. 8 for a magnetic flux tube perpendicular to the solar wind flow. For this orientation the effect is greatest.

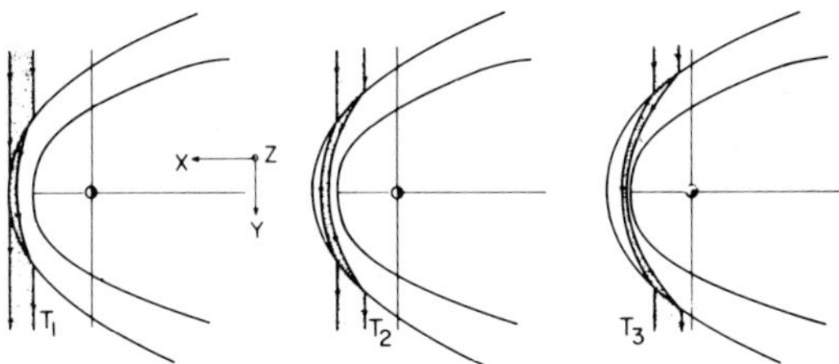

FIG. 8. Idealized sketches of the squeezing process at successive times T_1, T_2, and T_3. As plasma convects around the magnetosphere, flux tubes become draped around the nose of the magnetosphere (from ZWAN and WOLF, 1976).

ZWAN and WOLF (1976) solved the depletion problem by separating it into the plasma dynamics along a convected flux tube and the motion of a flux tube in the symmetry plane transverse to B. Conservation of mass and momentum along B provide differential equations for the flow speed parallel to B and for the cross-section of a given flux tube. The total pressure, $p_{tot} = p_{gas} + B^2/8\pi$, was taken from the gasdynamic model of SPREITER *et al.* (1966), and the magnetic pressure was subtracted from p_{tot} in accordance with the compression of the cross-section. The details of the calculations cannot be described in this context. However, a simple estimate yields an expression for the thickness, D, of the *depletion layer*, i.e. the region where ρ drops below 50% of the gasdynamic value:

$$D \cong 1.24 \, l \, M_A^{-2} \, . \tag{16}$$

l is the length along the magnetopause over which p_{tot} decreases by a factor of

2. For a geocentric distance of the stagnation point of 10 R_E, l turns out to be about 11 R_E. With an upstream Alfvénic Mach-number of 8 we get $D = 0.2$ R_E.

CROOKER *et al.* (1979) have searched for the depletion signature using IMP 6 plasma measurements. Figure 9 shows the expectations on the density variation between bow shock and magnetopause (a) along the stagnation streamline and (b) along a parallel line displaced by 10 R_E, according to the gasdynamic model (solid lines) and to the depletion theory of ZWAN and WOLF (1976) (dashed lines). The predictions of the gasdynamic model can be checked against Fig. 3. The observed densities drop typically by about 40% towards the magnetopause within distances of 0.2–0.4 R_E.

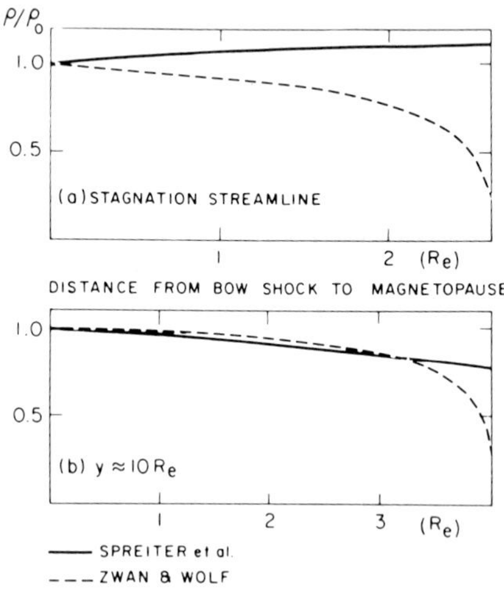

FIG. 9. Model density variations between the bow shock and the magnetopause, normalized to the value just inside the bow shock. The solid curves are taken from density contour maps of SPREITER *et al.* (1966). The dashed curves are calculated from model parameters of ZWAN and WOLF (1976). Variations are (a) along the stagnation streamline and (b) at a distance $y \approx 10$ R_E measured from the stagnation streamline along magnetic flux tubes (from CROOKER *et al.*, 1979).

One of the consequences of this depletion is the set up of a pressure anisotropy with $p_\perp/p_\parallel$ increasing. This has actually been confirmed. Figure 10 shows traces of n, B and $p_\perp/p_\parallel$ for a magnetopause crossing at 7 R_E from the stagnation streamline, exhibiting all predicted effects very clearly. Enhancements of $p_\perp/p_\parallel$ may trigger plasma waves by the mirror instability. Another possible consequence of the depletion may be a change of the

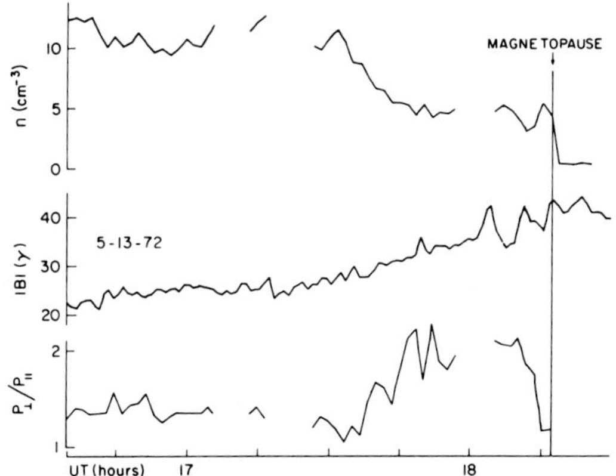

FIG. 10.   Average density n, magnetic field magnitude $|B|$, and pressure anisotropy ratio $p_\perp/p_\parallel$ variations along an IMP6 orbit that crossed the magnetopause nearest to the stagnation stream-line ($D \approx 7\ R_\mathrm{E}$). The vertical dashed lines mark the $180°$ phase difference between the peaks and troughs of the density and field waveforms (from CROOKER *et al.*, 1979).

reconnection rate at the stagnation point, since v_A is strongly affected and the reconnection rate seems to scale with v_A according to magnetohydrodynamic theories (see Subsection 5.1).

1.4   Structure of the magnetopause

At the magnetopause there is a balance between the total internal and external pressures. To the extent that the magnetopause can be described as a plane tangential discontinuity, i.e. with little or no normal component of the flow, this balance reads:

$$\frac{B_\mathrm{i}^2}{8\pi} + p_\mathrm{i} = \frac{B_0^2}{8\pi} + p_0 . \tag{17}$$

Before we turn to the microstructure and macroscopic thickness of this boundary layer, we want to warn the reader not to believe that necessarily $p_\mathrm{i} \ll p_0$ and $B_\mathrm{i}^2 \gg B_0^2$. Recently, rather complete plasma and field measurements have become available allowing a check of the pressure balance at the magnetopause. Figures 11 and 12 are two examples from the vicinity of the subsolar point and near the dawn flank, respectively. Whereas the pressure behaves normally in the former exmple, we have the reversed situation at the flank; the external magnetic pressure is substantially higher than the internal one. This not atypical example is the consequence of the

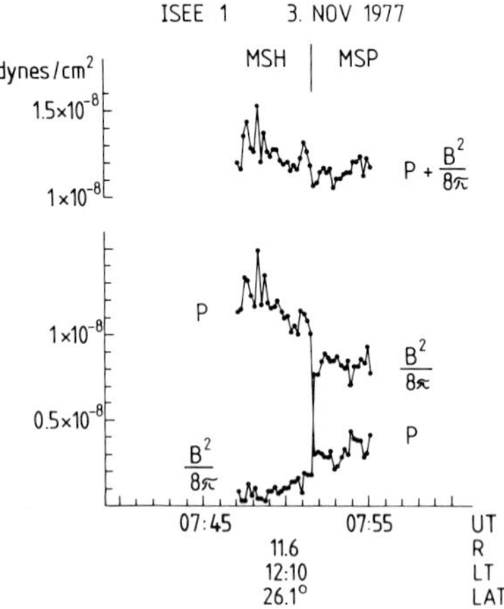

FIG. 11.   Pressure balance across the dayside magnetopause, showing the expected predominance of plasma pressure P and magnetic pressure $B^2/8\pi$ on the magnetosheath (left) and magnetosphere (right) sides, respectively. (MPE/UCLA)

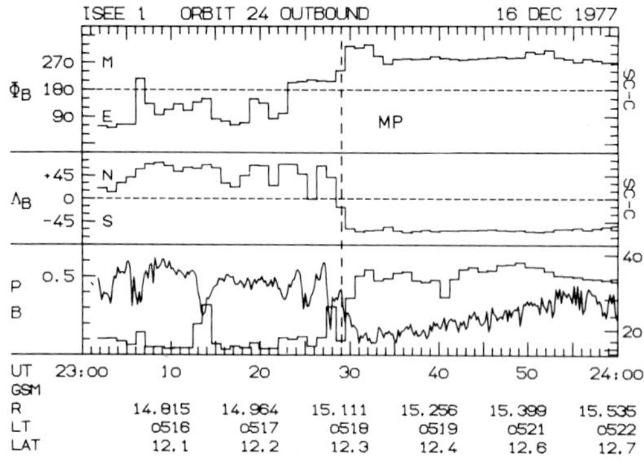

FIG. 12.   Pressure balance across a dawn magnetopause. Note the inverted role of plasma and field pressures: P dominates in the magnetosphere (left), while $B^2/8\pi$ dominates in the magnetosheath. (MPE/UCLA)

presence of the hot plasma sheet population extending along the flanks of the magnetopause. The magnetic signature of such "inverse" pressure equilibria has already been pointed out by HEPPNER *et al.* (1967). In both cases shown, the total pressure is continuous. It is not unusual to even encounter pressure peaks at the magnetopause which are a manifestation of the macroscopic dynamics of this surface (see Section 3.2).

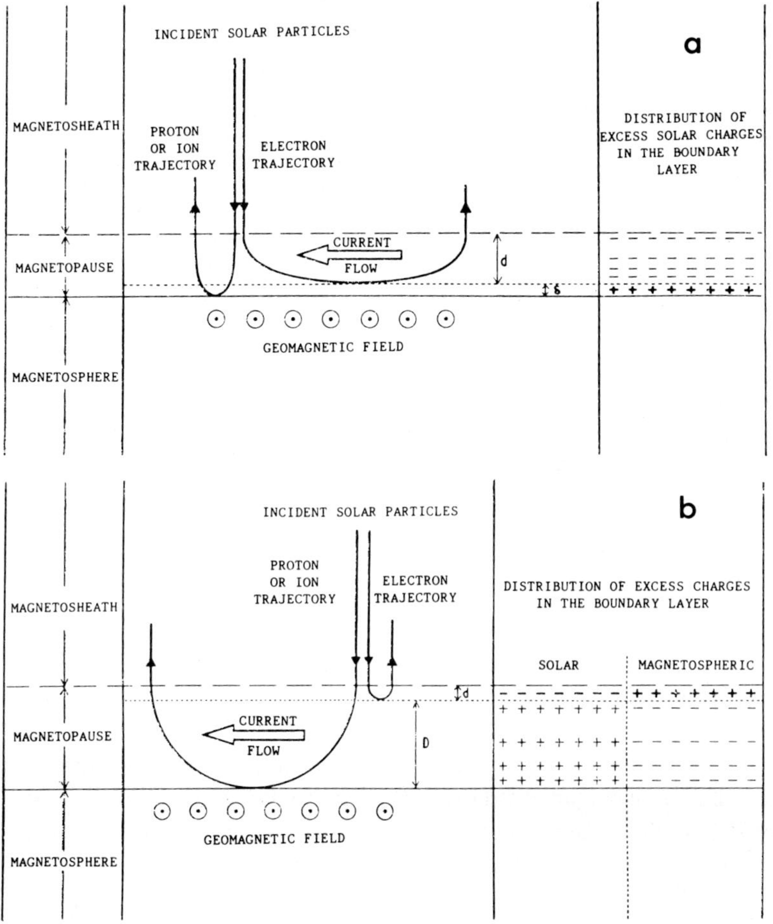

FIG. 13. (a) A schematic illustration of the trajectories of magnetosheath ions and electrons incident normally on a plane boundary layer when the polarization electric field due to charge separation is present; $d \sim 1\,\text{km}$, $\delta \sim 1\,\text{m}$. (b) A schematic illustration of the trajectories of magnetosheath ions and electrons incident normally on a plane boundary layer when the polarization electric field is completely neutralized by ambient magnetospheric charged particles; $D \sim 100\,\text{km}$, $d \sim 1\,\text{km}$ (from WILLIS, 1975).

The ideal magnetopause is an example of plasma-field boundaries which have been studied extensively, e.g. by FERRARO (1952), DUNGEY (1958), GRAD (1961) and ROSENBLUTH (1963). An oversimplified microscopic picture of the interface is given in Figs. 13a, b (WILLIS, 1975). The two charged particle species, ions and electrons, impinge at right angles on the boundary. The external space is considered as magnetic field-free. Because of their heavier mass, the ions tend to penetrate more deeply into the magnetic field than the electrons. This sets up a charge separation and outward pointing electric polarization field thus restraining the ions (Fig. 13a). Before they can be deflected by the magnetic field, they are returned by this polarization field with very little actual charge separation needed ($\delta n/n \approx v^2/c^2$). The electrons, however, experience the Lorentz force and gain energy in the polarization field. The transverse velocity component of the electrons accounts essentially for the electric current in the interface, which in case of the magnetopause is usually referred to as Chapman-Ferraro current. In the artificial case of equal external velocities, v_0, of the incident ions and electrons, the combined penetration depth, δ, is of the order of the geometric mean of the electron and ion gyroradii. If one uses $p = 2n(m_i + m_e)v_0^2$ in the pressure balance Eq. (17) and replaces v_0/B by the gyro-radius, one finds a thickness of the order of the plasma skin depth:

$$\delta = \frac{1}{\sqrt{2}} \frac{c}{\omega_{pe}} \tag{18}$$

where $\omega_{pe} = (4\pi e^2 n/m_e)^{1/2}$ is the plasma frequency. $c/\omega_{pe} = 5.3 \ \mathrm{km \cdot n}^{-1/2}$. Hence the minimum thickness of the magnetopause would be of the order of 1 km.

The situation changes if there is enough internal plasma present as to neutralize the incident ions (Fig. 13b). Any polarization field may become short-circuited by currents directed along the internal field lines and closing through the ionosphere (PARKER, 1967). In such a case, the ions penetrate unimpededly, and the thickness of the current layer would be of the order of the ion gyroradius or about 100 km. The closure of the short-circuiting currents poses, however, a problem when the solar wind parameters are changing faster than the polarization field propagates to the ionosphere. In addition, plasma turbulence driven by the same currents may limit the parallel electrical conductivity and reduce the depolarization. Furthermore, transverse diffusion of external plasma caused by turbulence in the magnetopause may scatter particles onto "trapped" orbits within the magnetopause and thus broaden the layer. Hence, the ion gyroradius, R_{gi}, is by no means an upper limit to the magnetopause. We shall return to the subject of microturbulence in the magnetopause in Section 5. Observations indicate that its actual width lies in the range of several 100 to $\approx 1{,}000 \ \mathrm{km}$

(RUSSELL and ELPHIC, 1978).

PARKER (1976) pointed out an interesting problem that arises when the external plasma has a substantial flow component parallel to the magnetic field. If there is enough cold internal plasma background as to neutralize the more deeply penetrating ions, we have the situation sketched in Fig. 14.

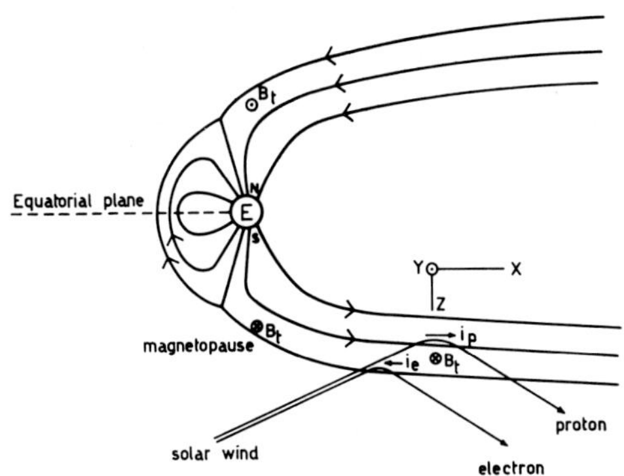

FIG. 14.   Schematic representation of the boundary between the streaming solar-wind plasma and the geomagnetic field. The currents i_p and i_e result from the different penetration depths into the magnetic field of the streaming protons and electrons. These currents induce a magnetic field along the y direction. This transverse field B_t has opposite directions above and below the equatorial plane (from SU and SONNERUP, 1971).

Electric currents are set up parallel to B, a thin layer of electron current $(O(c/\omega_{pe}))$ and a thick layer of ion current $(O(R_{gi}))$ opposing each other. These currents generate a transverse magnetic field between them whose magnitude can exceed that of the primary field when the flow velocity is supersonic. In this case, it is difficult to see how this new magnetic field could be confined by the external plasma. Hence Parker concluded that the magnetopause cannot be in an equilibrium state. However, this would exist only on a small scale. The discussed effect would probably generate a certain roughness of the surface field and lead to enhanced mixing with the external plasma, thus increasing its tangential drag on the field.

Several authors have pursued further this intriguing effect. SU and SONNERUP (1971) found that stabilization of the induced magnetic field is possible when the electron pressure exceeds a certain critical limit. However, this condition may not always be fulfilled in real situations. STOREY and CAIRO (1979) studied the problem for certain closed geometries and found a general stability condition which, unfortunately, is not met by a magnetosphere imbedded in a stellar wind. There may, however, be other mechanisms that relieve the situation. Trapped energetic particles, for instance, may alter the current distribution at the magnetopause completely and even suppress the transverse magnetic field (ALPERS, 1969). Another serious change of the idealized model would be introduced by the presence of strong plasma turbulence excited by the discussed currents.

SONNERUP and LEDLEY (1979) point out that a frequently observed rotation of the tangential magnetic field on the magnetospheric edge of the magnetopause current layer could be the manifestation of Parker's transverse field. An example will be shown in Section 4 (Fig. 54). If this is a common feature, its dynamical effect certainly needs further exploration.

2. The Bow Shock

2.1 Collisionless shocks

Because the solar wind approaches the magnetosphere at supersonic speed, a shock wave is formed which enables the flow to be deflected around the magnetospheric obstacle. As discussed in the previous section, the macroscopic aspects of the bow shock are reasonably well described in the gasdynamic analogy, in spite of the fact that the solar wind is not a collision-dominated gas but a very tenuous plasma (density $\sim 5\,\mathrm{cm}^{-3}$). Near the earth's orbit, ordinary binary (Coulomb) collisions among solar wind particles become so infrequent that they can be ignored. Consequently, the bow shock represents a collisionless shock wave, and belongs to a class of shocks which is of great importance in the entire universe.

In collision-dominated gases, changes in density and temperature across the shock are uniquely determined through the conservation of mass, momentum and energy, because the frequent collisions ensure Maxwellian velocity distributions. The shock thickness one obtains is a few collision mean free paths.

In a collision-free plasma, on the other hand, changes in state across the shock are achieved through collective interactions between particles and fields (magnetic and electric), caused by instabilities. The energy in these collective motions must be taken into account in the jump relations. Moreover, although the collective interactions play the role of 'effective' collisions, they will usually not ensure isotropic pressures. These anisotropies further change the jump relations. Another complication can arise from fast heat conduction along magnetic lines of force. Generally speaking, the instabilities, or the dissipation processes, determine the structure of the shock transition (e.g. TIDMAN and KRALL, 1971).

Still, it is useful to write down the solution of the jump relations (Eq. (9) in Section 1). Application of these relations implies that Maxwellian velocity distributions are maintained and that the turbulence dies out not too far behind the shock.

In a coordinate system where the shock is in the y, z-plane, the upstream flow v_1 is along the x-axis, and the upstream field B_1 is along the z-axis (perpendicular shock), one obtains for the jumps in mass density, normal flow velocity, and tangential magnetic field:

$$\frac{\rho_1}{\rho_2} = \frac{v_{2x}}{v_1} = \frac{B_{1z}}{B_{2z}}$$

$$= \frac{1}{8} \left\{ \frac{2h_1}{\rho_1 v_1^2} + 1 + \frac{B_{1z}^2}{4\pi\rho_1 v_1^2} + \left[\left(\frac{2h_1}{\rho_1 v_1^2} + 1 + \frac{B_{1z}^2}{4\pi\rho_1 v_1^2} \right)^2 + \frac{2B_{1z}^2}{\pi\rho_1 v_1^2} \right]^{1/2} \right\} \qquad (19a)$$

and for the jump in specific enthalpy:

$$\frac{h_2}{\rho_2} = \frac{h_1}{\rho_1} + \frac{1}{2}(v_1^2 - v_{2x}^2) + \left(1 - \frac{v_1}{v_{2x}}\right)\frac{B_{1z}^2}{4\pi\rho_1}. \qquad (19b)$$

The magnetosonic Mach-number is given by:

$$M = \frac{v_1}{(2h_1/3\rho_1 + B_{1z}^2/4\pi\rho_1)^{1/2}} > 1 .$$

For parallel propagation ($B_{1z} = 0$, $B_{1x} \neq 0$), the jump conditions are obtained from Eqs. (19) by setting B_z equal to zero.

From Eq. (19b) the jump in pressure or temperature is obtained by noting that $h = (5/2)p$ for $\gamma = 5/3$, where p is the total pressure related to the ion and electron temperatures by $p = nk(T_i + T_e)$.

In the limit of $M \to \infty$, Eqs. (19) reduce to:

$$\frac{\rho_1}{\rho_2} = \frac{v_{2x}}{v_1} = \frac{B_{1z}}{B_{2z}} = \frac{1}{4} \qquad (20a)$$

$$\frac{h_2}{\rho_2} = \frac{15}{32}v_1^2 . \qquad (20b)$$

In terms of the temperature, Eq. (20b) can be rewritten as:

$$\frac{k(T_e + T_i)}{m_i} = \frac{3}{16}v_1^2 \qquad (20c)$$

where m_i is the proton mass.

Another important distinction between collisionless shocks in magnetized plasmas and ordinary gasdynamic shocks arises from the number of different waves on which to base a shock solution. While in gases there is only one such wave (the sound wave), plasmas can support a large variety of wave types which can develop into shocks, e.g. magnetosonic, ion sound, and whistler waves. These waves have properties which strongly depend on frequency, direction of propagation with respect to the magnetic field, and plasma temperature. This explains the widely varying characteristics of collisionless shocks and their dependence on (a) the direction of propagation with respect to the magnetic field, (b) the plasma temperature, as measured by β (the ratio of plasma thermal pressure to magnetic field pressure), and (c) the Mach-number of the shock (the ratio of incident plasma flow speed in the shock frame to the characteristic wave velocity).

For any of the above waves to develop into a shock, wave steepening is required. This is obtained through nonlinear effects. But to maintain a stationary shock, steepening must be balanced somehow. In ordinary gasdynamic shocks this is achieved by collisions (viscosity, thermal conductivity). In collisionless shocks in plasmas there are several ways to balance steepening: dispersion, anomalous resistivity, and anomalous viscosity.

As an illustration for nonlinear wave steepening and its balance by dispersion let us consider a wave with the dispersion relation shown in Fig. 15: it is dispersionless ($\omega/k=$ const.) at low k (large wavelength). A wave packet which starts in the linear (dispersionless) region of the curve (ω_1, k_1 to ω_2, k_2) will, through second order coupling, obtain components $\omega_1 \pm \omega_2$, $k_1 \pm k_2$. As long as these values lie in the dispersionless region, this means that the pulse will steepen at its leading edge. It will continue to do so until components in the dispersive portion of the curve are generated. In this case, the shorter wavelengths (large k) will either run ahead of the main pulse (branch a), or fall behind it (branch b). As the dispersion curve of whistlers resemble curve (a), these waves could develop into a shock with a precursor, whereas magnetosonic waves with a curve like (b) would produce a shock with an oscillatory trailing edge.

Another possibility for limiting wave steepening is resistivity. The changing magnetic field across the shock implies the flow of electric currents. If the field gradient becomes so large that the current exceeds the limit for some current-driven instability (see Subsection 5.3) the resulting anomalous resistivity will limit further steepening. This effect is important for low

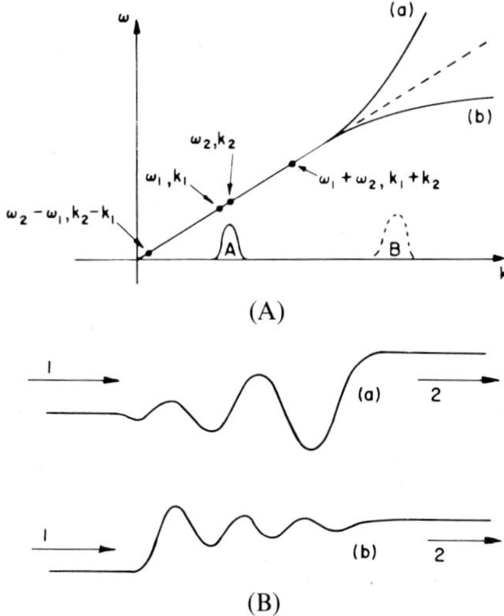

FIG. 15. Dispersion relation with two branches and corresponding shock profiles (from TIDMAN and KRALL, 1971).

Mach-number, low β shocks.

For higher Mach-numbers, resistivity is not sufficient any longer to allow a continuous ion flow: the wave is said to 'break'. This means that multi-velocity flows develop, which lead (through instability of these flows) to an effective viscosity which prevents further steepening. Large electric potentials and ion reflection are essential features of such shocks (e.g. BISKAMP, 1973).

For high β (~ 1) shocks, thermal velocities much exceed current flow velocities. Consequently, current micro-instabilities play a minor role and magnetic oscillations are therefore not damped. Many types of waves are excited and parametric instabilities are thought to develop (e.g. GALEEV, 1976). At present, such shocks are the least understood theoretically. As we will see in the next subsection, this category is also the one most frequently met at the earth's bow shock.

Electric potential jumps are a basic feature of shocks in plasmas. They occur because of the different motion of ions and electrons. Ions move through the shock nearly unaffected by the magnetic field since their gyroradii are much larger than the shock thickness. In short, they are "unmagnetized". The opposite is true for electrons which move adiabatically ('magnetized') and thus are slowed down (in the direction normal to the shock) by the increasing magnetic field. This causes a charge-separation electric field which slows down the ions so that quasi-neutrality is maintained.

Given a high enough potential jump and a finite temperature of the incoming plasma, ions which have insufficient velocity to cross the potential

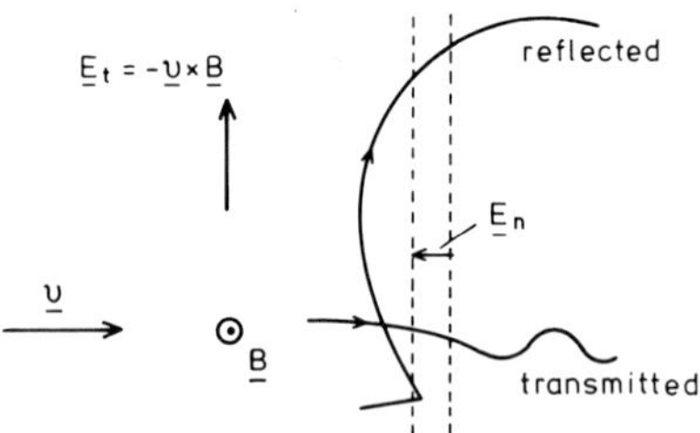

Fig. 16.   Ion orbits and electric fields for a perpendicular shock. $E_t = -v \times B$ is the interplanetary electric field; E_n is the charge-separation field across the shock (after CAIRNS, 1979).

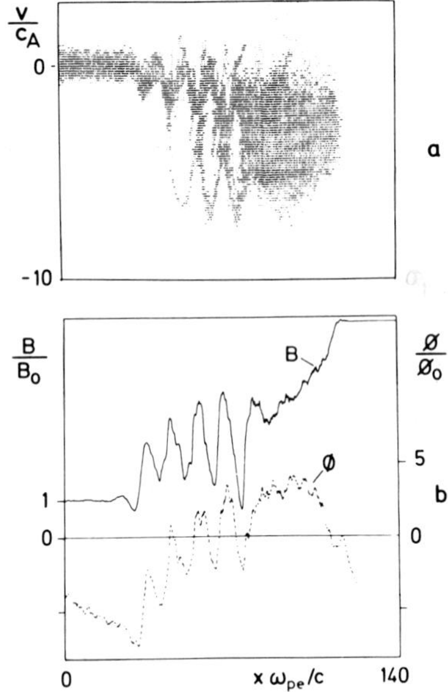

Fig. 17.   Numerical simulation of an oblique shock ($M_A \approx 5$) showing the non-linear interaction in front of the shock. (a) Distribution of ion velocities relative to the upstream velocity; (b) total magnetic field and electric potential profiles, as a function of distance across the shock. Note in (a) the very efficient ion dissipation leading to a broad almost thermalized downstream distribution (from BISKAMP, 1973).

barrier will be reflected. If the magnetic field forms a large angle with the shock normal, the gyromotion will carry the reflected ions back to the shock (Fig. 16). By this time, however, they have gained enough energy in the tangential electric field to cross the barrier. The result is a population of more energetic ions in the downstream region, which effectively constitutes a heating of the ions (e.g. CAIRNS, 1979).

If the magnetic field is directed more obliquely with respect to the shock, the relfected ions are able to escape upstream and become part of the upstream ion population discussed in Subsection 2.3.

The importance of potential jumps and ion reflection for high Mach-number shocks has been demonstrated by numerical simulations (e.g. BISKAMP, 1973). Figure 17 shows an example of an oblique shock wave some time after the shock was generated by a magnetic piston moving into the plasma. Reflected ions (negative v) are readily apparent, as are the steep and variable potentials ϕ, which appear to form 'subshocks'. The strong interaction of the reflected ions with the incoming plasma leads to a rapid thermalization of the ion distribution.

2.2   Observed bow shock structure

A schematic overview of the bow shock phenomenology is shown in Fig. 18 (GREENSTADT and FREDRICKS, 1979). It demonstrates the strong influence of the angle, θ_{nB}, between shock normal and interplanetary magnetic field (IMF) on the characteristic magnetic field profiles. For perpendicular ($\theta_{nB} \sim 90°$) or quasi-perpendicular shocks ($\theta_{nB} \gtrsim 50°$), transitions are fairly monotonic and smooth. For quasi-parallel ($\theta_{nB} \lesssim 50°$) and parallel shocks ($\theta_{nB} \approx 0°$), the transitions are characterized by strong fluctuations which extend over a large spatial range, both upstream and downstream. Because of the curvature of the bow shock, the entire range of θ_{nB} is simultaneously realized along the shock.

As was noted in the previous subsection, other parameters which control the nature of the shock are the plasma β and the Mach-number M. Empirically, a classification scheme has resulted, which is shown in Fig. 19: shocks are called laminar for $\beta \lesssim 0.1$, $M < M_c$; quasilaminar for $\beta \lesssim 0.1$, $M > M_c$; quasi-turbulent for $\beta \gtrsim 1$, $M < M_c$ and turbulent for $\beta \gtrsim 1$, $M > M_c$ (sometimes a fifth category, quasi-electrostatic, is introduced for $\beta \gg 1$, $M > M_c$). Here M_c denotes a loosely defined critical Mach-number which empirically seems to be ~ 3. Laminar shocks are also referred to as resistive shocks because of the decisive role of anomalous resistivity noted in the previous subsection. From statistical studies of the earth's bow shock it has

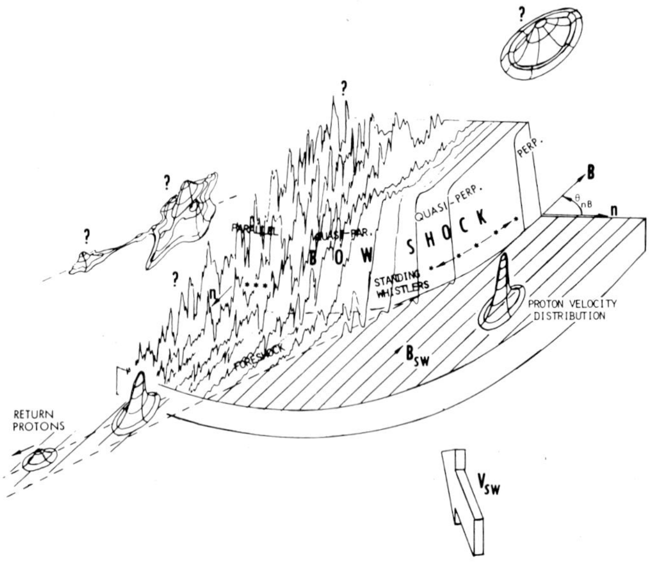

FIG. 18.   Schematic representation of magnetic field profiles at different locations of
the earth's bow shock, demonstrating the strong dependence on θ_{nB}. Situation is
drawn for an angle of 45° between the interplanetary magnetic field, B, and solar
wind velcoty, v_{sw}. Note, at lower left, the reference to the counterstreaming solar
wind and reflected (return) protons (from GREENSTADT and FREDRICKS, 1979).

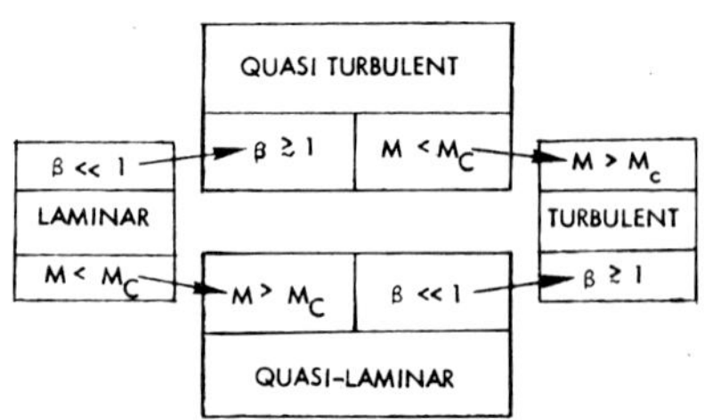

FIG. 19.   Shock classification scheme (from GREENSTADT, 1974).

been determined that the frequency of observation of the 4 classes are 1.9%,
1.6%, 4%, and 92.5%, respectively (FORMISANO, 1977). Thus 'turbulent'
shocks, which are the ones least understood theoretically, clearly dominate.
Bow shock observations can therefore serve as an important guide for
improving our understanding of these structures.

Figures 20 through 22 show examples of quasi-perpendicular shock
crossings for the four classes outlined above. The laminar shock (Fig. 20a) is
characterized by a monotonic magnetic field profile without turbulence.

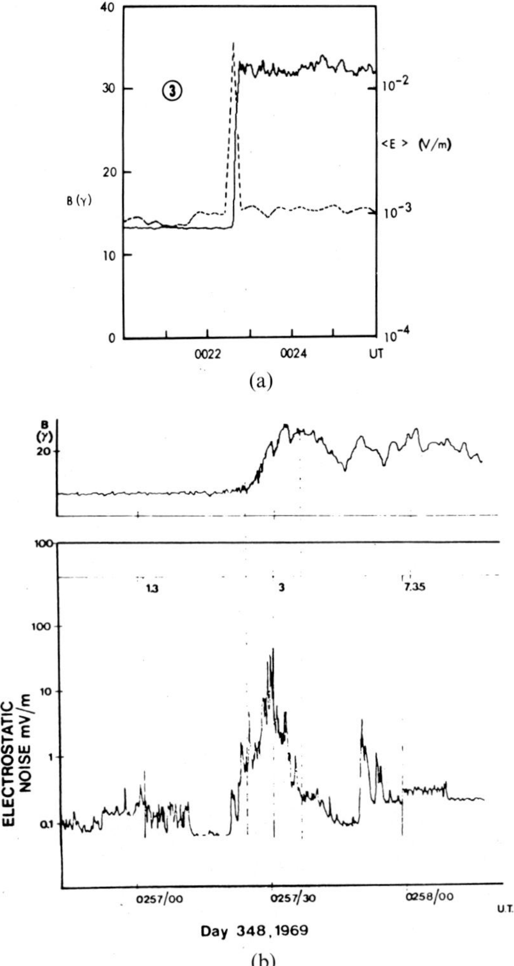

FIG. 20.   (a) Laminar shock profiles for $\theta_{nB} \approx 78°$. B is the magnetic field, $\langle E \rangle$ the electrostatic noise (from GREENSTADT *et al.*, 1975). (b) Profile of a quasi-laminar shock for $\theta_{nB} = 83°$, $\beta = 0.2$, $M = 3.7$ (from FORMISANO, 1977).

There is some electrostatic noise (a few mV/m) right in the ramp, as expected for resistive shocks. The quasi-laminar shock (Fig. 20b) still has a monotonic ramp, but damped oscillations appear, reminiscent of the dispersion effects schematically presented in Fig. 15. The electrostatic noise has become very intense.

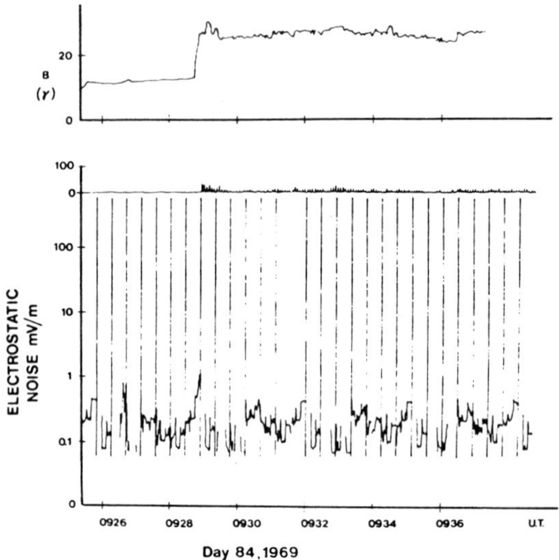

FIG. 21. Quasi-turbulent shock structure with $\theta_{nB} = 52°$, $\beta = 0.45$, $M = 2.8$ (from FORMISANO, 1977).

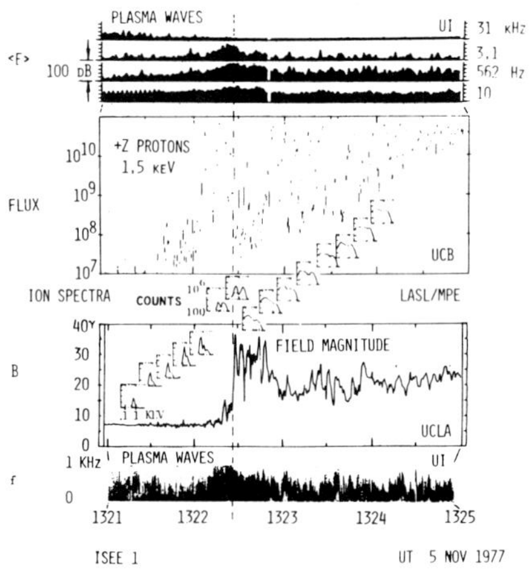

FIG. 22. Field and particle data for a turbulent shock with $\theta_{nB} = 68°$, $\beta = 2.2$, $M_A = 5.7$ (from GREENSTADT et al., 1980).

The quasi-turbulent case (Fig. 21) has a magnetic field profile almost indistinguishable from the laminar case. However, the electrostatic noise is more intense and especially remains so in the magnetosheath. In the turbulent case (Fig. 22), finally, the magnetic profile has lost its monotonic and smooth nature. A 'foot' has developed, as expected for substantial ion reflection (cf. Fig. 16), and turbulence is evident both up and downstream of the main transition. Plasma waves are intense and have a burst-like appearance. Ion energy spectra (inserts) have a bimodal shape, probably related to the multi-velocity streams or ion reflection, i.e. anomalous viscosity, as discussed in the previous subsection.

For oblique propagation one expects the shock to have a whistler precursor, as shown schematically in Fig. 15. Such wave structures are indeed observed, as shown in Fig. 23 for a pair of adjacent bow shock crossings.

Quasi-parallel shocks ($\theta_{nB} \lesssim 50°$) are by nature less well defined as illustrated by the oscillatory magnetic field profiles in Fig. 18. These fluctuations extend far upstream as well as downstream and are responsible for the turbulent nature of the magnetosheath behind the quasi-parallel

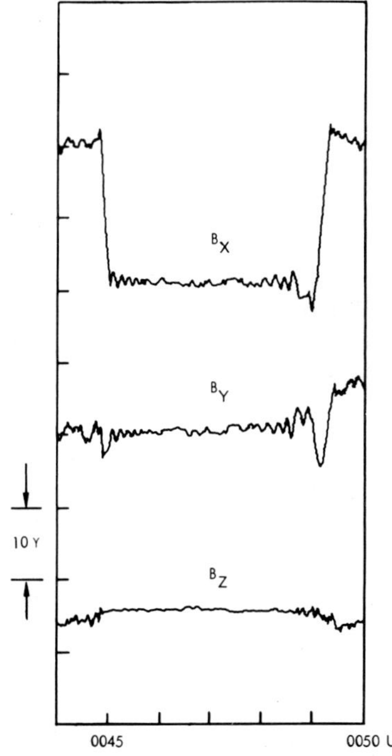

FIG. 23. Magnetic field components for a pair of adjacent shock crossings showing stationary upstream wave structure (from GREENSTADT et al., 1975).

portion of the bow shock. Similarly, the magnetosheath is turbulent behind a turbulent quasi-perpendicular shock. Note that in all those cases where the turbulence does not die out quickly enough behind the shock, application of the jump conditions (Eq. (19)) is not possible.

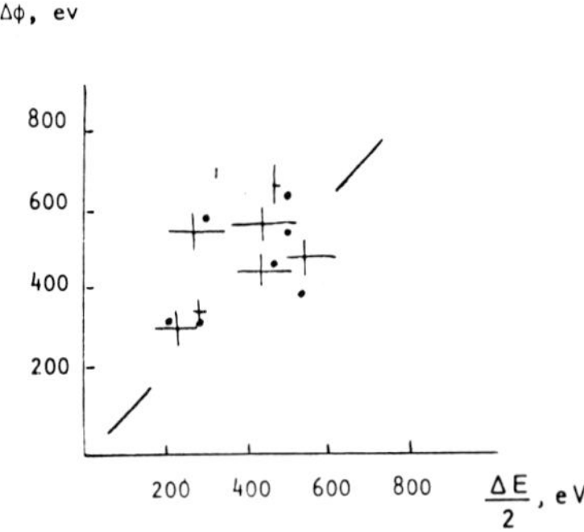

FIG. 24. Inferred potential barrier at the earth's bow shock as a function of the kinetic energy loss by protons (from GALEEV, 1976).

Measurements of the electrostatic potential within the shock play an important role for the understanding of the behaviour of ions and thus for the entire shock structure. Direct measurements have until now been reported only from laboratory experiments (e.g. PODGORNY, 1979). In case of the bow shock, potentials have so far been inferred indirectly. Figure 24 compares the potential jump, derived on a theoretical basis with the measured kinetic energy loss of protons (GALEEV, 1976). It is seen that only half the kinetic energy loss is spent in overcoming the potential barrier. The rest is dissipated and heats the ions.

2.3 Upstream phenomena

As illustrated in Fig. 18, there is a very extended region upstream of the shock where shock-associated phenomena are observed. These are (a) various types of waves and (b) accelerated particles. Both phenomena are closely related, since the most prominent type of upstream waves has been identified as a MHD wave, which cannot propagate upstream against the supersonic solar wind. These waves therefore must be generated locally by upstreaming protons (FAIRFIELD, 1969).

In the previous subsections it was demonstrated that ion reflection is a basic property of collisionless shocks. If the angle between IMF and shock normal is not too large, these ions can escape upstream, and such beams have indeed been observed (ASBRIDGE *et al.*, 1968).

An interesting feature of the reflected ion beams is their velocity which can be much higher than that of the incident solar wind (when measured in the shock frame). The reason for the acceleration is the interplanetary electric field, $\boldsymbol{E} = -\boldsymbol{v} \times \boldsymbol{B}$, along which particles are displaced in the reflection process. The energy gain can be easily calculated (SONNERUP, 1969) by transformation into a coordinate system which slides along the shock with such a velocity that the solar wind appears to flow along the IMF. In this coordinate system there can be no $\boldsymbol{v} \times \boldsymbol{B}$ electric field, and the particle energy must be preserved upon reflection. Transforming back into the shock frame gives the following expression for the ratio of reflected to incident particle energy:

$$\frac{E_{\mathrm{r}}}{E_{\mathrm{i}}} = 1 + 4 \frac{\cos^2 \phi - \cos\phi \cos\theta \cos\psi}{\cos^2 \theta} \tag{21}$$

where ϕ is the angle between $\boldsymbol{v}$ and the shock normal $\boldsymbol{n}$, θ the angle between $\boldsymbol{B}$ and $\boldsymbol{n}$, and ψ the angle between $\boldsymbol{v}$ and $\boldsymbol{B}$. Here it has been assumed that the reflection process conserves the magnetic moment of the particle motion. For the special case where the solar wind flows along the shock normal ($\phi = 0$; $\theta = \psi$), which is appropriate for the nose region of the shock, Eq. (21) reduces to

$$\frac{E_{\mathrm{r}}}{E_{\mathrm{i}}} = 1 + 4 \tan^2 \theta \tag{22}$$

which illustrates that substantial ion acceleration already occurs for $\theta \gtrsim 45°$, whereas for θ approaching $90°$ (perpendicular shock) the ratio approaches infinity in the non-relativistic treatment.

Figure 25 shows the good agreement between measured ion beam energies and those predicted by Eq. (21) (PASCHMANN *et al.*, 1980). Reflected ion energies in that study varied between ~ 1.5 and ~ 25 times the incident solar wind ion energy, which typically is $\sim 1\,\mathrm{keV}$.

For electrons, the same mechanism is applicable. Since, however, the solar wind bulk energy for electrons is very small ($< 1\,\mathrm{eV}$), electron energization does not occur until θ approaches $90°$. The observation of sheets of keV electrons on nearly tangent magnetic flux tubes (ANDERSON *et al.*, 1979) supports this view.

The solar wind and the reflected ions represent two counterstreaming ion beams which should quickly become unstable to the generation of MHD waves (e.g. BARNES, 1970). In this process the beam itself would be destroyed

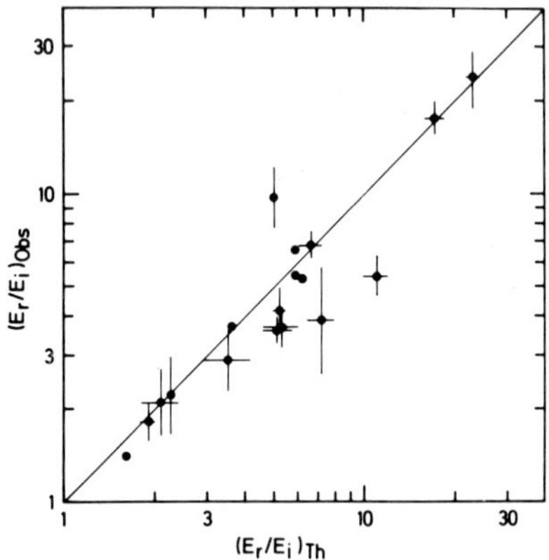

FIG. 25. Scatter diagram of the observed ratios $(E_r/E_i)_{obs}$ versus the predicted ratios $(E_r/E_i)_{th}$ of reflected beam and incident solar wind energies for 18 events observed with ISEE in 1977/78 (from PASCHMANN *et al.*, 1980).

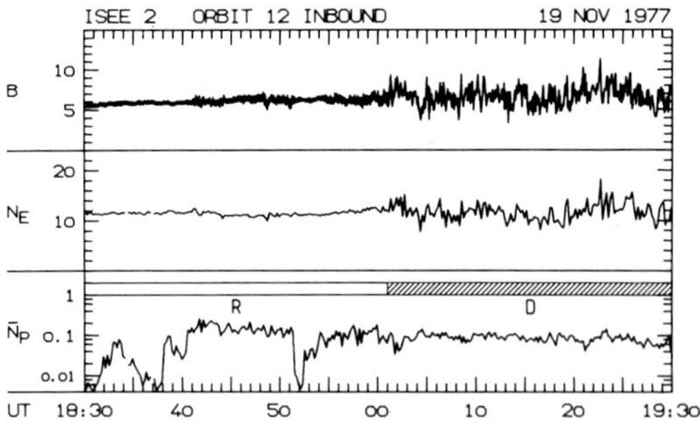

FIG. 26. Upstream wave event on 8 December 1977. The three panels show the field strength B (in gamma), the solar wind electron density N_E (in cm^{-3}) and the suprathermal ion densities $\bar{N}_p$ (in cm^{-3}) as observed on ISEE-1. The occurrence of reflected (R) and diffuse (D) ion populations is identified by the bars above the $\bar{N}_p$ plot (from PASCHMANN *et al.*, 1979a).

and converted into a more isotropic distribution. Indeed, such ion distributions are quite common in the upstream region (the 'diffuse' ion distributions of GOSLING *et al.*, 1978). Moreover, contrary to the reflected ions, these diffuse ions are always associated with low-frequency waves, as demonstrated by Fig. 26 (PASCHMANN *et al.*, 1979a).

In total one obtains an upstream particle situation as depicted in Fig. 27. Although this picture is drawn for the vicinity of the bow shock, it is interesting to note that some of the effects (notably the keV electrons and

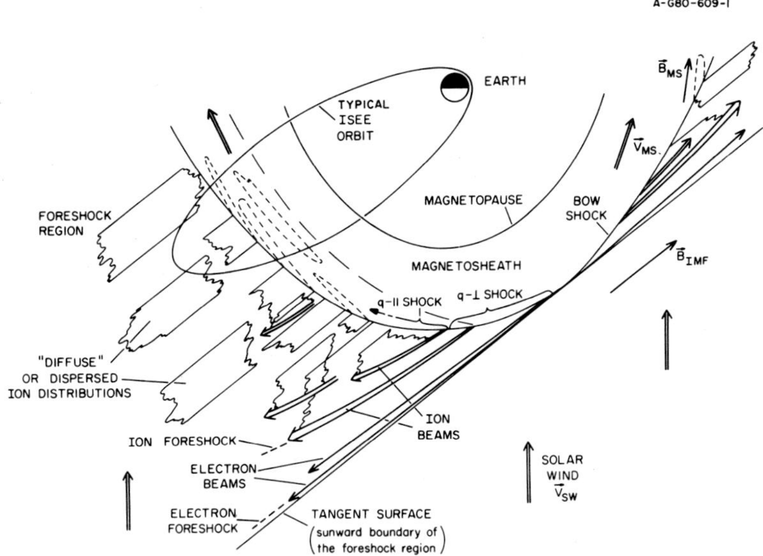

Fig. 27.   Schematic illustration of upstream particle morphology (from Eastman *et al.*, 1981).

high energy ions) are still observed $\sim 200\ R_{\rm E}$ upstream of the bow shock (e.g. Anderson, 1981; Scholer *et al.*, 1980).

3. The Boundary Layer

3.1   *Boundary layer morphology*
3.1.1   *Overview*

An important signature of the interaction between the solar wind and the magnetosphere is the existence of a boundary layer *inside* of and adjacent to the magnetopause, which consists of plasma with nearly the same temperature and often similar flow properties as the compressed solar wind plasma flowing around the magnetosphere. This layer is a manifestation of the fact, already noted in Section 1, that the magnetopause is not impermeable to the solar wind. (The boundary layer as it is understood here, should not be confused with the layers of energetic particles outside of the magnetopause, nor with the magnetopause current layer, nor with the plasma sheet boundary layer which is located at the interface between the magnetotail lobes and the plasma sheet.)

The boundary layer extends over the entire known portion of the

                    G. HAERENDEL AND G. PASCHMANN

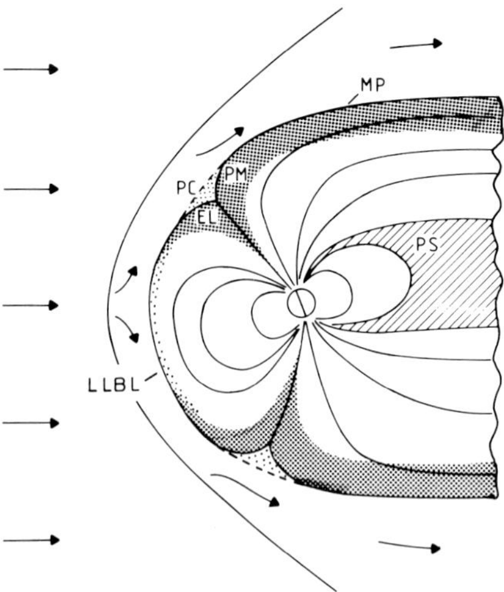

FIG. 28.   Noon-midnight meridian cross-section of the magnetosphere, illustrating schematically the boundary layer and polar cusp (PC) configuration. EL denotes the entry layer, PM the plasma mantle, MP the magnetopause, and PS the plasma sheet. The portion of the magnetosheath inside the dashed line is referred to as the exterior cusp.

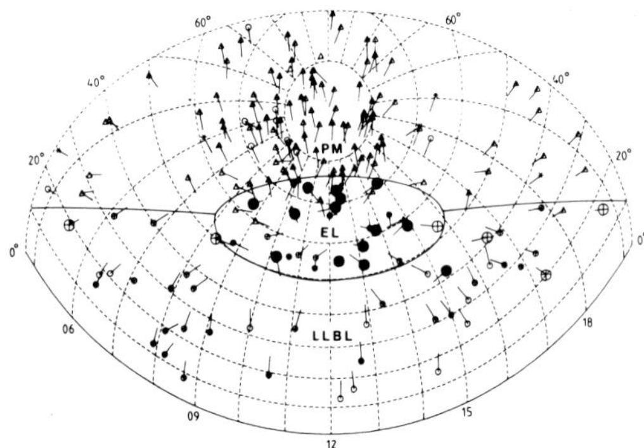

FIG. 29.   Front view of section of model magnetopause. The heavy solid lines delineate the three distinct regions observed, the entry layer (EL), the plasma mantle (PM), and the low-latitude boundary layer (LLBL). Locations of Heos 2 magnetopause crossings in solar magnetic (SM) coordinates are marked by various symbols. Short lines emanating from these symbols indicate the component of the magnetospheric magnetic field tangential to the model surface. Triangles represent the plasma mantle. The other symbols refer to EL and LLBL, and are distinguished according to density jump at the magnetopause and thickness of the layer: solid circles and crosses are used for jumps >4 and <4 respectively, large/small symbols for thicknesses >0.5 R_E and <0.5 R_E, respectively. Open circles denote crossings where no boundary layer was found (from HAERENDEL et al., 1978).

magnetopause, although its properties vary substantially from one region to the other. As a natural consequence, a variety of names are being used for different portions of the boundary layer. While these often do not appear very logical, they are convenient, at least for spatial reference.

Figure 28 shows the boundary layer as it appears in the noon-midnight meridional cross-section of the magnetosphere. Three regions have been identified: the *low-latitude boundary layer* (LLBL), the *entry layer* (EL), and the *plasma mantle* (PM) (HONES *et al.*, 1972; ROSENBAUER *et al.*, 1975; PASCHMANN *et al.*, 1976; HAERENDEL *et al.*, 1978; EASTMAN and HONES, 1979). The extent of these regions in the other dimension is illustrated in Fig. 29.

Some of the characteristic properties of the various boundary layer regions can be deduced from a comparison of the profiles of the macroscopic plasma parameters, presented in Fig. 30.

Figure 30a shows a plasma parameter profile during a plasma mantle crossing. It illustrates the typical gradual increase of plasma density by three orders of magnitude over a distance of several earth radii. At the same time the temperature also increases. The streaming of the plasma along the geomagnetic field towards the tail at speeds of ~ 100 km/sec is also readily apparent.

In contrast, the entry layer (Fig. 30b) shows a significantly different behavior. In particular, density and temperature reach their magnetosheath levels right at the inner edge of the layer and do not exhibit any significant change at the magnetopause. This absence of a density jump is strongly suggestive of local plasma entry. Moreover, the flat density profile argues for an effective inward transport (see Subsection 3.4).

The plasma β, i.e. the ratio of plasma thermal pressure to the magnetic field pressure, was larger than 1 throughout the entry layer crossing of Fig. 30b, whereas in the plasma mantle it generally is smaller than 1. The flow direction in the entry layer crossing of Fig. 30b is remarkable in that it is predominantly directed sunward, i.e. *opposite* to the external magnetosheath flow. As discussed below, different and rather irregular flows are observed in the entry layer at other times. Of course, there is another difference: whereas the plasma mantle occupies the otherwise nearly empty tail lobe region, the entry layer (and the low-latitude boundary layer) border on the hot plasma of the plasma sheet or ring current, allowing some mixing of the plasmas.

At medium latitudes on the dayside there seems to be a tendency for rather gradual density variations across the layer (EASTMAN and HONES, 1979). At low latitudes a variety of profiles is observed including one where the density changes abruptly at the magnetopause and again at the inner

                    G. HAERENDEL AND G. PASCHMANN

edge of the layer. Figure 30c shows an example observed with ISEE-1. Note
the pronounced variations of plasma flow speed and direction.

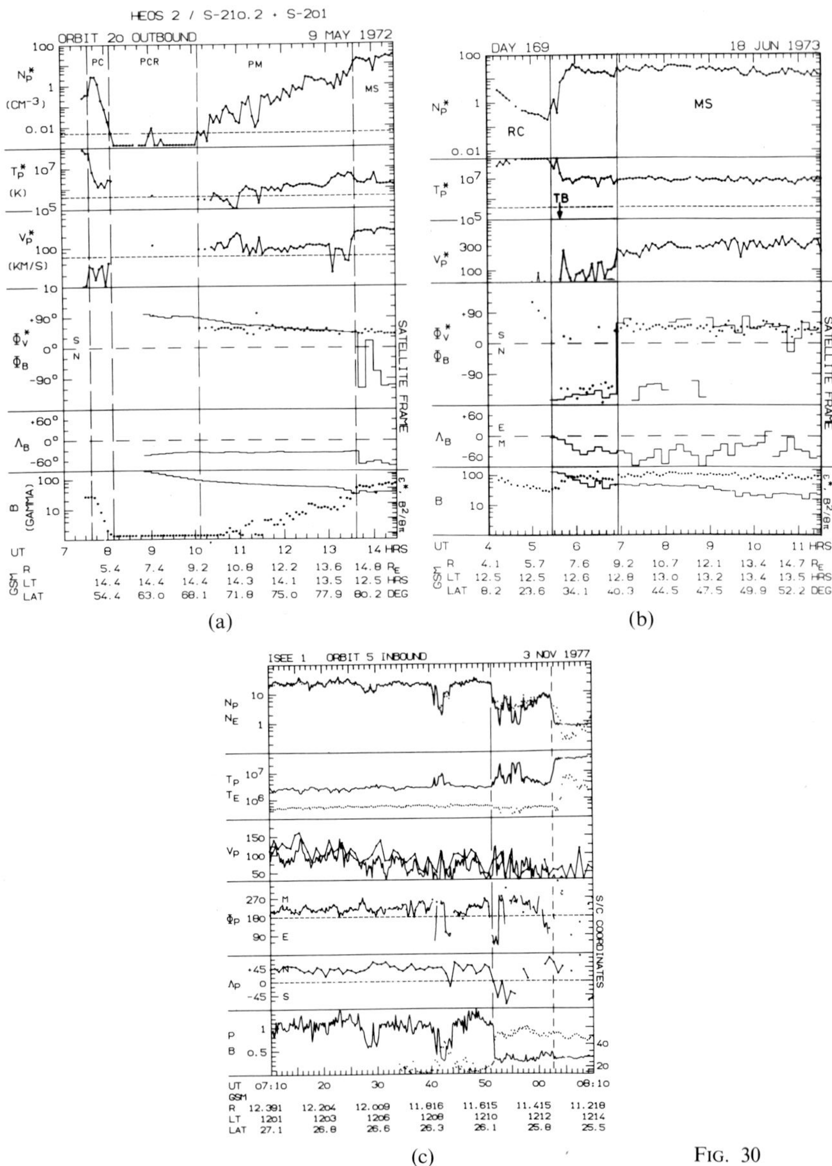

(a)

(b)

(c)

Fig. 30

From the density profiles measured at various locations it is evident that there are systematic thickness variations. The largest thicknesses (up to 10 R_E) are observed in the plasma mantle, the smallest (a fraction of an R_E) in the subsolar region. Apart from temporal variations, the boundary layer thickness increases with increasing distance from the subsolar point. When comparing the thickness of the boundary layer with that of the magnetopause, it is found that the former exceeds the latter by a large factor, except in the subsolar region where the two are often comparable (Fig. 56).

Plasma parameter profiles measured in any given region of the boundary layer show substantial variability which indicates that the boundary layers are not uniform and static. This fact is emphasized by recent ISEE magnetopause crossings where occasionally no indication of a boundary layer was found.

A pervading property of the plasma in the boundary layers is the occurrence of strong and, sometimes, organized flows. Some features of the flow were already discussed above. An attempt to summarize results obtained by various observers is presented in Fig. 31. The simplest situation is that found in the plasma mantle: here the plasma is streaming along the polar cap field lines into the geomagnetic tail, just as expected for an expansion of entry layer plasma into the tail lobe region. A fairly consistent picture also has been observed at low latitudes near the flanks of the magnetosphere: again the flow is tailward, this time, however, it is not directed parallel to the field lines but essentially parallel to the external magnetosheath flow. The consequently large flow component perpendicular to the magnetic field led EASTMAN *et al.* (1976) to propose this

FIG. 30.   (a) Plasma and magnetic field data for a polar magnetosphere crossing by Heos 2, from the polar cusp (PC) on the dayside, across the polar cap (PCR) and through the plasma mantle (PM) into the magnetosheath (MS). N_p, T_p, v_p are the proton density, temperature and flow speed, respectively. Φ_v (dots) is the flow direction and Φ_B the field direction, both in the noon-midnight meridian plane. A_B is the field elevation angle with respect to that plane. $\Phi_v = \Phi_B$ denotes tailward flow along the field. B is the field strength and ε the plasma energy density (from ROSENBAUER *et al.*, 1975). (b) Heos 2 outbound traversal from the ring current (RC) region through the entry layer (shaded region) into the magnetosheath (MS) (from PASCHMANN *et al.*, 1976). (c) ISSE-1 inbound crossing of the low-latitude boundary layer near local noon. The first dashed line indicates the magnetopause crossing, the second the inner edge of the boundary layer. N_p, N_E are the proton and electron densities, T_p, T_E the temperatures. v_p, Φ_p, and A_p are the speed, azimuth, and elevation angle of the proton flow velocity in ecliptic coordinates. P and B are the plasma and magnetic field pressures, respectively (from PASCHMANN, 1979).

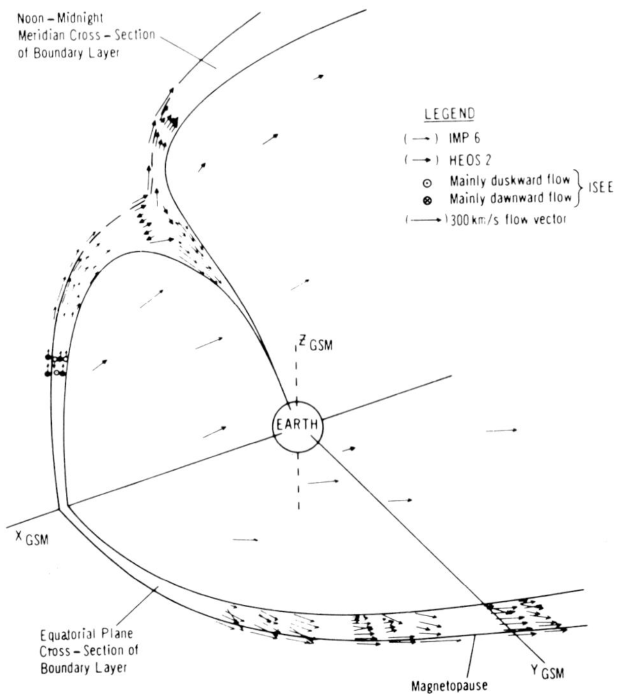

FIG. 31. Typical plasma flows in the boundary layers as observed by various satellites. Almost all flow vectors were derived from two-dimensional measurements (from EASTMAN, 1979).

region as the site for the dynamo driving field-aligned currents into the auroral ionosphere (see Subsection 3.3).

The flow pattern gets much more complicated near local noon. At high latitudes, in the entry layer, flow speeds are variable, and both sunward and anti-sunward directions occur (see Fig. 31), mostly along the magnetic field, but often with substantial cross field components (PASCHMANN et al., 1976). At mid-latitudes, there seems to be a persistence of poleward flow, but at low latitudes the flows become very irregular. This is illustrated by Fig. 32 which shows the results of three-dimensional flow measurements which first became available with ISEE-1 and -2. It is quite apparent from the figure that the flow velocity in the boundary layer is highly non-uniform. Directions vary from poleward to equatorward, duskward to dawnward. Such a behavior is reminiscent of eddy convection, which will be discussed in Subsection 3.4.

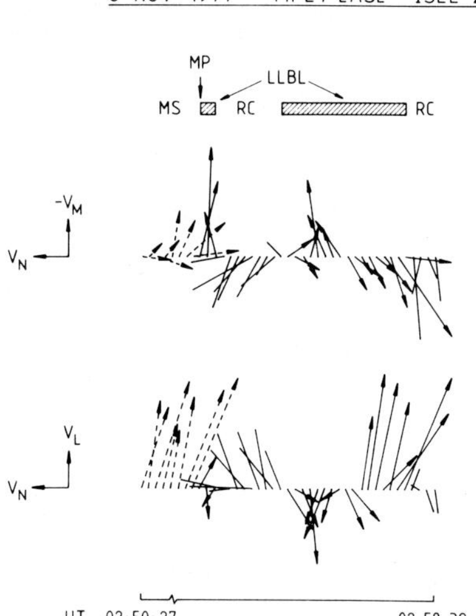

FIG. 32.   Three-dimensional flow velocities measured with ISEE-2 during a noon crossing of the low-latitude boundary layer. Flow vectors were projected into two orthogonal planes of the L, M, N boundary normal coordinate system (see Section 4.3).   Magnetosheath (MS) flow vectors are shown as dashed, boundary layer (BL) vectors as solid lines with arrows. The solid lines without arrows refer to ring current (RC) flows. The length of the velocity axes corresponds to 50 km/sec (from PASCHMANN, 1979).

3.1.2   Polar cusp region

The polar cusp (Fig. 28) is a particularly interesting region of the magnetosphere. First, it represents, as the name implies, a cusp-like indentation in the magnetopause, which severely perturbs the external flow. Second, the cusp possesses a unique magnetic field morphology, with field lines threading the cusp spreading, like an umbrella, over the entire magnetopause. Both aspects make the cusp a focal point of boundary layer dynamics.

The field geometry is illustrated in Fig. 33. It shows that magnetic field directions indeed converge to a point.

In Subsection 1.2 the flow in the polar cusp region was discussed and the possibility that a free-flow surface develops which separates a region of stagnating plasma (the exterior cusp in Fig. 28) from a free-flow region. Evidence for this division is presented in Fig. 34. Here the location of three kinds of boundary crossings in the cusp region have been mapped: crossings

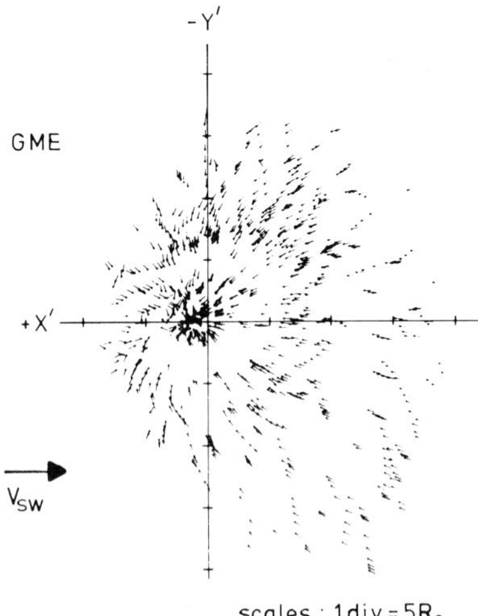

FIG. 33.   View from above the north pole of high-latitude Heos magnetosphere field vectors. Field lines appear to radiate from a point near the noon meridian and thus indicate a funnel type geometry near the dayside cusp (from FAIRFIELD, 1979, after Hedgecock and Thomas).

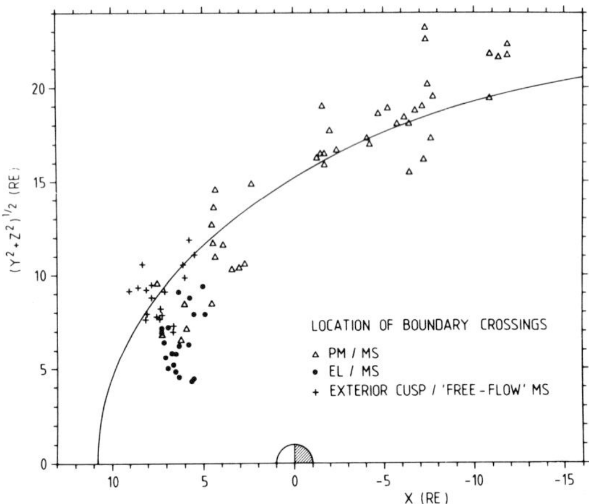

FIG. 34.   Positions of Heos 2 high-latitude magnetopause and "free-flow" magnetosheath crossings as compared to Fairfield's curve for the average magnetopause location (as observed near the ecliptic plane). The symbols PM, EL, and MS denote the plasma mantle, entry layer, and magnetosheath, respectively (from SCKOPKE, 1979).

of the magnetopause from the mantle (triangles) and from the entry layer (circles), and crossings from the exterior cusp to the regular "free-flow" magnetosheath (crosses). It is evident from the figure that the exterior cusp/free-flow crossings, as well as the magnetopause crossings from the plasma mantle occur along a line which represents the average location of the magnetopause in the equatorial plane, rotated by 90° about the earth sun line. This is good evidence, at least on a statistical basis, for the free-flow line concept discussed in Subsection 1.2. The positions of the entry layer/magnetosheath crossings, on the other hand, demonstrate the deep

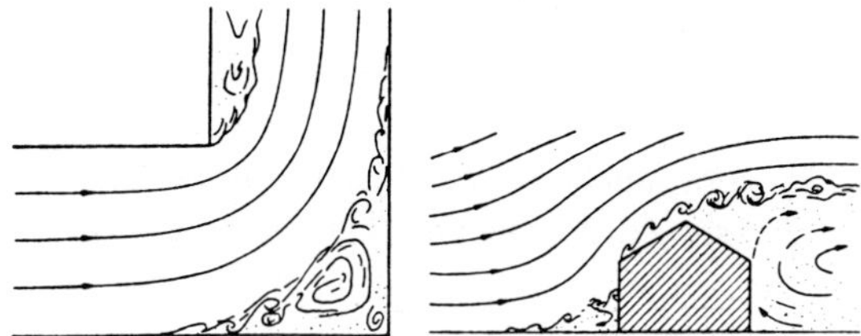

FIG. 35.   Sketches of vortex formations in flows around corners. This process seems to account for the turbulence found in the polar cusp regions.

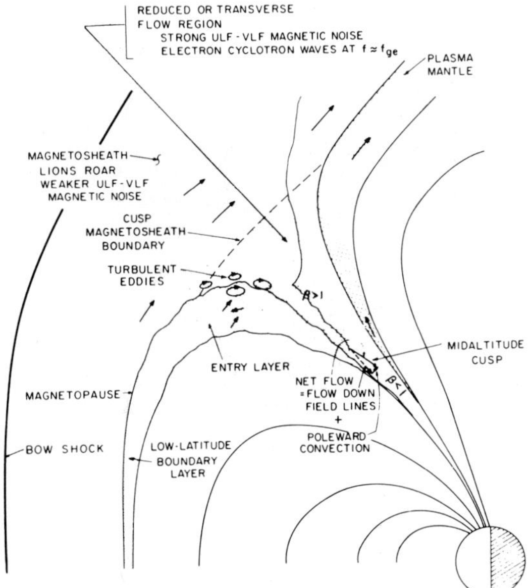

FIG. 36.   Schematic view of the cusp region (from FAIRFIELD and HONES, 1978).

indentation of the magnetopause in the cusp region.

The *exterior cusp* is characterized by elevated temperatures and irregular flows, both in direction as well as in magnitude. These features make it distinguishable from the "free-flow" magnetosheath. The irregular flow properties can be understood if one notes that the flow past the cusps resembles the hydrodynamic flow around a corner. This is illustrated in Fig. 35. As in the hydrodynamic analogue, one should expect some sign of vortex formation and separation in this region. The free-flow line would thus vary strongly with time. The turbulence resulting from these unsteady flows in the exterior cusp may have important implications for the entry and transport of boundary layer plasma (see Subsection 3.4).

The overall situation in the polar cusp region is summarized in Fig. 36.

3.1.3 *Plasma mantle*

As a particular consequence of the cusp topology, field lines which cross the equator on the sunward side and field lines extending far into the magnetotail join in the cusp region. This makes for a simple way of transporting plasma from the entry layer into the plasma mantle. The way this is achieved is illustrated schematically in Fig. 37. Magnetospheric convection (driven, for example, by reconnection) will transport magnetic flux from the sunward to the tailward side, taking its plasma with it. Once the field lines have become tail-like, the plasma can expand along the lines of force and acquire the ordered tailward flow characteristic of the plasma mantle. As illustrated by Fig. 37, particles having different parallel velocities

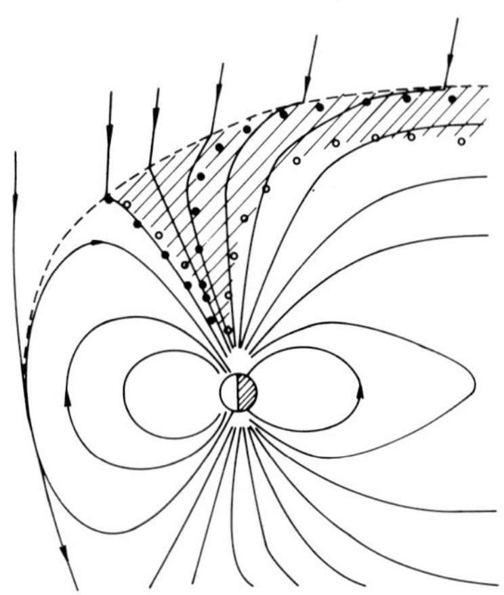

FIG. 37. Schematic representation of particle motion in the polar cusp and plasma mantle (from ROSENBAUER *et al.*, 1975).

along the field line will be affected differently by the convection velocity. As the latter is the same for all particles, particles with smaller parallel velocities will be transported deeper into the tail (open circles) than particles with large parallel speeds (closed circles). This *"velocity filter"* effect immediately explains why the characteristic energy in the plasma decreases with increasing distance from the magnetopause. At low altitudes this corresponds to a decrease in average energy with increasing latitude.

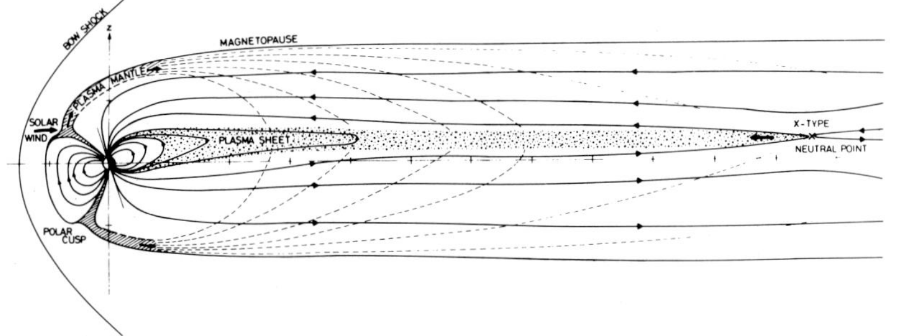

FIG. 38.   Schematic diagram of the transport of mantle plasma onto the plasma sheet under the influence of the large-scale convection electric field. The mantle particles disperse and move along the dashed lines. Mantle plasma entering the field reversal region within the "neutral point" may become trapped and form the plasma sheet. The magnetic field (solid lines) is depicted as an instantaneous snapshot of the quiet time configuration (from PILIPP and MORFILL, 1978).

Another consequence of the superpositon of field-aligned tailward flow and inward directed convective motion is that the inner edge of the plasma mantle should eventually reach the mid-plane of the tail, as illustrated in Fig. 38. This fact has caused some speculation about the plasma mantle as a possible source for the plasma sheet (PILIPP and MORFILL, 1978).

Figure 37 indicates that the boundary layer extends to very low altitudes along cusp field lines ("low-altitude cusp"). It is therefore plausible that plasma of ionospheric origin should become mixed with the solar wind plasma. Evidence for such an ionospheric contribution is the observation that at times the mantle plasma contains significant amounts of singly ionized oxygen (FRANK et al., 1977). Figure 39 shows a measured density profile of various ions in the plasma mantle (LUNDIN et al., 1981) and demonstrates the importance of ionospheric ions (He^+, O^+). The coupling of the boundary layer to the ionosphere will be discussed further in Subsection 3.3.

G. HAERENDEL AND G. PASCHMANN

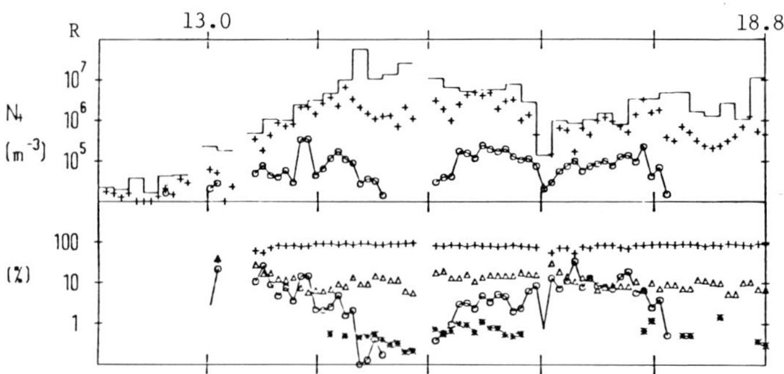

FIG. 39. Ion densities (cm^{-3}) and composition (%) during an outbound traversal of the plasma mantle. The different symbols refer to the following ion species: $+$(H$^+$); Δ(He^{2+}); o(O$^+$); $*$(He$^+$) (from LUNDIN et al., 1981).

Concerning the source region of the mantle plasma, it should be noted that it is probably not restricted to the entry layer as Fig. 37 seems to imply. Entry along the tail magnetopause may be an important contribution as well.

3.2 Plasma blobs, vortex flows and geomagnetic pulsations

The generation of surface waves at the magnetopause by the Kelvin-Helmholtz instability has received much attention in the literature (e.g. LANZEROTTI and SOUTHWOOD, 1979). They have been held responsible for the excitation of resonances of internal geomagnetic field lines in the Pc 3–5 range (SOUTHWOOD, 1974; CHEN and HASEGAWA, 1974). Recent observations of the temporal structure of the low latitude boundary layer (SCKOPKE et al., 1981) and of periodic flow patterns in the plasma sheet near the morning and evening flanks (HONES et al., 1981) show that the surface waves exist at the inner edge of the boundary layer and travel with speeds of a few hundred kilometers per second towards the tail.

Figure 40 presents plasma and field data from a 50-min encounter of the low latitude boundary layer at ≈ 0800 L.T. and $40°$ GSM latitude (SCKOPKE et al., 1981). The density and temperature are extremely variable, as is the flow velocity. A more detailed view at the density distribution in different energy bands, as given in Fig. 41 for the same event, is quite revealing. The overall density increases are caused by the low energy ranges which dominate in the "cool" magnetosheath. They are correlated with decreases of the higher energy bands for positive ions as well as electrons, which dominate the hot, but dilute magnetospheric plasma. SCKOPKE et al. (1981) distinguish 4 regions: (1) inner magnetosphere, (2) a "halo" region, which is an intermediate step towards typical boundary layer populations, (3) boundary

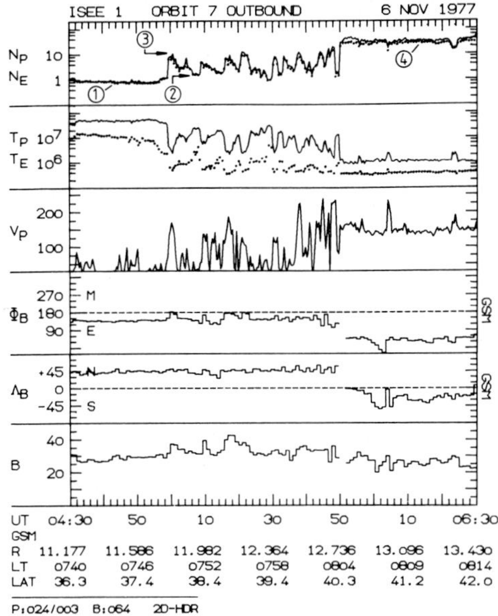

Fig. 40. Plasama and magnetic-field data from an ISEE outbound pass through the outer magnetosphere, low-latitude boundary layer, and magnetosheath near 0800 hours local time and 40° northern GSM latitude. The plasma parameters are, from top to bottom: proton (N_p) and electron (N_E) densities as solid and dotted curves, temperature T_p and T_e, and proton bulk speed (v_p). Magnetic-field data (lower three panels) are given as GSM azimuth (ϕ_B) and elevation (Λ_B) angles and field magnitude, B. Density levels 1–4 indicate characteristic values for the outer magnetosphere, boundary layer halo, boundary layer proper, and magnetosheath, respectively (from SCKOPKE et al., 1981).

layer and (4) magnetosheath. The transitions between regions (2) and (3) are rather abrupt. The density is fairly constant in the latter, but falls well below that of the magnetosheath by a factor 3–4. Characteristic variations of the flow speed through each encounter of boundary layer plasma have led to the interpretation sketched in Fig. 42. The boundary layer events of typically 2–5 min duration are manifestations of the layer thickness varying between zero (or a few 100 km) and ≈ 1 R_E. The boundary layer plasma is always in contact with the magnetopause and travels on its inside towards the tail with a speed of a few 100 km/sec (≈ 150 km/sec in the example shown). This corresponds to a length of the plasma "blobs" of 3–8 R_E along the magnetopause. The halo regions in between exhibit a small increase of

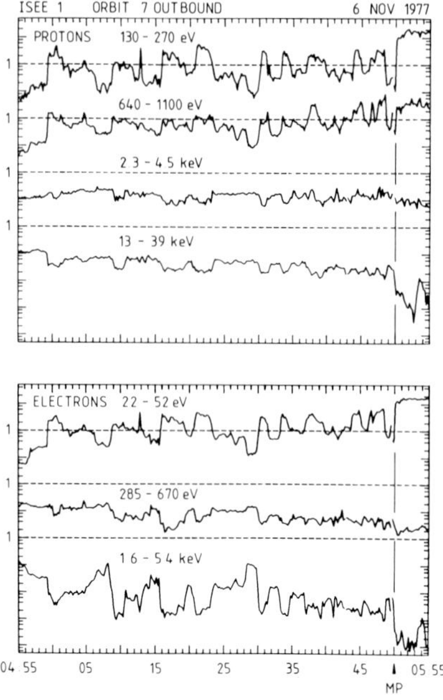

FIG. 41. Proton and electron partial densities, i.e. contributions from certain energy bands to the total densities shown in Fig. 40. Units are cm^{-3}, and curves are displaced by two decades each. The vertical line near 0550 UT marks the magnetopause crossing (from SCKOPKE *et al.*, 1981).

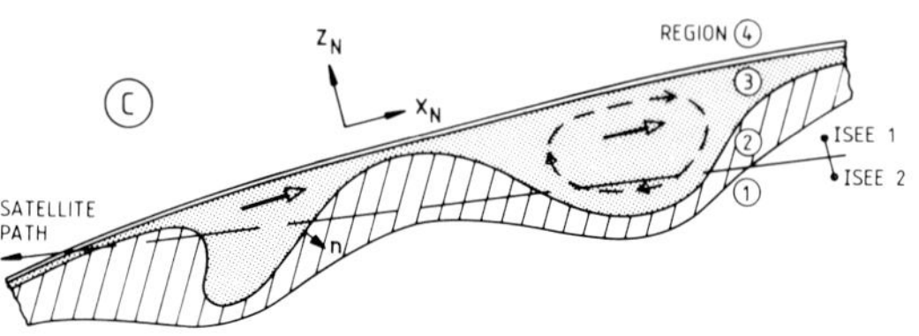

FIG. 42. Sketch illustrating explanation of the observations in Fig. 40 in terms of individual plasma elements moving along the magnetopause (from SCKOPKE *et al.*, 1981).

density over the magnetospheric value as well as of the temperature over the boundary layer level.

The oscillatory variation of the inner edge of the boundary layer which has been derived from the distribution of the dominant low energy plasma is also expressed in the energetic particle fluxes. WILLIAMS (1979a) has established an interesting technique of interpreting azimuthal anisotropies of

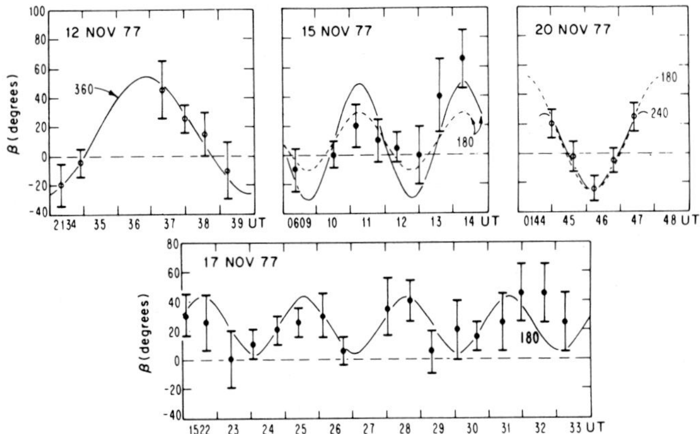

Fig. 43. Four time intervals for which nearly contiguous determinations of the magnetopause orientation angle β could be made. β is plotted versus time for consecutive 1-min intervals as indicated. Periodic variations with the indicated period in seconds are shown for reference (from WILLIAMS, 1980).

the energetic proton distributions above $\approx 20\,\text{keV}$ in terms of distance of the particle boundary from the observer and orientation of the normal vector. Figure 43 shows some examples of periodic variations of the normal vector of about 3 minutes period. That these boundary distortions might be associated with the inner edge of the boundary layer rather than the magnetopause, is suggested by the drops of the energetic proton fluxes shown in Fig. 41.

SCKOPKE et al. (1981) interpret the boundary layer modulation as being caused by a Kelvin-Helmholtz instability of the shear flow existing at the interface of boundary layer and inner magnetosphere. A theoretical analysis of the situation by LEE et al. (1981) confirms that this interface is more unstable than that between magnetosheath and boundary layer. The reasons are, on the one hand, the greater density jump at the inner interface and, on the other hand, a stabilization of the outer interface by sheared external and internal magnetic fields. Such shears are normally small between boundary layer and magnetospheric plasma.

Further towards the flanks of the magnetosphere, vortex motions are regularly encountered (HONES et al., 1981). They appear to be the tailward manifestations of the plasma blobs encountered at the dayside magnetopause. The whole plasma sheet is apparently participating in the vortex flow which is clockwise on the morning side and counterclockwise on

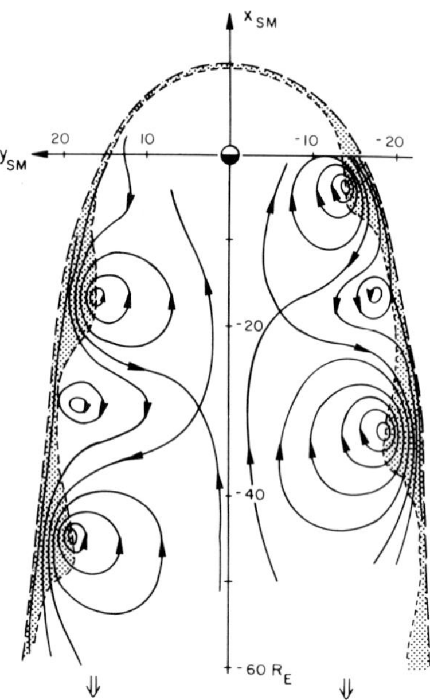

FIG. 44. Schematic interpretation of the plasma vortex measurements. The shaded area is the boundary layer, its wavy inner edge indicated by the dotted line. The dashed line represents the magnetopause. The pattern of flow represented by the solid lines moves tailward through the magnetosphere (as indicated by the white arrows at bottom) driven by Kelvin-Helmholtz instability at the boundary layer's inner edge (from HONES *et al.*, 1981).

the evening side (when viewed from above the ecliptic plane) thus revealing its likely origin via the Kelvin-Helmholtz instability. Figure 44 is an interpretation of the bulk of data. Comparison of the rotational sense of the flow in boundary layer and plasma sheet proper signifies that the axes of the vortices lie actually at the inner edge of the boundary layer. The periods have increased from frontside values of 2–5 min to 5–20 min and the flow velocities to several 100 km/sec. This yields wavelengths of 20–40 R_E and reveals a continued momentum transfer to the boundary layer plasma as it moves from late morning or early afternoon hours towards the dawn-dusk meridian and further into the tail. Such a tailward motion has also been deduced from ground data of geomagnetic pulsations in the Pc 3–5 range (OLSON and ROSTOKER, 1978).

The identification of the origin of long period magnetospheric pulsations is a significant step forward in the understanding of solar wind-magnetosphere interaction. Yet another step has to be made, namely to identify the processes which lead to the mass and momentum transfer through the magnetopause. After a discussion of the coupling between boundary layer and ionosphere, we will devote the rest of Section 3 as well as the following two sections to this question.

3.3 Coupling to the ionosphere

Whatever the process is that achieves the transfer of magnetosheath (i.e. shocked solar wind) plasma to the magnetosphere, it is necessarily accompanied by a momentum transfer. This is signified by the predominance of flow directions on the inside that are parallel to the external ones, although with somewhat lower magnitude. This momentum can be coupled via the magnetic field to the ionosphere and should manifest itself through a typical anti-sunward convection pattern at low altitudes. Before focussing our attention on the dyanmics of the coupling, we should like to demonstrate the pure geometrical connection of the various boundary layer domains with the ionosphere. This is done in Fig. 45. Because of the strong distortion of the field lines originating in the polar cap the whole low latitude boundary layer from noon till late evening or early morning hours projects onto a

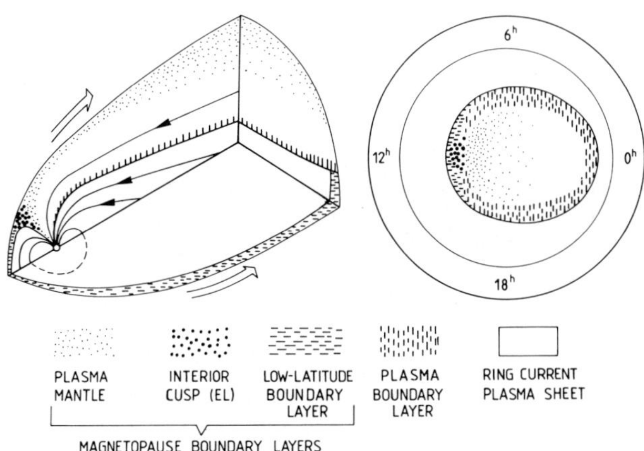

Fig. 45. (a) Schematic diagram of various observed magnetospheric boundary layers. The great extent of the low-latitude boundary layer into the nightside is speculative. Open arrows indicate the flow of the magnetosheath plasma. (b) Schematic diagram of the regions occupied by the various boundary layers, mapped down to the ionosphere along magnetic field lines. Hours indicate local time. The boundary between the low-latitude and the plasma sheet boundary layers is highly uncertain (from VASYLIUNAS, 1979).

narrow range of latitudes (poleward of $\approx 75°$) between 0900 and 1500 L.T. The cusp region projects onto an interval of ≈ 1 hour around noon, and the mantle to a region poleward of this with no clear termination at higher latitudes. The E-W elongation of the ionospheric trace of the boundary layer had initially led to the suspicion (HEIKKILA and WINNINGHAM, 1971) that also the high altitude geometry of the polar cusp was elongated. For several years the name "polar cleft" was en vogue. However, the HEOS-2 data led to a clarification of this point (HAERENDEL and PASCHMANN, 1975; PASCHMANN et al., 1976). The cusplike geometry manifests itself most strikingly by the orientation of magnetic vectors shown in Fig. 35. The ionospheric trace of the boundary layer is continued towards earlier and later local times by the outer parts of the plasma sheet, which have been named "plasma sheet boundary layer". This region is the site of processes that are discussed in Chapter 4 of this volume.

The predominance of anti-sunward flows in the ionospheric trace of the boundary layer, as measured on low orbiting satellites (HEPPNER, 1972; HEELIS et al., 1976), is obviously a consequence of momentum transfer through the magnetopause, and further via field-aligned currents into the ionosphere. They have been identified with satellite magnetometers (IIJIMA and POTEMRA, 1976). The currents close through transverse dissipative currents (Pedersen currents) in the ionosphere, which give rise to electric

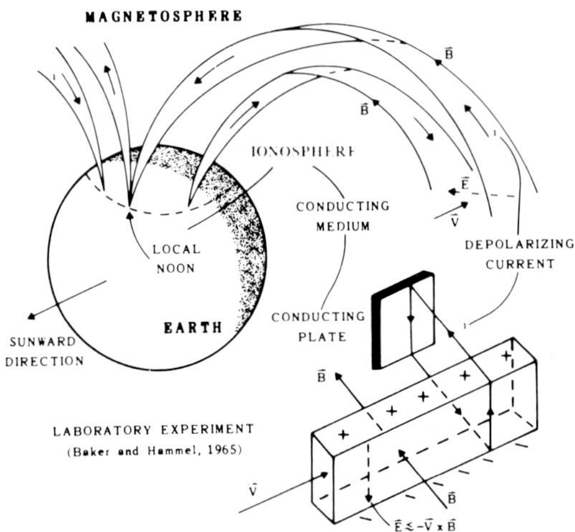

FIG. 46.   Illustration of polarization electric field, $\vec{E}$, and depolarizing currents, $\vec{i}$, generated by plasma moving with velocity, $\vec{v}$, transverse to a magnetic field, $\vec{B}$ (from EASTMAN et al., 1976).

polarization fields that are consistent with the observed ones.

For the dynamo driving the field-aligned currents, several, quite different, mechanisms have been proposed. EASTMAN *et al.* (1976) suggest a diffusive entry of magnetosheath plasma into the boundary layer under conservation of the tangential flow momentum and point to the analogy with an MHD generator (Fig. 46). SONNERUP (1980) pursues a viscous model of momentum transfer. The most effective way would be via reconnection. It allows the solar wind to communicate its momentum towards low altitudes directly through magnetic shear stresses. This process, which is of so great significance for astrophysical plasmas, will be discussed in much detail in Sections 4 and 5.

HAERENDEL *et al.* (1978) concluded that reconnection may occur

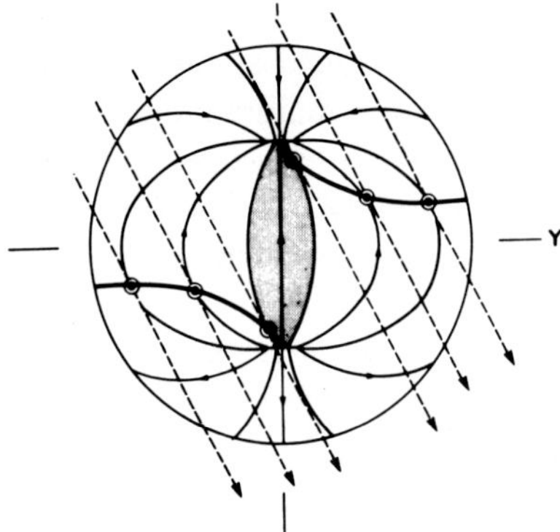

FIG. 47. View from sun of dayside magnetopause. Earth field lines (solid) converge in cusp-like pattern. Merging lines (heavy solid) are locus of points where IMF lines (dashed) are antiparallel to Earth field lines (from CROOKER, 1979).

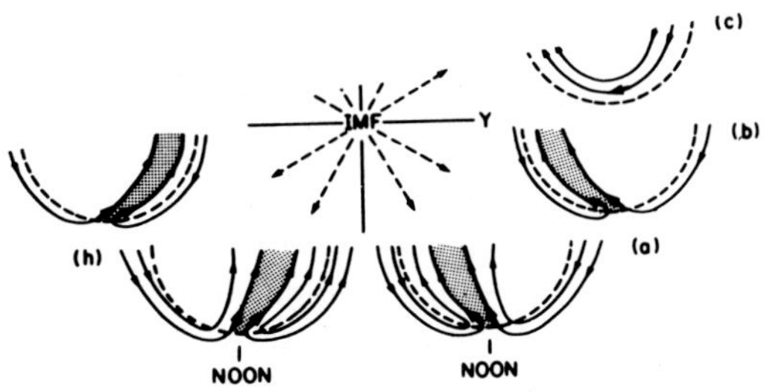

FIG. 48. Predicted convection patterns in dayside northern hemisphere polar cap for eight IMF orientations in y–z plane. Streamlines crossing polar cap boundary (heavy dashed line) effect flux transfer from closed dayside to open tail lobe field lines (from CROOKER, 1979).

dominantly in the polar cusp region, a concept which has yet to be confirmed by plasma and field measurements in that region which are sufficiently fast to resolve a transient reconnection process of only 10–30 sec duration (HAERENDEL, 1978). A simplified version of this concept is the hypothesis of CROOKER (1979) that reconnection prevails where external and internal fields are opposing each other (Fig. 47). This would yield a simple way to explain the concentration of the low altitude convection (Fig. 48) and Hall current to either morning or evening side of the polar cap, depending on the y-component of the interplanetary magnetic field (HEPPNER, 1972; FRIIS-CHRISTENSEN et al., 1972) (see also Subsection 4.1). However, this dependence can also be understood in the framework of the more conventional picture of low latitude reconnection (as sketched in Fig. 53b).

The field-aligned currents which achieve the long-range transfer of shear stresses from the boundary layer to the lower magnetosphere and ionosphere may be limited by the set-up of field-aligned potential drops (SONNERUP, 1980). The physics of the processes maintaining these voltages is dealt with in Chapter 5 of this volume. A consequence of the current limitation and closure of equipotential lines above the ionosphere is the set up of return flow in the boundary layer. Furthermore, the decoupling from the ionosphere, owing to the parallel potential drops, would facilitate interchange motions inside the boundary layer, as initiated, for instance, by the Kelvin-Helmholtz instability. SONNERUP (1980) estimates the total voltage between boundary layer and ionosphere to be of the order of 2 kV over a width of ≈ 100 km in the ionosphere.

One of the consequences of such a decoupling is the possibility of accelerating cold ions of ionospheric origin by the parallel electric fields. Wherever they are directed upward (predominantly on the evening side) they would lead to an injection of such ions into the boundary layer. The presence of substantial amounts of O^+- and He^+-ions (Subsection 3.1), which are indicative of an ionospheric origin, may be the by-product of this decoupling process.

3.4 *Transfer of mass, momentum and energy*

Density, temperature and composition of the boundary layer plasma strongly support the conclusion that it is of solar wind origin. A distinguishing feature of the entry layer is that frequently very little change of the density is observed across the magnetopause. This can be taken as evidence for a rather easy access of solar wind plasma to that region. Transfer processes which have been suggested in the literature comprise anomalous diffusion (EVIATAR and WOLF, 1968; EASTMAN *et al.*, 1976), direct influx along open field lines (LEVY *et al.*, 1964; HEIKKILA and WINNINGHAM, 1971; FRANK and GURNETT, 1977) as a consequence of reconnection, and an impulsive penetration of blobs of solar wind plasma with subsequent engulfment in the boundary layer (LEMAIRE and ROTH, 1978). Further transport in the boundary layer may be achieved by eddy transport (HAERENDEL, 1978).

Let us realize first how powerful the transport process has to be in order to account for the observed boundary layer properties. Much of the plasma mantle flow can be considered as an efflux from the entry layer into the tail. Total fluxes have been estimated by PILIPP and MORFILL (1978) to be of the order of 10^{26} sec^{-1}. The same figure was derived by EASTMAN *et al.* (1976) for the tailward transport of mass through the boundary layer at the flanks. This corresponds to an energy transport of about 10^{17} erg/sec. If we want to account for the above mass flux through the section of the magnetopause covering the entry layer, i.e. through an area of $\approx 10^{19}$ cm^2, and adopt a density of 20 cm^{-3}, we must postulate an average normal velocity $\langle u \rangle \cong 5$ km/sec.

Another way of expressing the normal mass flux is to refer to the voltage connected with the corresponding tangential electric field $\langle E_t \rangle = \dfrac{1}{c} \langle u \rangle B$. Taking $L \approx 5 \, R_E$ as a characteristic dimension and $B = 50 \, \gamma$, we find a voltage of the order of 10 kV, much less than the typically cited 50–60 kV across the polar cap (e.g. HEPPNER, 1972).

HAERENDEL (1978) proposed that the dominant transport mechanism inside the entry layer is some sort of *eddy diffusion*, initiated by the hydrodynamic turbulence in the exterior cusp region. A corresponding eddy transport coefficient should fulfill the condition:

$$D_{\text{eddy}} = \langle u \rangle d \qquad (23)$$

where d is the thickness of the layer. The entry layer can become as thick as 2 R_E. This means that D_{eddy} should have a magnitude of $5 \cdot 10^{14}$ cm^2/sec. Observations of transverse velocity fluctuations, $\delta v_\perp$, with amplitudes of up to 50 km/sec and time-scales, τ, of typically 20 sec allow an estimate of D_{eddy}, according to

$$D_{\text{eddy}} = \langle (v_\perp)^2 \rangle \tau . \qquad (24)$$

The numrical values lies in the postulated range.

We have little difficulty in understanding, at least qualitatively, the redistribution of the plasma, once it has penetrated the magnetopause, be it by eddy convection or growth of plasma blobs set up by the Kelvin-Helmholtz instability. Decoupling from the ionosphere, by field-aligned potential drops (SONNERUP, 1980) or by virtue of kinetic Alfvén-waves (HASEGAWA and MIMA, 1978), would allow flux tubes to be pushed around quite freely. What is much harder to account for is the initial loading of internal field lines with external plasma.

The easiest way of loading is probably by *intermittent reconnection.* This could account for the transient nature of the low latitude boundary layer as well as the highly disordered flow velocities observed in the polar cusp region (entry layer). However, the frequently observed large density jumps at the low latitude magnetopause, as well as the rather low flow velocities found in the cusp region, appear to be inconsistent with a rotational discontinuity (PASCHMANN *et al.,* 1976). On the other hand, as Section 4 will demonstrate, there is ample indirect and direct evidence for the frequent occurrence of reconnection. The problem is just whether it should be regarded as an additional independent mode of solar wind—magnetosphere interaction. It should be kept in mind that boundary layers with all their-transport effects, such as the vortex flows in the plasma sheet or the ionospheric soft particle precipitation, exist permanently and show little to no modulation with geomagnetic activity, which in turn depends strongly on the orientation of the interplanetary field and thus very likely on the occurrence of reconnection. Based on a rather limited data set, HAERENDEL *et al.* (1978) even found an anticorrelation of negative interplanetary B_z (favourable for reconnection) and the thickness of the low latitude boundary layer.

An *impulsive penetration* of whole plasma blobs as discussed by LEMAIRE and ROTH (1978) poses severe problems when the external magnetic field is not strictly aligned with the internal one. This limits the applicability of the concept decisively (SCHINDLER, 1979).

Finally, some kind of *anomalous diffusion* may be responsible for the initial transfer of plasma. The diffusivity, D_{an}, would arise from plasma turbulence in the magnetopause, probably initiated by instabilities of the Chapman-Ferraro currents. SCKOPKE *et al.* (1981) adopt this concept and estimate from the mass transport carried by the blobs of boundary layer plasma, how effective the diffusion would have to be if the whole magnetopause from the subsolar point to the point of observation (0800 L.T.) contributed to the diffusive entry. Figures like $D_{an} = 10^{13} \, \mathrm{cm}^2 \, \mathrm{sec}^{-1}$ over a thickness of $\delta = 400 \, \mathrm{km}$ would satisfy the requirements.

A general expression for D_{an} is:

$$D_{an} \cong R_{ge}^2 v^* \cong \left(\frac{c}{\omega_{pe}} \right)^2 v^* . \tag{25}$$

Hence $D_{an} = 10^{13} \, \mathrm{cm^2 \, sec^{-1}}$ implies an effective collision frequency of $v^* = 4 \cdot 10^3 \, \mathrm{sec^{-1}}$. This is close to the electron-cyclotron frequency, Ω_e, in a 20 γ field. In Section 5, we shall look into such processes from the point of view of microscopic theory and shall realize that it is not easy to account for such a high value of v^*, unless the layer of unstable current is much thinner than 400 km. Observed wave amplitudes at the magnetopause are as well substantially lower than the quoted value of v^* would imply.

Nevertheless, by comparing the magnitudes postulated for D_{eddy} and D_{an}, we can get an easy estimate of the boundary layer thickness, d, that would arise from further redistribution by turbulent convection. $d/\delta = (D_{eddy}/D_{an})^{1/2} \approx 10$. Hence, the observed thickness of the low latitude boundary layer is consistent with such a two-step process, which is suggested by the often plateau-like density profile in the boundary layer at a level well below the external plasma density.

In conclusion, we must say that the fundamental process that establishes the existence of the boundary layer has not yet been identified beyond doubt. It is likely that a combination of the entry process proper, which loads thin layers with solar wind plasma, and a subsequent irregular distribution process, some sort of eddy convection, may explain the origin of the layer.

4. Evidence for Reconnection at the Magnetopause

In Section 3, frequent reference was made to reconnection as a process which would permit a direct transfer of mass, momentum, and energy from the solar wind into the magnetosphere. In this section we will describe the available observational evidence for the occurrence of this process at the magnetopause.

Reconnection (or merging) of magnetic field lines is a process which changes the *topology* of magnetic fields. While for a closed magnetosphere one can distinguish only two classes of topologically different magnetic field lines (interplanetary and terrestrial), reconnection implies the existence of three classes, as illustrated in Fig. 49: interplanetary (I), terrestrial (III), and 'open' field lines (II). The latter have one end on the earth and the other in interplanetary space. These three regions are separated by surfaces called separatrices. The separatrices intersect along x-lines (or neutral lines) which

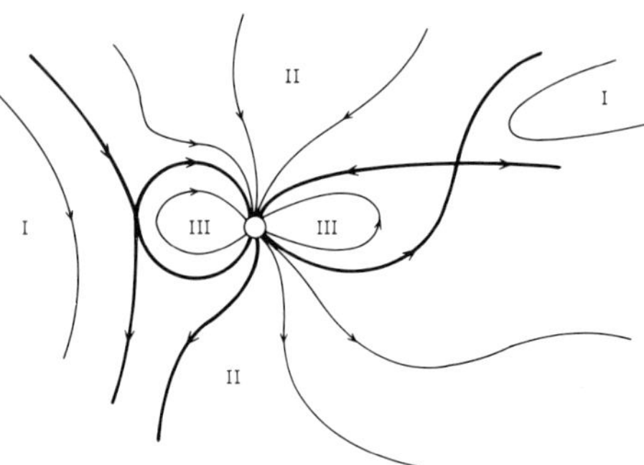

FIG. 49.   Magnetic field topology for magnetosphere with reconnection occurring at
the subsolar magnetopause and in the geomagnetic tail. Regions I, II, and III refer to
interplanetary, "open" and 'closed' terrestrial field lines, respectively. The regions are
bounded by separatrices (heavy lines).

encircle the earth (perpendicular to the plane of Fig. 49). The x-lines are the
site of the "diffusion region", which is the central part of a reconnection
situation (see Subsection 5.1). As pointed out already in Section 1,
reconnection may be defined as a "process whereby plasma flows across a
surface containing topologically different magnetic fields", i.e. across a
separatrix (VASYLIUNAS, 1975).

Figure 49 implies that reconnection occurs both on the frontside as well
as in the magnetotail. Here we are only concerned with the former, while the
latter is the subject of Chapter 4 of this volume.

Figure 49 shows the intersection (with the plane of the figure) of an x-
line near the subsolar point of the magnetosphere, as expected for
reconnection with a southward pointing interplanetary magnetic field (IMF).
Other configurations, including those where reconnection occurs with
northward IMF, are conceivable. Differences will also result from the y-
component of the IMF, which on average is much larger than the z-
component.

The indirect evidence for frontside reconnection, to be discussed in
Section 4.1, exploits the two basic features of Fig. 49: the existence of *open
field lines*, and the controlling influence of the *IMF orientation*.

The direct evidence (Subsection 4.2) is based on investigations of local
magnetopause properties, using a two-dimensional, steady-state description
of reconnection. In Subsection 4.3 the evidence for transient, small-scale
reconnection will be discussed.

4.1   Indirect evidence

4.1.1   Open field lines

Strong support for an open magnetosphere comes from observations of direct access of energetic solar electrons to the earth's polar caps. Such electrons are emitted from the sun during times of solar disturbances. Figure 50 shows an intensity profile of 376 keV electrons measured by a polar-orbiting satellite. The important feature is the flat profile near the central polar cap, indicating a spatially uniform electron bombardment. This is the situation one would expect if all polar cap field lines were open, i.e. extend into interplanetary space (cf. Fig. 49). By contrast, in a closed magnetosphere the field lines emerging from the central polar cap would stay far away from the magnetopause. Solar electrons could therefore gain access to these field lines only by cross-field diffusion. This would lead to an intensity profile with a depleted central polar cap (e.g. MORFILL and SCHOLER, 1973).

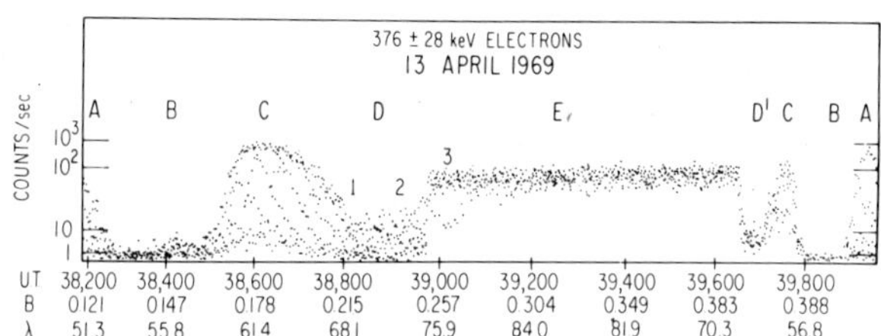

FIG. 50.   Uniform bombardment of the polar cap by solar electrons, in an analog record of 376 keV electron intensities measured along a satellite pass across the polar region. UT, magnetic field intensity, invariant latitude are shown along the time axis (from VAMPOLA, 1971).

In the above example, energetic particles *entering* the magnetosphere were used to trace out the open magnetic field topology. In much the same way, one could conceivably also use energetic particles *escaping* from the magnetosphere. It is indeed tempting to interpret the layers of energetic electrons and ions observed outside the magnetopause (e.g. BAKER and STONE, 1978; WILLIAMS, 1979b) in such terms. However, it is difficult to differentiate particle exit along open field lines from other mechanisms, such as outward diffusion, or losses by particle drift orbits which encounter the magnetopause.

4.1.2   IMF Correlations

The other type of indirect evidence is derived from the observed relationship between variations of certain magnetospheric phenomena and

          G. HAERENDEL AND G. PASCHMANN

the changing properties of the solar wind, in particular the orientation of the IMF.

ARNOLDY (1971) gave an empirical relation between the auroral electrojet (AE) index and B_z, the IMF north-south component. The AE index is a measure of the auroral electrojet current intensity, based on magnetic records from a distribution of auroral zone stations (see, for example, AKASOFU, 1977, p. 249). Figure 51 compares the measured AE index with that deduced from the equation given at the top of the figure, for a 6-day period in 1967. On such a coarse time-scale the correspondence is quite good. On the scale of the individual auroral events ($\lesssim 1$ hour), however, large discrepancies can appear. In the empirical relation of Fig. 51 each summation is extended over 1 hour of 10-min averages of $B_s(=B_z)$, multiplied with the time increment τ. The indices 0, 1, 2 indicate the lead-time in hours. As can be seen from the weight factors, the sum preceding the time of interest by 1 hour dominates the expression.

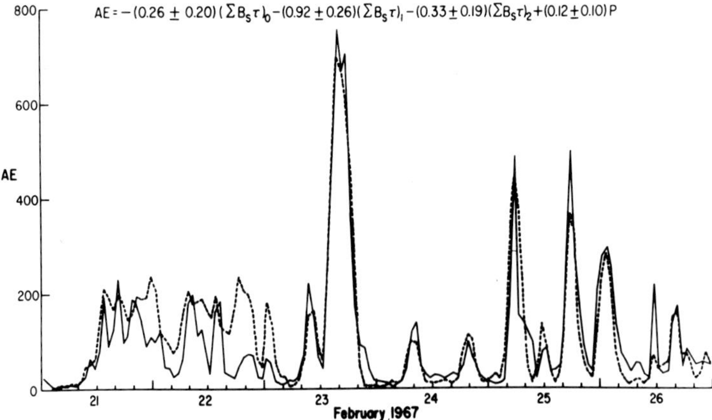

FIG. 51. Comparison of the empirical AE index deduced from the IMF B_z component and solar wind pressure (dashed curve), and the actual AE index (from ARNOLDY, 1971).

Another example in this category is the prediction of the magnetic activity index, D_{st}, on the basis of interplanetary parameters. D_{st}, which is computed from mid-latitude magnetic records, is a measure of the worldwide deviation of the magnetic H-component from its quiet-day values, an effect produced by magnetic storm-induced ring currents. BURTON et $al.$ (1975) have presented an algorithm for D_{st} based on (a) changes in solar wind kinetic pressure, $p = \rho v^2$, (b) a plasma injection rate into the ring current proportional to the solar wind electric field component E_y, (c) an exponential decay of the current with a time constant of ~ 8 hours. E_y is related to B_z via $E_y = -vB_z$, and is positive for southward B_z. A 'rectifier' effect was introduced by requiring that the injection becomes zero for $E_y \leq 0$, i.e. for northward B_z. The solar wind kinetic pressure enters as the square root to account for the (positive) contribution of magnetopause currents to ΔH measured at the earth's surface.

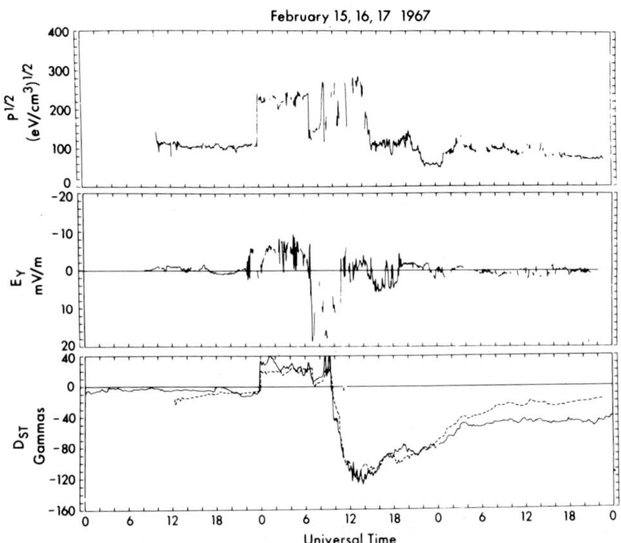

FIG. 52.   (From top to botom) The square root of the solar wind dynamic pressure, the dawn-to-dusk component of the interplanetary electric field, and the predicted (dashed line) and observed (solid line) D_{st} (from BURTON et al., 1975).

Figure 52 shows measured and predicted D_{st}'s, together with the quantities E_y and $p^{1/2}$, on which the prediction is based, for a magnetic storm in February 1967. The agreement is quite remarkable.

Another relationship, intended to describe the energy coupling from the solar wind to the magnetosphere, has been introduced by Perreault and AKASOFU (1978):

$$\varepsilon(t) = vB^2(\sin^4 \theta/2)l_0^2 \tag{26}$$

where v is again the solar wind velocity, B the IMF magnitude, θ is the azimuth angle (measured from the z-axis) of the IMF in the solar-magneto-spheric y, z plane, and l_0 is a constant length chosen as 7 R_E. ε has the dimensions of energy/time. Relation (26) has been quite successful in predicting the energy dissipated during magnetic storms, as well as the AE-index (AKASOFU, 1979). Contrary to the foregoing D_{st} algorithm, Equation (26) does not include a rectifier effect: $\varepsilon(t)$ is different from zero even for a northward B_z, i.e. $\theta < 90°$. However, the fact that $\sin\theta/2$ appears as the fourth power, makes the contribution for $B_z > 0$, i.e. $\theta < \pi/2$, rather small.

In none of the correlations discussed so far the sign of the IMF y-component plays a role. That it has a significant effect was shown by HEPPNER (1972). He demonstrated that E_y, the dawn-dusk component of the electric field measured over the earth's polar caps, shows a pronounced dawn-dusk asymmetry (E_y being stronger on one side than the other), with the asymmetry reversed when B_y is reversed (compare Subsection 3.3).

A strong influence of B_y has also been established for the distribution of the ionospheric currents over the polar cap. Daily variations of the horizontal magnetic field component on the ground have a markedly different pattern for days with positive and negative B_y (e.g. FRIIS-CHRISTENSEN et al., 1972).

4.2 In situ observations

From the material presented in Subsection 4.1 it seems unavoidable to conclude that reconnection occurs, in as much as no other mechanism is known, which could produce the basic features, namely open field lines and IMF control. In this subsection we will investigate whether plasma and field measurements at the subsolar magnetopause corroborate this view.

The configuration most appropriate for reconnection at the subsolar magnetopause is illustrated in Fig. 53a. It is based on the steady, two-dimensional model of LEVY et al. (1964) which describes a situation with a cool, dense plasma on one (the magnetosheath) side, and a hot, dilute plasma on the other (magnetospheric) side. Because of this, the configuration is referred to as the '*asymmetric*' case, as opposed to the symmetric case discussed by PETSCHEK (1964), where the conditions on both sides are assumed to be the same (compare Fig. 63b). The outstanding features in Fig. 53a are: the *diffusion region* around the x-line, where the change of magnetic field topology actually takes place (see Subsection 5.1), and two standing waves emanating from the x-line.

The outer is an Alfvén wave, or *rotational discontinuity*, which serves to rotate the field from its arbitrary orientation in the magnetosheath to the

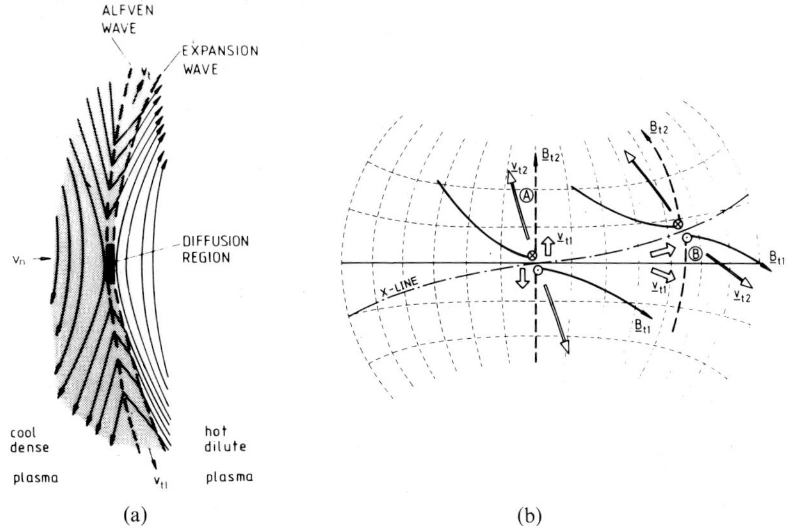

FIG. 53.   (a) Asymmetric reconnection configuration for antiparallel interplanetary
and terrestrial magnetic fields. Cool, dense plasma of the magnetosheath (left)
expands into the magnetosphere (right), which is filled with a hot, dilute plasma (after
LEVY *et al.*, 1964). (b) Front view of the magnetopause for reconnection with an
interplanetary field (solid lines) having a strong positive B_y component, in addition to
a southward component B_z. As a result, the x-line (dot-dash line) is tilted.
Magnetospheric portions of field lines are dashed. Arrows indicate plasma flow
velocities.

well-defined magnetospheric orientation. For an isotropic plasma, no change
in field strength is associated with the rotational discontinuity. The second,
innermost standing wave is an *expansion wave* where the magnetic field
magnitude changes to its magnetospheric value and where the
magnetosheath plasma terminates. In applying this model, one would
identify the magnetopause with the rotational discontinuity, and the region
between the two standing waves with the boundary layer (cf. Section 3). In
the reconnection model the boundary layer is entirely on open field lines.

In pictures like Fig. 53a the existence of an IMF y-component is
necessarily ignored. As Fig. 53b shows, B_y introduces an asymmetry, most
prominently a tilt of the reconnection line. This tilt is reversed if the sign of
B_y is reversed.

The diffusion region is the subject of Section 5. Here we are only
concerned with the magnetopause outside the diffusion region. What

features does one expect to observe if the above description is valid?

a)   There must be a normal component, B_n, of the magnetic field which is directed inward north of the x-line and outward south of it.

b)   There must be a tangential electric field, E_t, along the magnetopause, consistent with the motion of plasma towards and across the magnetopause and its direction must be such that $E_t \cdot J > 0$ (where J is the magnetopause current), implying conversion of electromagnetic into kinetic energy of the plasma.

c)   There is an inward directed plasma flow.

d)   Plasma which has crossed the magnetopause must exhibit a substantial change of the tangential flow velocity.

4.2.1   Normal magnetic field

To reliably determine B_n is a formidable task, since, for reasonable reconnection rates, B_n is only a few γ. This means that the local magnetopause normal must be known to within a few degrees. As the magnetopause constantly changes its shape as a result of surface waves, a model surface, such as determined in Subsection 1.1, is far from satisfactory. A method which has had remarkable success in determining B_n is the so-called minimum variance technique (SONNERUP and CAHILL, 1967).

For a one-dimensional magnetopause, i.e. one where the variations along the surface occur on length scales much larger than its thickness, div $B = 0$ requires that B_n should be constant across the magnetopause. Since in practice one cannot find a direction along which B_n is strictly constant, one chooses as an approximation the direction which yields minimum variance in the corresponding field component. Mathematically, this amounts to solving an eigenvalue problem. The normal vector is identified with the eigenvector belonging to the smallest eigenvalue. One important restriction on the applicability of the technique lies in the fact that for strictly unidirectional magnetopause currents the two smallest eigenvalues become equal, with the result that the normal direction is undetermined.

Figure 54 shows an example where the minimum variance technique yielded a finite B_n with high confidence level (SONNERUP and LEDLEY, 1979). The data for a 3-sec interval spanning the magnetopause transition are presented as hodograms, in the coordinates obtained from minimum variance analysis of that same interval. B_1 corresponds to the axis of maximum variance, B_3 to that of minimum variance. The latter is therefore identified with B_n. The hodogram on the left (in the plane tangent to the magnetopause) shows the semi-circular shape characteristic for a rotational discontinuity. At the same time the field has a consistent normal component

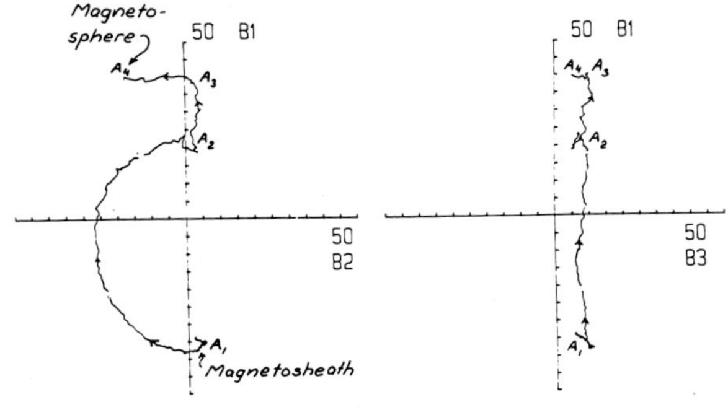

FIG. 54. Polar plots of magnetic field at the magnetopause. Left-hand figure shows the field components B_1 and B_2 tangential to the magnetopause during an OGO-5 crossing; right-hand figure shows the nearly constant magnetic-field component B_3 normal to the magnetopause. The field is given in units of γ. Intermediate wave or rotational discontinuity is segment A_1-A_2 of the left-hand trace: the slow expansion fan is segment A_2-A_3. The segment A_3-A_4 may be caused by a finite gyro-radius effect not contained in the MHD model (from SONNERUP and LEDLEY, 1979).

of (8 ± 0.4) γ, as shown by the hodogram on the right. Towards the magnetospheric end of the B_1, B_2 trace there are indications of the expansion wave, and also of Parker's transverse field (cf. Section 1.4).

4.2.2 Tangential electric field

Measurements of the electric field at the magnetopause have only recently become available. Moreover, a reliable determination of E_t is difficult. Not only are the fields expected to be small (proportional to B_n), they must also be transformed into a reference frame fixed with respect to the magnetopause, because only in such a frame E_t is required to be constant across the magnetopause.

Figure 55 shows the first example of a magnetopause crossing for which a persistent positive E_t was reported (MOZER *et al.*, 1979). From its magnitude (a few mV/m) and direction with respect to the local magnetopause current, an energy dissipation $E_t \cdot J$ of ~ 70 Watts km^{-2} was derived.

It should be noted that these measurements are always interpreted in terms of a quasi-stationary reconnection model for which a frame of reference can be found in which curl $E=0$. In reality this need not be the case. Induction electric fields may not only exist, but play a dominant role in the reconnection process. We shall return to this question in Subsection 5.4.

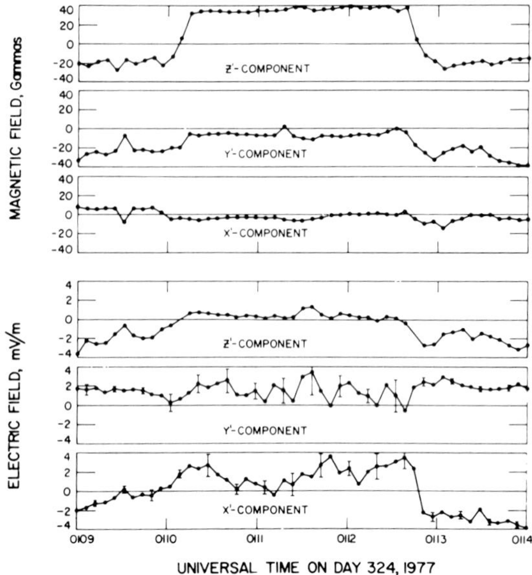

FIG. 55. Five minutes of electric and magnetic field data in a frame of reference oriented along the average magnetopause, during the passage of the magnetopause back and forth over the ISEE-1 spacecraft. The error bars in the electric field data are the standard deviations of a single point in the sine wave least squares fit of six seconds of data. E_y' is the tangential component in the dawn-to-dusk direction. The magnetopause motion has not been removed from the data (from MOZER et al., 1979).

4.2.3 Trangential plasma flows

As we have seen, the quantities, B_n and E_t, which are directly related to the reconnection rate, are difficult to measure. This is even more true for v_n, the inward plasma flow velocity.

The situation is quite different for the tangential plasma flows. These are expected to be quite large. Contrary to the case of B_n and v_n, they also are not sensitive to the choice of the magnetopause normal, and they do not require any transformation into a magnetopause frame, as in the case of E_t. On the other hand, the flows can assume arbitrary directions and occur over narrow length scales. Only the advent of three-dimensional flow measurement with high time resolution made the analysis possible.

The changes in plasma flow velocity at the magnetopause result from the Maxwell stresses. Picturing the sharply bent field lines in Fig. 53 as elastic bands acting on the plasma, the process could be described as a magnetic slingshot, which ejects the plasma at high speeds in poleward directed jets.

Quantitatively, one can approach the problem from two different directions (HUDSON, 1971).

First, one can start out from tangential momentum balance across a discontinuity. This implies that:

$$\left[F_n v_t - \frac{B_n}{4\pi} B_t \right] = 0$$

or

$$F_n [v_t] = \frac{B_n}{4\pi} [B_t] \,. \tag{27}$$

Here v_t and B_t are the tangential velocity and magnetic field, respectively, F_n is the mass flux across the discontinuity:

$$F_n = \rho(v_n - U_n) \,, \tag{28}$$

where v_n is the normal component of the plasma flow velocity and U_n is the magnetopause speed, both measured in the satellite frame. The brackets denote the differences of the enclosed quantities across the discontinuity. Conservation of mass implies $F_n = \text{constant}$.

One can also start out from the constancy of the tangential electric field, E_t, across a discontinuity in a quasi-stationary situation, and combine it with the conservation of mass flow. Assuming that other terms in Ohm's law are negligible, E_t can be expressed as

$$E_t = -(v \times B)_t$$

or

$$[B_n v_t - v_n B_t] = 0 \,.$$

Using

$$[B_n] = 0 \quad \text{and} \quad v_{ni} = \frac{F_n}{\rho_i}$$

one obtains

$$B_n [v_t] = F_n \left[\frac{B_t}{\rho} \right] \,. \tag{29}$$

The index $i = 1, 2$ refers to the different sides of the discontinuity.

Elimination of $[v_t]$ from Eqs. (27) and (29) yields

$$F_n^2\left[\frac{B_t}{\rho}\right]=\frac{B_n^2}{4\pi}[B_t]$$

or

$$B_{t1}\left(\frac{F_n^2}{\rho_1}-\frac{B_n^2}{4\pi}\right)=B_{t2}\left(\frac{F_n^2}{\rho_2}-\frac{B_n^2}{4\pi}\right). \tag{30}$$

We are interested in the only discontinuity, which is able to rotate the magnetic field by an arbitary angle, the rotational discontinuity. If B_{t1} and B_{t2} are not parallel, it follows from Eq. (30) that the factors in parentheses must vanish; in other words $\rho_1=\rho_2$, and

$$(v_n-U_n)=\pm\frac{B_n}{(4\pi\rho)^{1/2}}. \tag{31}$$

This means that the normal flow speed through the rotational discontinuity is equal to the Alfvén speed based on the normal field component.

With Eq. (30) expressions (29) and (30) take on the identical form

$$[v_t]=\pm\frac{[B_t]}{(4\pi\rho)^{1/2}}. \tag{32}$$

Here the positive and negative signs apply to the cases with $B_n<0$, $B_n>0$, respectively.

Equation (32) indicates that the change in tangential velocity is equal to the change in tangential Alfvén velocity.

For the case of an anisotropic plasma pressure, the equations are somewhat modified (e.g. HUDSON, 1971). Instead of Eq. (32), one obtains:

$$[v_t]=\left[\frac{(1-\alpha)^{1/2}B_t}{(4\pi\rho)^{1/2}}\right] \tag{33}$$

where α is the pressure anisotropy factor, defined as

$$\alpha\equiv\frac{(p_\parallel-p_\perp)}{B^2/4\pi}.$$

In the anisotropic case, the densities on both sides of the discontinuity need not be equal, but are related via

$$\rho_1(1-\alpha_1)=\rho_2(1-\alpha_2).$$

Equation (31) for the normal flow, finally, is replaced by

$$v_n - U_n = \pm \frac{(1-\alpha)^{1/2} B_n}{(4\pi\rho)^{1/2}}. \tag{34}$$

In the extreme case where $\boldsymbol{B}_t$ rotates through $180°$, Eq. (32) yields:

$$|\Delta v_t| = 2\frac{|\boldsymbol{B}_t|}{(4\pi\rho)^{1/2}} \tag{35}$$

which for typical magnetopause conditions is of the order of $300\,\mathrm{km\ sec^{-1}}$.

Now we will discuss the observational tests of the relationship between tangential plasma flow and magnetic field components described by Eqs. (32) and (33) (PASCHMANN *et al.*, 1979b; NISHIDA *et al.*, 1979; SONNERUP *et al.*, 1981).

Figure 56 shows plasma and magnetic field data from an outbound traversal of the subsolar magnetopause region at a latitude of $\sim 25°$. From

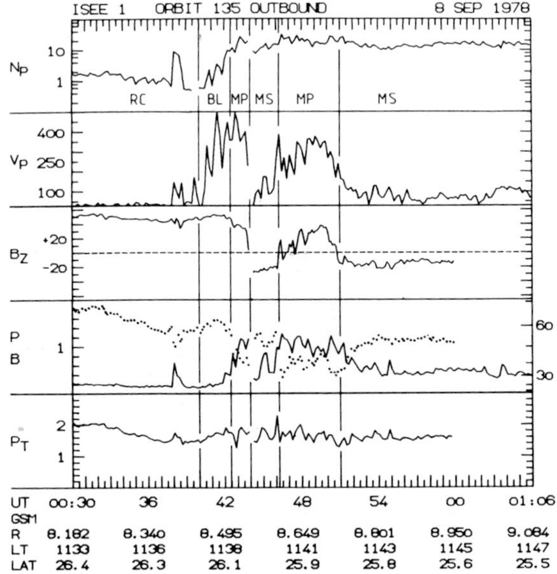

FIG. 56.   Plasma and magnetic field data for the ISEE-1 magnetopause crossings of 8 September 1978. From top to bottom, the figure shows the plasma density ($\mathrm{cm^{-3}}$) and bulk speed ($\mathrm{km\,s^{-1}}$), the north-south component of the magnetic field (γ), the magnetic field (dotted line) and plasma (solid line) pressures, and the total pressure P_t, all in dynes $\mathrm{cm^{-2}}$. GSM coordinates (radial distance, local time and latitude) are indicated along the UT axis. Multiple magnetopause (MP) crossings are identified by the transitions in B_z from northward to southward (from PASCHMANN *et al.*, 1979b).

the panel showing the plasma flow speed, v_p, the qualitative signature of the rotational discontinuity is immediately apparent: every time the magnetopause is encountered, the flow speed is much enhanced over the speed in the magnetosheath. In the magnetosheath, speed are $\sim 50\,\mathrm{km\,s^{-1}}$, consistent with the satellite being located fairly close to the stagnation region of the flow (cf. Section 1). Within the magnetopause and boundary layer, the flow speed increases to $\lesssim 500\,\mathrm{km\,s^{-1}}$, an increase of almost a factor of 10.

For the quantitative analysis, the full magnetic field and plasma flow vectors obtained during the first magnetopause crossings of Fig. 56 were converted into a normal coordinate system, and then plotted as hodograms as shown in Fig. 57.

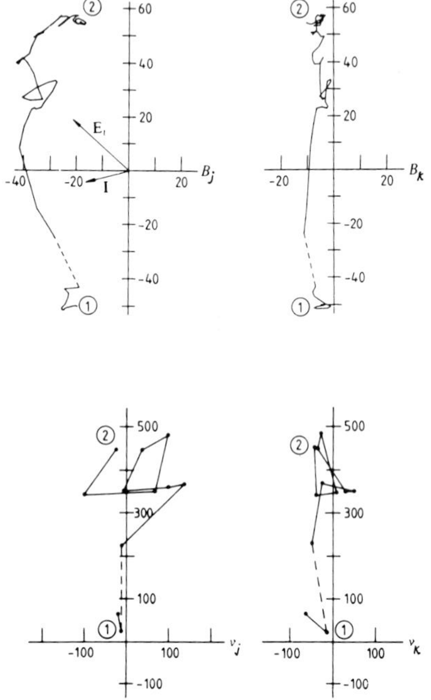

FIG. 57. Polar plots of the magnetic field $\boldsymbol{B}$ (top) and plasma flow velocity $\boldsymbol{v}$, for the first magnetopause crossing of Fig. 56, in the coordinate system obtained from minimum variance analysis. The k direction is along the outward directed magnetopause normal, the i, j plane is tangential to the magnetopause, with the i and j axes approximately due north and west, respectively. The points marked (1) and (2) refer to the magnetosheath and magnetosphere sides of the magnetopause. Also shown are vectors representing the magnetopause current $\boldsymbol{I}$ and the inferred electric field $\mathbf{E}_t$ (from PASCHMANN et al., 1979b).

Inspection of the tangential hodograms (left) immediately shows that the changes in velocity and magnetic field across the magnetopause are nearly parallel, as required by Eq. (32). Taking the change in B_t between points 1 and 2 in the tangential hodogram, and the measured plasma mass density, and inserting them into Eq. (32), one obtains a predicted change in plasma velocity of $580\,\mathrm{km\,s^{-1}}$. As can be seen from Fig. 57 the actually measured change is $\sim 425\,\mathrm{km\,s^{-1}}$. This represents a fairly good agreement, if one considers the difficulty of making accurate three-dimensional flow measurements.

A detailed analysis of 11 such cases (including the one already presented) has confirmed the conclusion that the magnetopause can have the structure of a rotational discontinuity, as required by reconnection (SONNERUP *et al.*, 1981). The analysis also included the contributions of plasma pressure anisotropy according to Eq. (33), which turned out to be rather small. Figure 58 shows the comparison of the measured velocity changes (arrows) with the predicted changes, after appropriate normalization. Vectors of unit length along the horizontal would indicate perfect agreement. The figure shows that the agreement is quite good, both in direction and magnitude.

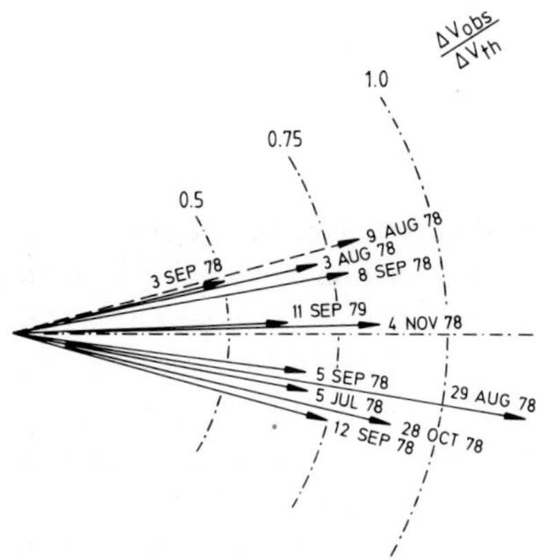

FIG. 58.   Result of the comparison between measured and predicted tangential velocity enhancements across the magnetopause for 11 cases observed by ISEE-1 and -2. Each vector represents the measured velocity change, rotated around the normal in such a way that the predicted velocity change lies along the horizontal (dotted line) in each case. Vectors have been normalized such that the predicted vector has unit length. In case of perfect agreement, the vectors should be horizontal and of length 1.0 (from SONNERUP *et al.*, 1981).

Whereas it is satisfying to find the concept of collisionless reconnection confirmed by measurements of the local plasma and field properties, it should not be overlooked that during a significant number of magnetopause crossings, no reconnection signatures were observed (HAERENDEL et al., 1978; SONNERUP and LEDLEY, 1979; SONNERUP et al., 1981), although the IMF had a southward component. We must conclude that either southward pointing IMF is not sufficient for reconnection to occur, or that the process is of small-scale and pulsating nature with extended phases where it is effectively switched off.

4.3 Flux transfer events

In the previous subsection we discussed the evidence for the structure of the magnetopause predicted by the two-dimensional, steady-state model of reconnection. There is, however, also ample evidence for a non-stationary nature of the process. Certain characteristic variations of the magnetic field near the magnetopause have been interpreted in terms of small-scale, transient erosion of magnetic flux.

Figure 59 shows magnetic field data for an inbound crossing of the magnetopause near local noon at a magnetic latitude of $\sim 35°$. The features to note here are the large spikes in the field strength after ~ 01 UT. These spikes occur *inside* the magnetosphere and have a duration of a few minutes. HAERENDEL et al. (1978) have interpreted these spikes as events of *impulsive erosion* of magnetic flux tubes. Figure 60 illustrates the proposed mechanism. An interplanetary flux tube is swept over the polar cusp region and reconnects with the terrestrial field at a location where the two fields are anti-parallel (position ① in Fig. 60). The dashed portion of this field line is then connected to the northern hemisphere, whereas the solid portion is connected with a field line from the southern hemisphere. It is the solid field line we are interested in. The sharp kink in that field line implies magnetic stresses which point in the sunward direction, and thus oppose the solar wind flow, which carries the external portion of the field line tailwards. This way the solar wind can do work on the magnetic field and "erode" magnetic flux that formerly belonged to the magnetosphere, in the fashion illustrated in Fig. 60 by the temporal progression from ① to ③. When mechanical work is done on the magnetic field against the magnetic tension, the result will be an increase in field strength. Through "extension of the lines of force" the density of the field increases (COWLING, 1957). This is exactly what is observed in Fig. 59.

The situation just described is the opposite to that experienced in

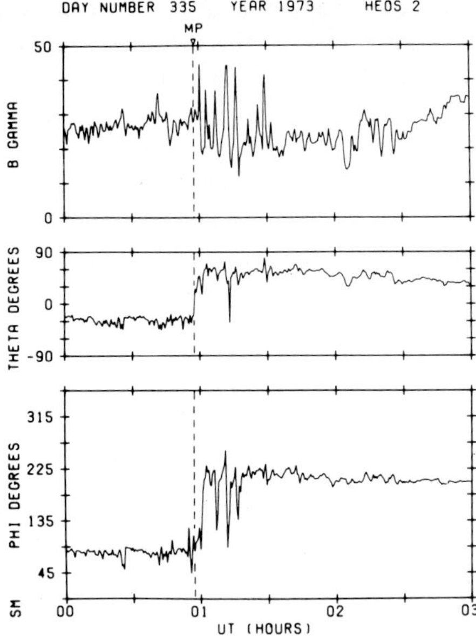

FIG. 59.   Magnetic field data for an inbound magnetopause crossing by Heos 2 near local noon at mid-latitudes showing flux erosion events inside the magnetopause (from HAERENDEL *et al.*, 1978).

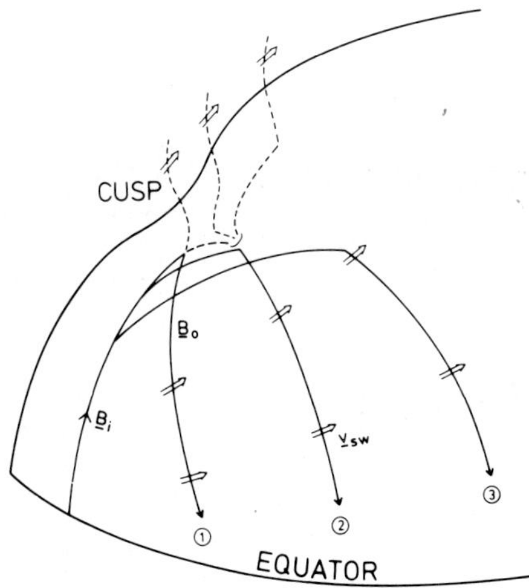

FIG. 60.   Erosion of magnetic flux from the frontside of the magnetosphere if merging is initiated in the cusp region. The external parts of the magnetic lines of force are draped around the magnetopause and carried in the antisolar direction. The magnetosheath plasma does work against the magnetic tension (from HAERENDEL *et al.*, 1978).

reconnection at the subsolar magnetopause (cf. Subsection 4.2 and Fig. 53). There the magnetic tensions are in the direction of the plasma flow and therefore do work on the plasma, which consequently is accelerated.

From the duration of the events one would conclude that the reconnection process for these flux tubes in the polar cusps is transient, with a duration of the order of a minute. Such a time-scale does not necessarily invalidate the applicability of the stationary jump relations, if they are applied sufficiently close to the x-line and one does not demand perfect quantitative agreement.

The example just discussed referred to a transient flux erosion as observed inside the magnetosphere. Similar events *outside* the magnetosphere have been reported by Russell and Elphic (1979), who referred to them as *flux transfer events*.

Figure 61 shows magnetic field measurements for ~ 1 hour during an inbound traversal of the magnetosheath into the magnetosphere at low latitudes near local noon. The data are displayed in L, M, N boundary normal coordinates, a cartesian coordinate system, where N is along the local magnetopause normal, L is in the plane defined by N and the z-axis of the solar magnetospheric (GSM) system, M completes the right-handed

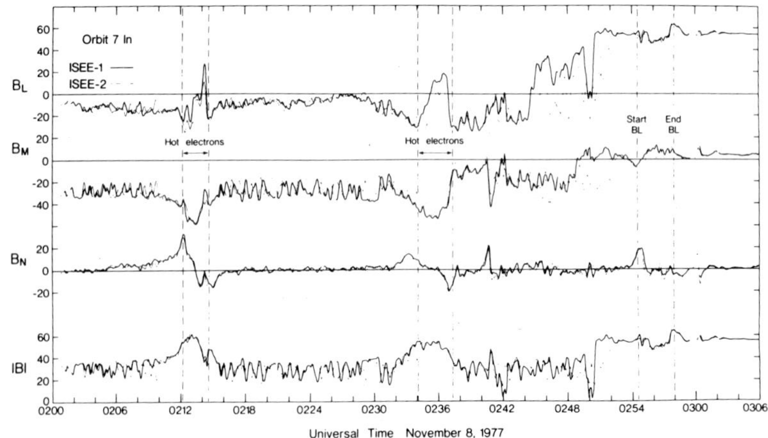

FIG. 61.   Magnetic field at ISEE-1 (heavy line) and ISEE-2 (light line) on November 8, 1977, in boundary normal coordinates. The spacecraft were at $(10.16, -1.77, 5.07)$ R_e in solar magnetospheric coordinates and separated by 299 km along the model boundary normal. The main flux transfer events are observed near 0212 and 0236 UT. Entry in the magnetosphere occurs at 0250 UT (from Russell and Elphic, 1979).

system. The periods of interest in Fig. 61 are the two intervals marked by the vertical dashed lines. In both events one notices a characteristic bipolar signature of the B_N component. The interpretation is illustrated in Fig. 62: a flux tube of magnetosheath field lines has reconnected with magnetospheric field lines, probably some time earlier and in the subsolar region; and while being transported by the magnetosheath plasma flow in the direction of the large arrow, the flux tube passes over the spacecraft. The draping of magnetosheath field lines over the flux tube then produce the noted B_N variation.

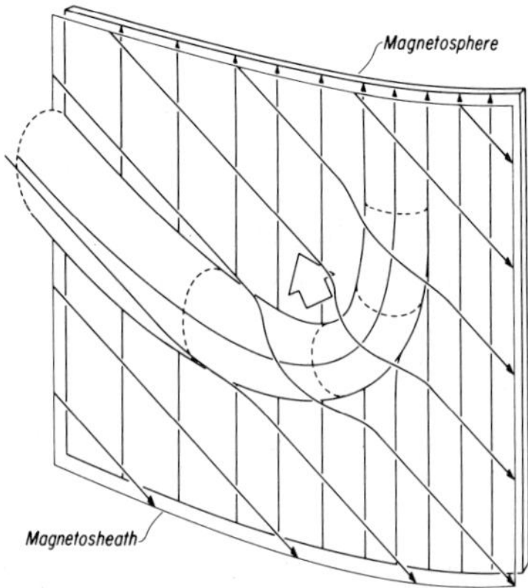

FIG. 62. Qualitative sketch of a flux transfer event. Magnetosheath field lines, slanted arrows, have connected with magnetospheric field line, vertical arrows, possibly off the lower edge of the figure. The connected flux tube is carried by the magnetosheath flow in the direction of the large arrow. Magnetosheath field lines not connected to the magnetosphere drape over the connected flux tube and are swept up by its motion relative to the magnetosheath flow (from RUSSELL and ELPHIC, 1979).

The situation depicted in Fig. 62 is characterized by the magnetic stresses of the flux tube being in the same direction as the external flow. This is in contrast to the previously discussed case of transient cusp reconnection. Consequently, the explanations for the increased field strength in both events cannot be the same. In the present case, pressure balance may be maintained by the magnetic tension in the external magnetosheath field lines which are draped around the flux tube (cf. Fig. 62).

The concept of identifiable flux tubes is often employed in discussions of astrophysical plasmas, e.g. in the solar corona or in cometary tails, where flux tubes appear visibly by differences in the emitted light. Now we find increasing reason to talk of finite size flux tubes which distinguish themselves from their environment by differences of the field topology.

5. Microprocesses Related to Reconnection

The previous section demonstrated that reconnection indeed occurs at the frontside magnetopause and plays a crucial role in the interaction of the solar wind with the magnetosphere. Since the medium is extremely collision-free, the gas-kinetic collision lengths being orders of magnitude larger than the size of the magnetosphere, not to speak of the width of the magnetopause, it is an interesting question which processes take over the role of collisions in ordinary fluids or gases. The interaction cannot be entirely collision-free; somewhere Ohmic dissipation must exist and enable the change of topology that the magnetic field undergoes in the reconnection process.

Besides the interest in identifiing the most relevant microscopic processes that create the plasma turbulence responsible for collision-free Ohmic dissipation, there exists some uncertainty about the control that these processes exert on the efficiency and dynamics of the reconnection process. Quite diverging opinions have been expressed on this question in the literature. They range from formulations of the reconnection rate in terms of the effective collision frequency, v^*, or anomalous resistiveity, η_{an}, (PARKER, 1957), over a weak logarithmic dependence (PETSCHEK, 1964) to complete independence (YEH and AXFORD, 1979). This divergence of opinions is a consequence of the different macroscopic geometries and boundary conditions adopted for the reconnection process and also of the consideration or neglect of any feed-back processes. Even if there are always powerful instabilities available to cope with the maximum reconnection rate that is allowed in any given situation, it is not clear how long such a process may last. All instabilities have the tendency to change the macroscopic parameters in a way that reduces the source of free energy driving them. Only if the external forces and transport processes are such that the free energy is replenished at the same rate at which the dissipated energy is removed, can the process be stationary. A stationary model is, of course, based on the hypothesis that this is the case. The present status of theoretical understanding is such that we are unable to predict whether, and under which circumstances, a quasi-stationary reconnection can exist at the frontside magnetopause, what the external driving forces are, or whether it can occur spontaneously and maintain itself for prolonged periods.

It is therefore an important task to study the microscopic processes and their dynamics in regions where reconnection occurs. Although the magnetopause regions are exhibiting plenty of plasma wave activity, there has been no report of unambiguous evidence for an encounter of the very neighborhood of the x-line, mostly referred to as *diffusion region*, in which the magnetic topology is changed by anomalous Ohmic dissipation.

This chapter will deal with the subject in four steps. First, we deduce quantitative limitations on the microprocesses from macroscopic paramerers and their observed values. This will be followed by an inspection of the candidate instabilities for producing the necessary plasma turbulence. Then we will compare the present knowledge of wave activity near the magnetopause with the theoretical predictions. Finally, we will return to the cause-and-effect questions about reconnection.

5.1   *Macroscopic constraints*

The two extreme geometries for which reconnection has been discussed

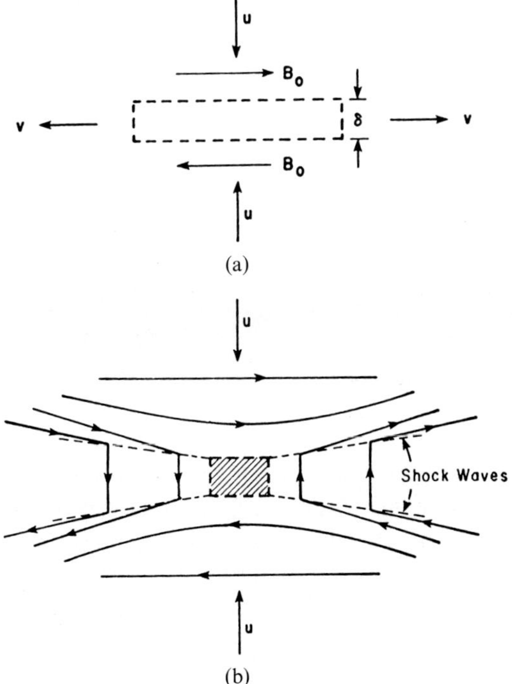

FIG. 63.   (a) Sweet-Parker mechanism of magnetic annihilation. (b) Petschek mechanism of X point reconnection.

in the literature are sketched in Fig. 63. They differ essentially with respect to the ratio of length and width of the diffusion region, l/δ. In the Sweet-Parker annihilation model (PARKER, 1957), l is determined by the size of the system under consideration, whereas in PETSCHEK's (1964) reconnection model, l adjusts to the reconnection rate and the microprocesses, i.e. the effective magnetic diffusion coefficient. The two models provide quite different answers for the rate of reconnection and its dependence on the electrical resistivity. The reason for this lies essentially in the efficiency with which the matter is being ejected from the diffusion region.

In both models, plasma flows with normal velocity u, into a layer of length, l, and is ejected with the speed, v, over a width, δ. We assume incompressibility and obtain from *conservation of mass*:

$$ul = v\delta \ . \tag{36}$$

Inside the layer, the magnetic field strength assumes low values, even drops to zero since the external fields are assumed to be anti-parallel. *Momentum balance* between the external magnetized and the internal (essentially) field-free region requires:

$$p = p_0 + \frac{B^2}{8\pi} \tag{37}$$

where p is the interior pressure. The pressure difference between the low field region and the distant sheet, where the magnetic field has recovered to its normal value, drives the efflux with velocity, v:

$$p - p_0 = \frac{\rho}{2}v^2 \approx \frac{B_0^2}{8\pi} \tag{38}$$

$$v = v_A = \frac{B_0}{\sqrt{4\pi\rho}} \tag{39}$$

(PARKER, 1957). Thus, irrespective of the magnitude of l and of magnetic configuration,   all models yield an efflux velocity of the order of the Alfvén velocity, as calculated with the unperturbed value of B.

The magnetic field must be regarded as frozen in the plasma outside the diffusion region. The fluid moving toward the diffusion layer carries magnetic flux with it which is annihilated by Ohmic dissipation. The energy goes partially into heating ($p > p_0$) and partially into acceleration. In the real world, a substantial amount of the heat generated may be lost by electron heat conduction and not be available for hydrodynamic acceleration.

In steady state, a balance exists between the magnetic field convection by the inward moving fluid and magnetic diffusion due to finite resistivity in the layer. The *diffusive equilibrium* can be expressed by:

$$u = \frac{D_{\mathrm{m}}}{\delta} \qquad (40)$$

where the magnetic diffusivity, D_{m}, is related to the (anomalous) electrical resistivity, η^*, by:

$$D_{\mathrm{m}} = \frac{\eta^* c^2}{4\pi} . \qquad (41)$$

Combining Eqs. (36), (39)–(41) yields an expression for the merging rate, u, in terms of electrical resistivity and the length, l:

$$u = \left(\frac{\eta^* c^2 v_{\mathrm{A}}}{4\pi l} \right)^{1/2} . \qquad (42)$$

Introducing the *magnetic Reynolds* number, R_{m}, by:

$$R_{\mathrm{M}} = \frac{v_{\mathrm{A}} l}{D_{\mathrm{m}}} \qquad (43)$$

we can express the merging rate by the Alfvén velocity, v_{A}, and R_{M}:

$$u = \frac{v_{\mathrm{A}}}{\sqrt{R_{\mathrm{M}}}} . \qquad (44)$$

Even with high values of the anomalous resistivity (i.e. D_{m}), R_{M} tends to be a large quantity for dilute astrophysical plasmas. Hence, magnetic annihilation in a thin extended neutral sheet is too slow to account for many observations (PARKER, 1957).

PETSCHEK (1964) showed a way out of this dilemma by matching the diffusion region to a region of standing waves, slow shocks, which result in coversion of magnetic energy into kinetic energy of flow. Thereby, l becomes an adjustable quantity and is effectively shortened. The merging rate, u, becomes strongly dependent on the boundary conditions and only weakly on R_{M} as defined with the external length, L. Other models (YEH and AXFORD, 1970) yield complete independence. As VASYLIUNAS (1975) has shown, increase of entropy is only consistent with an upper limit of the merging rate near v_{A}. The influx velocity far away from the diffusion region is therefore:

$$u = \varepsilon v_{\mathrm{A}} \qquad \text{with} \quad \varepsilon \lesssim 1 . \qquad (45)$$

The various reconnection models yield values of ε between 0.1 and ≈ 1, whereby the lower values seem to be more appropriate for free space situations and are also consistent with determinations of the normal component of $\boldsymbol{B}$ or tangential component of $\boldsymbol{E}$ (see Subsection 4.2).

With these preparations we can derive a few useful relations characterizing the range of macroscopic quantities which has to be accounted for by the microprocesses. Using the relation,

$$\eta^* = \frac{4\pi v^*}{\omega_{pe}^2} \tag{46}$$

between resistivity and effective collision frequency, v^*, and Eqs. (40) and (41), we can relate the thickness of the stationary diffusion region to v^*:

$$\delta = \frac{D_m}{u} = \left(\frac{c}{\omega_{pe}}\right)^2 \frac{v^*}{\varepsilon v_A}. \tag{47}$$

The inadequacy of Coulomb collisions is impressively demonstrated by inserting for v^* the corresponding collision rate. With $\varepsilon = 0.1$ we find $\delta = 10^{-5}$ cm, an absurd answer in view of a plasma skin depth, c/ω_{pe}, and electron gyroradius of about 1 km. c/ω_{pe} is certainly a lower limit for δ, as discussed in Subsection 1.4. This implies that the effective momentum exchange rate between ions and electrons must be at least 10 orders of magnitude greater than v_{coul} in the diffusion region.

What is the most likely source of free energy for the plasma instabilities creating the anomalous resistivity? Without doubt it must be the current in the layer separating the opposing magnetic fields. A characteristic of current-driven instabilities is the existence of a threshold which the relative velocity between ions and electrons has to exceed before instability sets in:

$$|v_e - v_i| \geq v_{crit} \tag{48}$$

Ampère's law applied to the sheet current, J, in the diffusion region,

$$J = \frac{c\Delta B}{4\pi} = en(v_i - v_e)\delta \tag{49}$$

then establishes an upper limit for δ in terms of the magnetic field jump, ΔB, through the current sheet and v_{crit}:

$$\delta \leq \frac{c|\Delta B|}{4\pi\, en\, v_{crit}}. \tag{50}$$

The lower the velocity threshold of an instability is, the broader the diffusion region can be. From Eq. (47) we see, however, that for a given reconnection rate, higher values of δ imply higher effective collision frequencies.

In order to prepare some numerical estimates, we rewrite Eqs. (47) and (50) by calibrating the actual relative velocity in the current layer in units of the ion thermal velocity:

$$|v_i - v_e| = r\, v_{th,i} \tag{51}$$

with $\beta_i = 8\pi p_i / B_0^2$ we then obtain:

$$v^* = \frac{\varepsilon}{r\sqrt{\beta_i}}\, \Omega_e \tag{52a}$$

$$\delta = \frac{1}{r\sqrt{\beta_i}}\, \frac{c}{\omega_{pi}} \tag{52b}$$

$$D_m = \frac{\varepsilon}{r\sqrt{\beta_i}} \left(\frac{c}{\omega_{pe}}\right)^2 \Omega_e \tag{52c}$$

In pressure equilibrium, β_i lies between 0.5 and 1.0 depending on the ion-electron temperature ratio. If we adopt $B_0 = 30\,\gamma$ and $n_0 = 20\,\mathrm{cm}^{-3}$, which implies an ion temperature of nearly $10^6\,{}^\circ\mathrm{K}$, we estimate:

$$v^* \cong \frac{\varepsilon}{r} \cdot 7.5\,\mathrm{kHz} \tag{53a}$$

$$\delta \cong \frac{70\,\mathrm{km}}{r} \tag{53b}$$

$$D_m \cong \frac{\varepsilon}{r}\, 10^{14}\,\mathrm{cm}^2/\mathrm{sec}\,. \tag{53c}$$

Merging rates of $\varepsilon = 0.2$ are consistent with the observations. The thickness of the magnetopause as determined by the magnetometers on ISEE- 1 and -2 turned out between 500 and $\sim 1{,}000\,\mathrm{km}$ (RUSSELL and ELPHIC, 1979). It is, however, unlikely that a diffusion region was encountered during the measurements reported. At present we do not know what the actual thickness of such a region should be. It could be much narrower than $500\,\mathrm{km}$. So, we can only conclude that $D_m \approx r^{-1} \cdot 2 \times 10^{13}\,\mathrm{cm}^2/\mathrm{sec}$, or $v^* = r^{-1} \cdot 1.5 \times 10^3\,\mathrm{Hz}$ is required by the chosen macroscopic parameters.

5.2   Relevant instabilities

Current-driven instabilities which may set up plasma turbulence of high enough intensity to support the magnetic diffusion in the reconnection process, have been investigated by SYROVATSKY (1972), CORONITI and EVIATAR (1977), HAERENDEL (1978), HUBA *et al.* (1978), and PAPADOPOULOS (1979). The last reference is a comprehensive review of six candidate instabilities. They are: (1) Buneman, (2) ion acoustic, (3) electron-cyclotron drift, (4) lower hybrid drift, (5) modified two-stream, and (6) ion-cyclotron drift instabilities. Their velocity thresholds range from $v_{\mathrm{th,e}}$ (Buneman) to $(m_{\mathrm{e}}/m_{\mathrm{i}})^{1/2} \cdot v_{\mathrm{th,i}}$ (ion-cyclotron drift). Since in this context it is not possible to review the linear and non-linear properties of these instabilities, we will comment only briefly on each of them, essentially following PAPADOPOULOS (1979), and pick the lower hybrid drift instability as the most likely and at present most favoured candidate for detailed discussion.

The source of free energy is the current necessary to support the pressure gradient. It is given by:

$$j = \frac{nc(T_{\mathrm{i}} + T_{\mathrm{e}})}{B} \frac{\partial lnn_0}{\partial z} \tag{54}$$

Instead of speaking of a current threshold, en v_{crit}, we can refer to a critical gradient scale, L_{crit}, below which an instability may be initiated:

$$L_{\mathrm{crit}} = \frac{1}{2} \frac{v_{\mathrm{th,i}}^2}{\Omega_{\mathrm{i}} v_{\mathrm{crit}}} \tag{55}$$

1)   *Buneman instability.* Its velocity threshold is very high, $v_{\mathrm{crit}} = v_{\mathrm{th,e}}$, implying a layer thickness of the order of 1–2 km, according the Eq. (53b). Before the magnetopause reaches such a low width, it would be subject to some of the other instabilities.

2)   *Ion-acoustic instability.* This instability is well known for situations with $T_{\mathrm{e}} \gg T_{\mathrm{i}}$. However, at the magnetopause we rather find $T_{\mathrm{e}} \leq T_{\mathrm{i}}$. In this case, the threshold is raised to $v_{\mathrm{th,e}}$ and implies as narrow current sheets as the Buneman instability does (CORONITI and EVIATAR, 1977).

3)   *Electron-cyclotron drift instability.* This instability results from an interaction of ion-acoustic and electron Bernstein waves. It is insensitive to the ratio $T_{\mathrm{e}}/T_{\mathrm{i}}$. For $\Omega_{\mathrm{e}}/\omega_{\mathrm{pe}} \ll 1$, as applies to the magnetopause, the critical velocity is:

$$v_{\mathrm{crit}} \cong \frac{\Omega_{\mathrm{e}}}{\omega_{\mathrm{pe}}} v_{\mathrm{th,e}} . \tag{56}$$

For the parameters chosen above, this is close to or below the ion thermal velocity. The instability has, however, the disadvantage of a rather low nonlinear saturation level, when the actual drift velocity, v_d, is close to v_{crit}. Saturation is determined by turbulent scattering of the electrons by the waves and destruction of the necessary phase coherence. The effective collision frequency at saturation is:

$$v^* = 2 \times 10^{-2} \left(\frac{v_d}{v_{th,e}} \right)^3 \Omega_e . \tag{57}$$

HAERENDEL (1978), therefore, concluded that this instability would be effective only in thin current layers, such as needed to drive the two former instabilities, and that the ion-acoustic mode would probably be more efficient.

4) *Lower hybrid drift instability*. This instability is characterized by flute-type perturbations with $k \cdot B = 0$ and exists even for $T_i > T_e$. It is not purely electrostatic, but becomes more electromagnetic with increasing β_i. Depending on the ratio of $v_d/v_{th,i}$, there are somewhat different regimes with wave phase velocities above and below the ion thermal velocity. For $v_d \ll v_{th,i}$, the instability arises from a resonance between ions and a drift wave. Maximum growth with rate:

$$\gamma_m = \frac{\sqrt{2\pi}}{8} \left(\frac{v_d}{v_{th,i}} \right)^2 \Omega_{LH} \tag{58a}$$

occurs at ω- and κ-values of:

$$\omega_m = \frac{1}{\sqrt{2}} \left(\frac{v_d}{v_{th,i}} \right) \cdot \Omega_{LH} \tag{58b}$$

$$k_m = 2 \frac{\Omega_{LH}}{v_{th,i}} \tag{58c}$$

(HUBA et al., 1978).

Different non-linear saturation processes have been discussed, such as plateau formation, ion trapping, electron resonance broadening, or wave-wave interactions. Galeev (Chapter 3 of this volume) estimates a saturation level expressed by the effective collision frequency:

$$v^* = \left(\frac{R_{gi}}{2} \left| \frac{\partial \ln n_0}{\partial z} \right| \right)^3 \left(\frac{T_i}{T_e} \right)^2 \Omega_{LH} . \tag{59}$$

The results of HUBA *et al.* (1978) are presented in Fig. 64 in terms of the dimensionless wave energy, ε_F/nT_i, where $\varepsilon_F = |\delta E_w|^2/8\pi$ is the field energy density of the electrostatic oscillations. The mean amplitude, δE_w, is, of

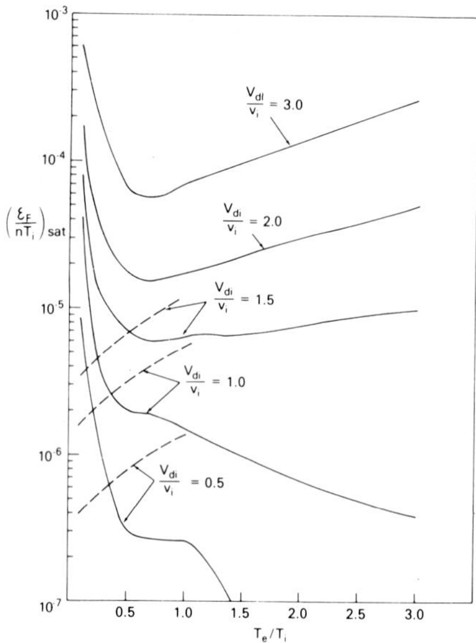

FIG. 64. Saturation energies due to electron resonance broadening (solid curve) and current relaxation (dashed curve) as a function of T_e/T_i and $v_d/v_{th,i}$, for $\beta = 1.0$ and $\omega_{pe}^2/\Omega_e^2 = 100$ (from HUBA et al., 1978).

course, an integral over the range of k-values with strong growth. For comparison with wave-data, a relation between ε_F and the effective collision frequency for ion-electron momentum transfer (see Eq. (46)) is useful. Galeev (this volume) derives:

$$v^* = \frac{\sqrt{\pi}}{4}\frac{m_i}{m_e}\Omega_{LH}\frac{\varepsilon_F}{n_0(T_i + T_e)}. \tag{60}$$

For the application to the diffusion region, there is a difficulty arising from the fact that the instability is stabilized for $\beta \gtrsim 3$.

5) *Modified two-stream instability.* It is similar to 4), but with $k_{\parallel} = (m_e/m_i)^{1/2}k_{\perp}$, and applies to homogeneous media in which a relative electron-ion drift is set up by other causes, as for instance by a sudden switch-on of a transverse electric field.

6) *Ion-cyclotron drift instability.* Of all the considered instabilities this has the lowest threshold. It applies to density gradient lengths, L_n, in the range:

$$\left(\frac{m_i}{m_e}\right)^{1/4} R_{gi} < L_n < \left(\frac{m_i}{m_e}\right)^{1/2} R_{gi} \tag{61}$$

The instability is flute-like $|\mathbf{k} \cdot \mathbf{B}_0 = 0|$ and independent of T_e/T_i. The waves are ion gyro-harmonics with $\gamma/\omega \lesssim 1$. The effective collision frequency provided by such waves lies typically below the ion gyro-frequency and is thus too weak to account for significant anomalous resistivity.

5.3 Wave observations

In order to enable us to judge the potential dynamic effects of the observed wave fields, we combine the macroscopic expectations as expressed by Eqs. (52a) with the microscopic relation (60). To be specific, we assume a merging rate of $u = 0.2\ v_A$, i.e. $\varepsilon = 0.2$, and a magnetic field value $B_0 = 30\ \gamma$. Thus $8\pi n_0\ (T_i + T_e) = 10^{-7}\ \mathrm{erg\,cm}^{-3}$. This leads to a mean amplitude, δE_w, of the electric wave field of:

$$\delta E_w \cong 10^3 r^{-1/2}\ (\mathrm{mV/m})\ . \tag{62}$$

The saturation levels of lower hybrid waves shown in Fig. 64 (after Huba et al., 1978) can be interpreted in terms of δE_w with help of the following relation:

$$\delta E_w \cong 10^4 \left(\frac{\varepsilon_F}{n_0(T_i + T_e)}\right)^{1/2} \quad (\mathrm{mV/m})\ . \tag{63}$$

Hence a realtive saturation energy of 10^{-4} as reached for $v_d/v_{\mathrm{th,i}} = 3$ (i.e. $r = 3$) corresponds to integrated electric wave amplitudes of $100\ \mathrm{mV/m}$, not too far below what is needed for reconnection, according to Eq. (62). It should be noted, however, that $r = 3$ implies a rather thin diffusion region, namely $\delta \cong 20\ \mathrm{km}$ (see Eq. (53)).

Observations of plasma waves near the magnetopause have been reported by Gurnett and Frank (1978), Perraut et al. (1979) and Gurnett et al. (1979). The magnetopause and boundary layer show up clearly in the wave data by enhanced broad-band low frequency noise with both electric and magnetic components (Fig. 65). The electric vector is transverse to $\mathbf{B}_0$. As seen in Fig. 66, most of the power resides at low frequencies ($< 100\ \mathrm{Hz}$), i.e. mostly at and below Ω_{LH}. Gurnett et al. (1979) quote as typical integrated wave amplitudes: $\delta E = 5.2\ \mathrm{mV/m}$ and $\delta B = 1.3\ \gamma$. Perraut et al. (1979) find peak amplitudes of δB of $10\ \gamma$ in the range from 0.5–$10\ \mathrm{Hz}$.

In a pure electromagnetic wave, the ratio $\delta B/\delta E$ is given by the index of refraction $n = ck/\omega$. This sets a lower limit for δE of the mixed

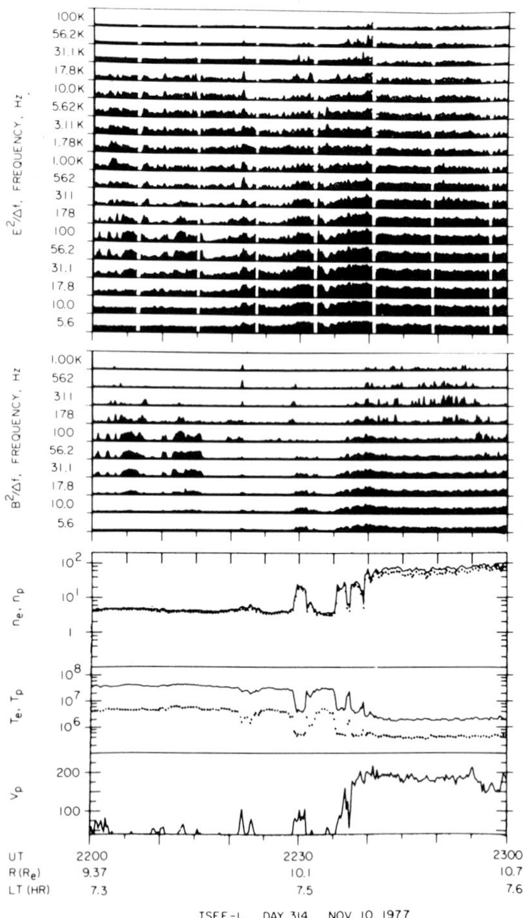

FIG. 65.   A detailed comparison of the plasma wave electric and magnetic field intensities and the plasma parameters for an outbound magnetopause crossing. Several distinct encounters with the plasma boundary layer can be seen before the spacecraft crosses the magnetopause current layer at about 2239 UT. Each of these encounters with the boundary layer is associated with enhanced plasma wave electric and magnetic field intensities (from GURNETT *et al.*, 1979).

electrostatic/electromagnetic lower hybrid waves:

$$\delta E_{(\mathrm{mV/m})} \geq 0.1 \; \delta B_{(\gamma)}$$

with $v_d \approx v_{\mathrm{th,i}}$ assumed. The data are consistent with this estimate and show that the electrostatic contribution is at least one order of magnitude higher

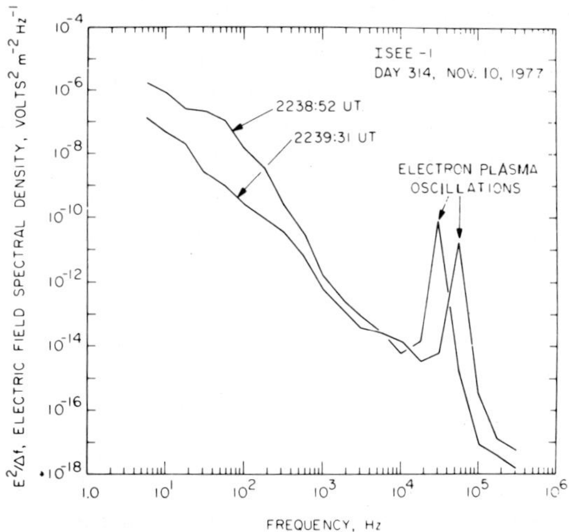

Fig. 66.   Selected electric field spectra from the magnetopause crossing in Fig. 65. showing the occurrence of intense narrowband emissions near the local electron plasma frequency (from Gurnett *et al.*, 1979).

than the electromagnetic (by amplitude). Thus there is good observational evidence for the existence of lower hybrid noise (probably strongly Doppler shifted) at the magnetopause. However, the electric wave amplitudes reported so far are almost two orders of magnitude too small to account for the resistivity needed in a diffusion region of the reconnection process. This may be due to incomplete data analysis, but is more likely a consequence of the rareness of encounters with such regions. Their latitudinal extent, $l = r^{-1}$ (Eq. (36)) should be of the order of 1,000 km only. All other magnetopause encounters would correspond to either tangential or rotational discontinuities with considerably lower noise. The signature of a true diffusion region would be the existence of electric wave amplitudes well exceeding 100 mV/m.

The relatively low noise level at the magnetopause makes it difficult to support the proposal that the low latitude boundary layer is filled by anomalous diffusion through the magnetopause over its entire extension on the frontside (Subsection 3.4). This needs values of D_{an} of the order of 10^{13} cm² sec⁻¹ or wave amplitudes as postulated for the diffusion region. This may be taken as evidence for a mode of mass transfer other than diffusion.

5.4 *Origin and maintenance of reconnection*

Most of the theories of reconnection dealt with stationary situations. The rotational discontinuities identified by direct observations had a duration of at most several minutes (see Subsection 4.2). There is much evidence of a rather transient nature of the process (PASCHMANN *et al.*, 1976; HAERENDEL *et al.*, 1978) and the term 'patchy' reconnection has been introduced. The understanding of this process has not advanced far enough as to explain why in the real three-dimensional world reconnection should exist at all. There is always the solution of a mere diversion of the flow by the magnetopause, which could be a pure tangential discontinuity and is, in fact, often enough found to show all the signs of that. It is not clear why a turbulent magnetopause current layer allowing for rapid magnetic diffusion, once established, should maintain itself. This is in contrast with collisionless shock waves for which we have a good mental picture of the process of wave steepening. Magnetic diffusion and heating are acting in a sense as to stabilize the magnetopause current. Their effect has to be balanced exactly by the convective transport of magnetic field into and kinetic energy out of the region in order to create a stationary situation. No simple mechanism of self-adjustment seems to exist. In view of this and of the sensitivity of stationary

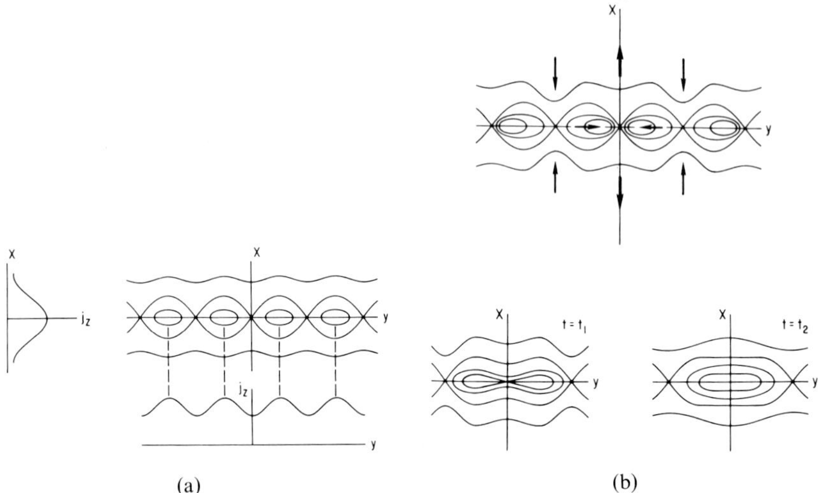

FIG. 67. (a) Equilibrium magnetic flux surfaces (magnetic islands) and current distributions in slab geometry. (b) Perturbed flux surfaces with zero resistivity. (c) Coalescence of neighboring islands with finite resistivity at times t_1, $t_2 > t_1$ (from FINN and KAW, 1977).

theories to the assumed boundary conditions, one is forced to conclude that not only in laboratory experiments (BAUM and BRATENAHL, 1980; STENZEL and GEKELMAN, 1979), but also in space, reconnection is driven by *external forces.*

One of the consequences of such a proposition is the importance of electric induction fields (CORONITI *et al.*, 1977; PELLINEN and HEIKKILA, 1978). At times this component may even be so strong that the electrostatic potential field tangential to the current sheet can even oppose the induction field and the direction of the current (STENZEL and GEKELMAN, 1979). External forces driving reconnection may be provided by magnetic tensions due to the draping of field lines around the frontside magnetopause, by MHD instabilities, or pressure fluctuations accompanying hydrodynamic turbulence, such as found in the cusps.

A large-scale reconnection situation may develop via the tearing mode instability of a thin current layer and grow through coalescence of magnetic islands (FINN and KAW, 1977). Figure 67 is a sketch of the latter process. The most important implication of such a picture would be that the microprocesses indeed exert a control on the merging rate. BISKAMP and WELTER (1980), for example, find that this rate depends on the resistivity as $u \sim \eta^{1/3}$ and that island coalescence is subject to a sloshing motion. The coalescence proceeds mainly during the phases of magnetic field compression. The transient nature of the boundary layer at low latitudes (SCKOPKE *et al.*, 1981) may well have its origin in such a process.

REFERENCES

AKASOFU, S.-I., *Physics of Magnetospheric Substorms*, D. Reidel Publishing Company, Dordrecht, Holland, 1977.

AKASOFU, S.-I., Interplanetary energy flux associated with magnetospheric substorms, *Planet. Space Sci.*, **27**, 425, 1979.

ALPERS, W., Steady state charge neutral models of the magnetopause, *Astrophys. Space Sci.*, **5**, 425, 1969.

ANDERSON, K. A., Measurements of bow shock particles far upstream from the earth, *J. Geophys. Res.*, **86**, 4445, 1981.

ANDERSON, K. A., R. P. LIN, F. MARTEL, C. S. LIN, G. K. PARKS, and H. RÈME, Thin sheets of energetic elecrons upstream from the earth's bow shock, *Geophys. Res. Lett.*, **6**, 401, 1979.

ARNOLDY, R. L., Signature in the interplanetary medium for substorms, *J. Geophys. Res.*, **76**, 5189, 1971.

ASBRIDGE, J. R., S. J. BAME, and I. B. STRONG, Outward flow of protons from the earth's bow shock, *J. Geophys. Res.*, **73**, 5777, 1968.

BAKER, D. N. and E. C. STONE, The magnetopause energetic electron layer, 1. Observations along the distant magnetotail, *J. Geophys. Res.*, **83**, 4327, 1978.

BARNES, A., Theory of generation of bow-shock-associated waves in the upstream interplanetary medium, *Cosmic Elec.*, **1**, 90, 1970.

BAUM, P. and J. A. BRATENAHL, Magnetic reconnection experiments, in *Advances in Electronics and Electron Physics,* Vol. 54, edited by C. Marton, p. 1, Academic Press, New York, 1980.

BEARD, D. B. and J. Y. CHOE, The magnetospheric boundary, in *Correlated Interplanetary and Magnetospheric Observations*, edited by D. E. Page, pp. 97–114, D. Reidel Publishing Company, Dordrecht, Holland, 1974.

BISKAMP, D., Collisionless shock waves in plasmas, *Nuclear Fusion*, **13**, 719, 1973.

BISKAMP, D., and H. WELTER, Coalescence of magnetic islands, *Phys. Rev. Lett.*, **44**, 1069, 1980.

BURTON, R. K., R. L. MCPHERRON, and C. T. RUSSELL, An empirical relationship between interplanetary conditions and D_{st}, *J. Geophys. Res.*, **80**, 4204, 1975.

CAIRNS, R. A., Kinetic models of shocks in collisionless plasmas, *Nuovo Cimento*, **2c**, 712, 1979.

CHEN, L. and A. HASEGAWA, A theory of long-period magnetic pulsations, 1. steady state excitation of field line resonance, *J. Geophys. Res.*, **79**, 1024, 1974.

CORONITI, F. V. and A. EVIATAR, Magnetic field reconnection in a collisionless plasma, *Ap. J. Suppl. Ser.*, **33**, 189, 1977.

CORONITI, F. V., F. L. SCARF, L. A. FRANK, and R. P. LEPPING, Microstructure of a magnetotail fireball, *Geophys. Res. Lett.*, **4**, 219, 1977.

COWLING, T. G. *Magnetohydrodynamics*, p. 9, Interscience Publ., New York, 1957.

CROOKER, N. U., Dayside merging and cusp geometry, *J. Geophys. Res.*, **84**, 951, 1979.

CROOKER, N. U., T. E. EASTMAN, and G. S. STILES, Observations of plasma depletion in the magnetosheath at the dayside magnetopause, *J. Geophys. Res.*, **84**, 869, 1979.

DRYER, M., Solar wind interactions—hypersonic analogue, in *Cosmic Electrodynamics,* **1**, 115, D. Reidel Publishing Company, Dordrecht, Holland, 1970.

DUNGEY, J. W., *Cosmic Electrodynamics*, Cambridge University Press, London, 1958.

EASTMAN, T. E., The plasma boundary layer and magnetopause layer of the Earth's magnetosphere (Thesis), The University of Alaska, Geophysical Institute, May 1979.

EASTMAN, T. E., E. W. HONES, Jr., S. J. BAME, and J. R. ASBRIDGE, The magnetospheric boundary layer: Site of plasma momentum and energy transfer from the magnetosheath into the magnetosphere, *Geophys. Res. Lett.*, **3**, 685, 1976.

EASTMAN, T. E. and E. W. HONES, Jr., Characteristics of the low latitude boundary layer and magnetopause layer at high time resolution, *J. Geophys. Res.*, **84**, 2019, 1979.

EASTMAN, T. E., R. R. ANDERSON, L. A. FRANK, and G. K. PARKS, Upstream particles observed in the earth's foreshock region, *J. Geophys. Res.*, **86**, 4379, 1981.

EVIATAR, A. and R. A. WOLF, Transfer processes in the magnetopause, *J. Geophys. Res.*, **73**, 5561, 1968.

FAIRFIELD, D. H., Bow shock associated waves observed in the far upstream interplanetary medium, *J. Geophys. Res.*, **74**, 3541, 1969.

FAIRFIELD, D. H., Average and unusual locations of the earth's magnetopause and bow shock, *J. Geophys. Res.*, **76**, 6700, 1971.

FAIRFIELD, D. H., Global aspects of the magnetopause, in *Magnetospheric Boundary Layers*, edited by B. Battrick, p. 5, Paris: ESA SP-148, 1979.

FAIRFIELD, D. H. and E. W. HONES, Jr., IMP 6 measurements in the distant polar cusp during substorms, *J. Geophys. Res.*, **83**, 4273, 1978.

FERRARO, V. C. A., Theory of a plane model, *J. Geophys. Res.*, **57**, 15, 1952.

FINN, J. M. and P. K. KAW, Coalescence instability of magnetic islands, *Phys. Fluids*, **20**, 72, 1977.

FORMISANO, V., The physics of the earth's collisionless shock wave, *J. Physiq.*, **38**, C6 65, 1977.

FRANK, L. A. and D. A. GURNETT, Distributions of plasmas and electric fields over the auroral zones and polar caps, *J. Geophys. Res.*, **76**, 6829, 1971.

FRANK, L. A., K. L. ACKERSON, and D. M. YEAGER, Observations of atomic oxygen (O^+) in the earth's magnetotail, *J. Geophys. Res.*, **82**, 129, 1977.

FRIIS-CHRISTENSEN, E., K. LASSEN, J. WILHJELM, J. M. WILCOX, W. GONZALEZ, and D. S. COLBURN, Critical component of the interplanetary magnetic field responsible for large geomagnetic effects in the polar cap, *J. Geophys. Res.*, **77**, 3371, 1972.

GALEEV, A. A., Collisionless shocks, in *Physics of Solar Planetary Environment*, edited by D. J. Williams, p. 464, American Geophysical Union, 1976.

GOSLING, J. T., J. R. ASBRIDGE, S. J. BAME, G. PASCHMANN, and N. SCKOPKE, Observations of two distinct populations of bow shock ions in the upstream solar wind, *Geophys. Res. Lett.*, **5**, 957, 1978.

GRAD, H., Boundary layer between a plasma and a magnetic field, *Phys. Fluids*, **4**, 1366, 1961.

GREENSTADT, E. W., Structure of the terrestrial bow shock, in *Solar Wind Three*, edited by C. T. Russell, p. 440, Institute of Geophysics and Planetary Physics, UCLA, 1974.

GREENSTADT, E. W., C. T. RUSSELL, F. L. SCARF, V. FORMISANO, and M. NEUGEBAUER, Structure of the quasi-perpendicular, laminar bow shock, *J. Geophys. Res.*, **80**, 502, 1975.

GREENSTADT, E. W. and R. W. FREDRICKS, Shock systems in collisionless space plasmas, in *Solar System Plasma Physics*, Vol. 3, edited by L. J. Lanzerotti, C. F. Kennel, and E. N. Parker, North-Holland Publishing Company, Amsterdam, 1979.

GREENSTADT, E. W., C. T. RUSSELL, J. T. GOSLING, S. J. BAME, G. PASCHMANN, G. K. PARKS, K. A. ANDERSON, F. L. SCARF, R. R. ANDERSON, D. A. GURNETT, R. P. LIN,

C. S. Lin, and H. Rème, A macroscopic profile of the typical quasi-perpendicular bow shock: ISEE 1 and 2, *J. Geophys. Res.*, **85**, 2124, 1980.

Gurnett, D. A. and L. A. Frank, Plasma waves in the polar cusp: observations from Hawkeye I, *J. Geophys. Res.*, **83**, 1447, 1978.

Gurnett, D. A., R. R. Anderson, B. T. Tsurutani, E. J. Smith, G. Paschmann, G. Haerendel, S. J. Bame, and C. T. Russell, Plasma wave turbulence at the magnetopause: observations from ISEE 1 and 2, *J. Geophys. Res.*, **84**, 7043, 1979.

Haerendel, G., Microscopic plasma processes related to reconnection, *J. Atmos. Terr. Phys.*, **40**, 343, 1978.

Haerendel, G. and G. Paschmann, Entry of solar wind plasma into the magnetosphere, in *Physics of the Hot Plasma in the Magnetosphere*, edited by B. Hultqvist and L. Stenflo, pp. 23–43, Plenum, New York, 1975.

Haerendel, G., G. Paschmann, N. Sckopke, H. Rosenbauer, and P. C. Hedgecock, The frontside boundary layer and the problem of reconnection, *J. Geophys. Res.*, **83**, 3195, 1978.

Hasegawa, A. and K. Mima, Anomalous transport produced by kinetic Alfvén wave turbulence, *J. Geophys. Res.*, **83**, 1117, 1978.

Heelis, R. A., W. B. Hanson, and J. L. Burch, Ion convection velocity reversals in the dayside cleft, *J. Geophys. Res.*, **81**, 3803, 1976.

Heikkila, W. J. and J. D. Winningham, Penetration of magnetosheath plasma to low altitude through the dayside magnetospheric cusps, *J. Geophys. Res.*, **76**, 883, 1971.

Heppner, J. P., Polar cap electric field distributions related to the interplanetary magnetic field direction, *J. Geophys. Res.*, **77**, 4877, 1972.

Heppner, J. P., M. Sugiura, T. L. Skillman, B. G. Ledley, and M. Campbell, OGO-A magnetic field observations, *J. Geophys. Res.*, **72**, 5417, 1967.

Hones, E. W., Jr., J. R. Asbridge, S. J. Bame, M. D. Montgomery, S. Singer, and S.-I. Akasofu, Measurements of magnetotail plasma flow made with Vela 4B, *J. Geophys. Res.*, **77**, 5503, 1972.

Hones, E. W., J. Birn, S. J. Bame, J. R. Asbridge, G. Paschmann, N. Sckopke, and G. Haerendel, Further determination of the characteristics of magnetospheric plasma vortices with ISEE1 and 2, *J. Geophys. Res.*, **86**, 814, 1981.

Huba, J. T., N. T. Gladd, and K. Papadopoulos, Lower-hybrid-drift wave turbulence in the distant magnetotail, *J. Geophys. Res.*, **83**, 5217, 1978.

Hudson, P. D., Rotational discontinuities in an anisotropic plasma, *Planet. Space Sci.*, **21**, 475, 1971.

Iijima, T. and T. A. Potemra, Field-aligned currents in the dayside cusp observed by Triad, *J. Geophys. Res.*, **81**, 5971, 1976.

Lanzerotti, L. J. and D. J. Southwood, Hydromagnetic waves, in *Solar System Plasma Physics*, Vol. 3, edited by L. J. Lanzerotti, C. F. Kennel, and E. N. Parker, p. 109, North-Holland Publishing Company, Amsterdam, 1979.

LEE, L. C., R. K. ALBANO, and J. R. KAN, Kelvin-Helmholtz instability in the magnetopause-boundary layer region, *J. Geophys. Res.*, **86**, 54, 1981.

LEMAIRE, J. and M. ROTH, Penetration of solar wind plasma elements into the magnetosphere, *J. Atm. Terr. Phys.*, **40**, 331, 1978.

LEVY, R. H., H. E. PETSCHEK, and G. L. SISCOE, Aerodynamic aspects of the magnetospheric flow, *AIAA J.*, **2**, 2065, 1964.

LUNDIN, R., B. HULTQVIST, N. PISSARENKO, and A. ZACKAROV, The plasma mantle: composition and other characteristics observed by means of the Prognoz-7 satellite, preprint Kiruna Geophys. Institute, March, 1981.

MIDGLEY, J. E. and L. DAVIS, Jr., Calculation by a moment technique of the perturbation of the geomagnetic field by the solar wind, *J. Geophys. Res.*, **68**, 5111, 1963.

MORFILL, G. and M. SCHOLER, Study of the magnetosphere using energetic solar particles, *Space Sci. Rev.*, **15**, 267, 1973.

MOZER, F. S., R. B. TORBERT, U. V. FAHLESON, C.-G. FÄLTHAMMAR, A. GONFALONE, A. PEDERSEN, and C. T. RUSSELL, Direct observation of a tangential electric field component at the magnetopause, *Geophys. Res. Lett.*, **6**, 305, 1979.

NISHIDA, A., T. E. EASTMAN, and E. W. HONES, Jr., Comparison of the magnetopause current layer with Alfvén wave, in *Magnetospheric Study 1979, Proc. Internat. Workshop on Selected Topics of Magnetospheric Physcis*, p. 344, Tokyo, March 1979.

OLSON, J. V. and G. ROSTOKER, Longitudinal phase variations of Pc 4–5 micropulsations, *J. Geophys. Res.*, **83**, 2481, 1978.

PAPADOPOULOS, K., The role of microturbulence on collisionless reconnection, in *Dynamics of the Magnetosphere*, edited by S.-I. Akasofu, p. 289, D. Reidel Publishing Co., Dordrecht, Holland, 1979.

PARKER, E. N., Sweet's mechanism for merging magnetic fields in conducting fluids, *J. Geophys. Res.*, **62**, 509, 1957.

PARKER, E. N., Confinement of a magnetic field by a beam of ions, *J. Geophys. Res.*, **72**, 2315, 1967.

PASCHMANN, G., Plasma structure of the magnetopause and boundary layer, in *Magnetospheric Boundary Layers*, edited by B. Battrick, p. 25, Paris: ESA SP-148, 1979.

PASCHMANN, G., G. HAERENDEL, N. SCKOPKE, H. ROSENBAUER, and P. C. HEDGECOCK, Plasma and magnetic field characteristics of the distant polar cusp near local noon: the entry layer, *J. Geophys. Res.*, **81**, 2883, 1976.

PASCHMANN, G., N. SCKOPKE, J. R. ASBRIDGE, S. J. BAME, and J. T. GOSLING, Energization of solar wind ions by reflection from the earth's bow shock, *J. Geophys. Res.*, **85**, 4689, 1980.

PASCHMANN, G., N. SCKOPKE, S. J. BAME, J. R. ASBRIDGE, J. T. GOSLING, C. T. RUSSELL, and E. W. GREENSTADT, Association of low-frequency waves with suprathermal ions in the upstream solar wind, *Geophys. Res. Lett.*, **6**, 209, 1979a.

PASCHMANN, G., B. U. Ö. SONNERUP, I. PAPAMASTORAKIS, N. SCKOPKE, G. HAERENDEL, S. J. BAME, J. R. ASBRIDGE, J. T. GOSLING, C. T. RUSSELL, and R. C. ELPHIC, Plasma acceleration at the earth's magnetopause: evidence for reconnection, *Nature*, **282**, 243, 1979b.

PELLINEN, R. J. and W. J. HEIKKILA, Energization of charged particles to high energies by an induced substorm electric field within the magnetotail, *J. Geophys. Res.*, **83**, 1544, 1978.

PERRAUT, S., R. GENDRIN, P. ROBERT, and A. ROUX, Magnetic pulsations observed on board Geos-2 in the ULF range during multiple magnetopause crossings, in *Magnetospheric Boundary Layers,* edited by B. Battrick, p. 113, Paris: ESA SP-148, 1979.

PERREAULT, P. and S.-I. AKASOFU, A study of geomagnetic storms, *Geophys. J. R. Astr. Soc.*, **54**, 547, 1978.

PETSCHEK, H. E., Magnetic field annihilation, AAS-NASA Symposium on the Physics of Solar Flares, *NASA Spec. Publ.*, SP-50, 425, 1964.

PILIPP, W. G. and G. MORFILL, The formation of the plasma sheet resulting from plasma mantle dynamics, *J. Geophys. Res.*, **83**, 5670, 1978.

PODGORNY, I. M., Collisionless shocks: simulation and laboratory experiments, *Nuovo Cimento*, **2C**, 834, 1979.

ROSENBAUER, H., H. GRÜNWALDT, M. D. MONTGOMERY, G. PASCHMANN, and N. SCKOPKE, Heos-2 plasma observations in the distant polar magnetosphere: the plasma mantle, *J. Geophys. Res.*, **80**, 2723, 1975.

ROSENBLUTH, M. N., Infinite conductivity theory of the pinch V. Surface layer model in the limit of no collisions, in *Plasma Physics and Thermonuclear Research*, Vol. 2, edited by C. L. Longmire, J. L. Tuck, and W. B. Thompson, p. 271, Pergamon, London, 1963.

Russell, C. T. and R. C. ELPHIC, Initial ISEE magnetometer results: magnetopause observations, *Space Sci. Rev.*, **22**, 681, 1978.

RUSSELL, C. T. and R. C. ELPHIC, ISEE observations of flux transfer events at the dayside magnetopause, *Geophys. Res. Lett.*, **6**, 33, 1979.

SCHINDLER, K., On the role of irregularities in plasma entry into the magnetosphere, *J. Geophys. Res.*, **84**, 7257, 1979.

SCHOLER, M., F. M. IPAVICH, G. GLOECKLER, D. HOVESTADT, and B. KLECKER, Upstream particle events close to the bow shock and 200 R_E upstream: ISEE-1 and ISEE-3 observations, *Geophys. Res. Lett.*, **7**, 73, 1980.

SCKOPKE, N., External plasma flow, in *Magnetospheric Boundary Layers,* edited by B. Battrick, p. 37, Paris: ESA SP-148, 1979.

SCKOPKE, N., G. PASCHMANN, G. HAERENDEL, B. U. Ö. SONNERUP, S. J. BAME, T. G. FORBES, E. W. HONES, Jr., and C. T. RUSSELL, Structure of the low latitude boundary layer, *J. Geophys. Res.*, **86**, 2099, 1981.

SLUTZ, R. J., The shape of the geomagnetic field boundary under uniform external pressure, *J. Geophys. Res.*, **67**, 505, 1962.

SONNERUP, B. U. Ö., Acceleration of particles reflected at a shock front, *J. Geophys. Res.*, **74**, 1301, 1969.

SONNERUP, B. U. Ö., Magnetic field reconnection, in *Solar System Plasma Physics*, Vol. 3, edited by L. J. Lanzerotti, C. F. Kennel, and E. N. Parker, pp. 45–108, North-Holland Publishing Company, Amsterdam, 1979.

SONNERUP, B. U. Ö., Theory of the low-latitude boundary layer, *J. Geophys. Res.*, **85**, 2017, 1980.

SONNERUP, B. U. Ö. and L. J. CAHILL, Jr., Magnetopause structure and attitude from Explorer 12 observations, *J. Geophys. Res.*, **72**, 171, 1967.

SONNERUP, B. U. Ö. and B. G. Ledley, Electromagnetic structure of the magnetopause and boundary layer, in *Magnetospheric Boundary Layers*, edited by B. Battrick, pp. 401–411, paris: ESA SP-148, 1979.

SONNERUP, B. U. Ö., G. PASCHMANN, I. PAPAMASTORAKIS, N. SCKOPKE, G. HAERENDEL, S. J. BAME, J. R. ASBRIDGE, J. T. GOSLING, and C. T. RUSSELL, Evidence for magnetic field reconnection at the earth's magnetopause, *J. Geophys. Res.*, 1981 (submitted).

SOUTHWOOD, D. J., Some features of field line resonances in the magnetosphere, *Planet. Space Sci.*, **22**, 483, 1974.

SPREITER, J.R., A. L. SUMMERS, and A. Y. ALKSNE, Hydromagnetic flow around the magnetosphere, *Planet. Space Sci.*, **14**, 223, 1966.

SPREITER, J. R., A. Y. ALKSNE, and A. L. SUMMERS, External aerodynamics of the magnetosphere, in *Physics of the Magnetosphere*, edited by R. L. Carovillano, D. Reidel Publishing Co., Dordrecht, Holland, 1968.

STENZEL, R. L. and W. GEKELMAN, Experiments on magnetic-field line reconnection, *Phys. Rev. Lett.*, **42**, 1055, 1979.

STOREY, L. R. O. and L. CAIRO, Kinetic theory of the boundary layer between a flowing isotropic plasma and a magnetic field, in *Magnetospheric Boundary Layers*, edited by B. Battrick, p. 289, Paris: ESA SP-148, 1979.

SU, S.-Y. and B. U. Ö. SONNERUP, On the equilibrium of the magnetopause current layer, *J. Geophys. Res.*, **76**, 5181, 1971.

SYROVATSKY, S. I., Origin of the geomagnetic tail and neutral sheet, in *Critical Problems of Magnetospheric Physics, Proc. of the Joint COSPAR/IAGA/URSI Symposium, Madrid, 11–13 May, 1972*, edited by E. R. Dyer, p. 35, 1972.

TIDMAN, D. A. and N. A. KRALL, *Shock Waves in Collisionless Plasmas*, John Wiley-Interscience, New York, 1971.

VAMPOLA, A. L., Access of solar electrons to closed field lines, *J. Geophys. Res.*, **76**, 36, 1971.

VASYLIUNAS, V. M., Theoretical models of magnetic field line merging, 1, *Revs. Geophys. Space Phys.*, **13**, 303, 1975.

VASYLIUNAS, V. M., Interaction between the magnetospheric boundary layers and the ionosphere, in *Magnetospheric Boundary Layers*, edited by B. Battrick, p. 387, Paris: ESA SP-148, 1979.

WALTERS, G. K., On the existence of a second standing shock wave attached to the magnetopause, *J. Geophys. Res.*, **71**, 1341, 1966.

WILLIAMS, D. J., Magnetopause characteristics inferred from three-dimensional energetic particle distributions, *J. Geophys. Res.*, **84**, 101, 1979a.

WILLIAMS, D. J., Observations of significant magnetosheath antisolar energy flow, *J. Geophys. Res.*, **84**, 2105, 1979b.

WILLIAMS, D. J., Magnetopause characteristics at 0840–1040 hours local time, *J. Geophys. Res.*, **85**, 3387, 1980.

WILLIS, D. M., The microstructure of the magnetopause, *Geophys. J. R. Astr. Soc.*, **41**, 355, 1975.

WU, C. C., R. J. WALKER, and J. M. DAWSON, A three-dimensional MHD model of the earth's magnetosphere, UCLA preprint PPG-528, 1980.

YEH, T. and W. I. AXFORD, On the re-connexion of magnetic field lines in conducting fluids, *J. Plasma Phys.*, **4**, 207, 1970.

ZWAN, B. J. and R. A. WOLF, Depletion of solar wind plasma near a planetary boundary, *J. Geophys. Res.*, **81**, 1636, 1976.

Chapter 3

MAGNETOSPHERIC TAIL DYNAMICS

A. A. GALEEV

Space Research Institute of U.S.S.R. Academy of Sciences, Moscow, U.S.S.R.

1. Introduction: Model of the Equilibrium Neutral Sheet

The plasma and magnetic field distribution within the magnetotail (see Chapter 1) happens to be thermodynamically nonequilibrium and therefore it is unstable against excitation of a number of plasma collective modes. If the wave length of the excited oscillations is much shorter than the characteristic scale of the plasma configuration, then one speaks about microinstabilities. The microinstabilities do not change the global plasma and field distribution, causing however a substantial enhancement of transport coefficients in comparison with their classical values. It is important to calculate these transport coefficients both to construct a model for the plasma and field distribution in the quiet magnetospheric tail and to account quantitatively for the corresponding dissipative processes in the global stability analysis. Therefore we begin our consideration of the magnetotail dynamics with the analysis of the microinstabilities resulting in anomalous resistivity and pitch angle scattering within the magnetotail plasma sheet. The theoretically estimated level of the excited plasma oscillations could be compared here with the observational data obtained by satellites.

However here we are interested most of all in the macroinstabilities resulting in a substantial change of the magnetotail configuration and an explosive dissipation of the energy stored in it, i.e. in a magnetospheric substorm. We will show that the tearing instability of the magnetotail neutral sheet is an appropriate mechanism for substorms and it naturally explains the accompanying phenomena. Moreover there is not yet any other

                               A. A. GALEEV

alternative substorm mechanism which has been advanced theoretically to
such details and would permit to make such reliable quantitative estimates
for the accompanying phenomena, as it is in the case of the tearing mode
substorm mechanism.

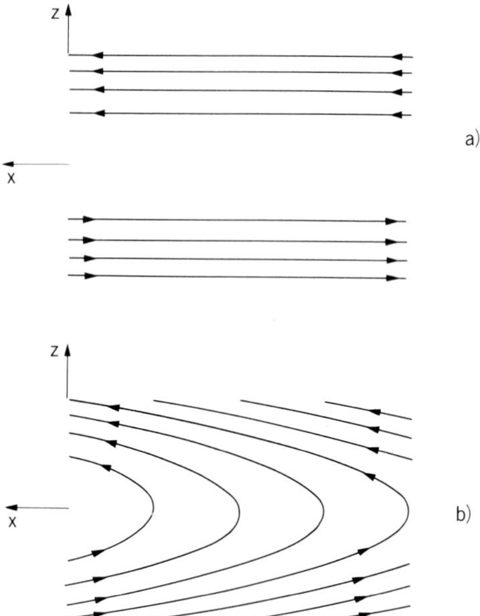

FIG. 1.   One-dimensional (a) and two-dimensional (b) magnetic field models for
the magnetospheric plasma sheet.

In the magnetotail plasma sheet stability analysis we will use either the
Harris model for the one-dimensional equilibrium plasma and magnetic
field distribution, or the modification of this model to the weakly two-
dimensional case. For the case of the plane neutral sheet shown in Fig. 1(a), the
particle distribution function is assumed to be a local Maxwellian shifted
by the constant velocity u_j along the current direction. Such a distribution, a
function of constants of motion only, can be easily written:

$$f_{0j}(z, \boldsymbol{v}) = n_0 \left(\frac{m_j}{2\pi T_j} \right)^{3/2} \exp \left(-\frac{m_j u_j^2}{2T_j} \right) \exp \left[-\frac{\varepsilon}{T_j} + \frac{u_j P_y}{T_j} \right] \tag{1}$$

where two constants of motion are: energy $\varepsilon = m_j v^2/2 + e_j \varphi_0(z)$ and
generalized momentum $P_y = m_j v_y + e_j A_{0y}(z)/c$, $\varphi_0(z)$ and $A_{0y}(z)$ are equili-

brium electrostatic and vector potentials; e_j, m_j are the charge and the mass of particles of species j; T_j is the temperature which is assumed to be constant. We can maintain plasma quasineutrality if we assume

$$\varphi_0 \equiv 0\,, \qquad u_j = 2cT_j/e_jB_0\varDelta \tag{2}$$

where B_0 and $\varDelta$ are constants having dimension of magnetic field strength and length, respectively. Solving now Maxwell's equation

$$\text{rot } \boldsymbol{B} = (4\pi/c)\sum_j e_j \int v f_{0j}\, \mathrm{d}v\,, \tag{3}$$

we find the equilibrium magnetic field and plasma density profiles (HARRIS, 1962)

$$\begin{aligned}
\boldsymbol{B} &= B_0\text{th}(z/\varDelta)\hat{e}_x \\
n(z) &= n_0/\text{ch}^2(z/\varDelta)\,.
\end{aligned} \tag{4}$$

Pressure balance requires the following

$$n_0(T_e + T_i) = B_0^2/8\pi\,. \tag{5}$$

We can show that the current velocity u_j coincides with the diamagnetic drift velocity, i.e.

$$u_j = -\frac{cT_j}{e_jB_xn}\frac{\mathrm{d}n}{\mathrm{d}z}\,. \tag{6}$$

The Harris model is usually a satisfactory approximation for the plasma and magnetic field configuration within the magnetotail plasma sheet. However for a number of cases it is essential to account for the two-dimensionality of the real field configuration (see Fig. 1(b)). The simplest approximate generalization of the Harris configuration for the two-dimensional case can be written as

$$\begin{aligned}
\boldsymbol{B} &= B_0\text{th}(z/\varDelta)\hat{e}_x + B_n\hat{e}_z \\
n(x,\, z) &= n_0(x)/\text{ch}^2(z/\varDelta)
\end{aligned} \tag{7}$$

where the magnetic field component perpendicular to the neutral sheet is assumed to be constant.

For $B_n \ll B_0$, the deviation from the Harris configuration is small and the magnetic field-line tension along x-axis $(B_n/4\pi)\cdot(\partial B_x/\partial z)$ could be easily balanced by the weak pressure gradient $(\partial n(T_e + T_i)/\partial x)$. When considering plasma sheet sections with the length shorter than the characteristic nonuniformity scale $L_x \simeq (B_0/B_n)\varDelta$ we can neglect dependences of density and magnetic field on the x-coordinate.

In our discussion of the quantitative effects we will assume that the typical plasma and field parameters in the quiet plasma sheet are:

$$B_0 = 2 \times 10^{-4} \mathrm{G}, \quad T_i = 5 \text{ keV}, \quad T_e = 0.25 \, T_i \,;$$
$$n_0 = 0.36 \text{ cm}^{-3}, \quad \varDelta = 1 R_E = 6.4 \times 10^8 \text{ cm} \,.$$

2. Microinstabilities of the Quiet Magnetospheric Tail

2.1   The lower hybrid-drift instability of the plasma sheet and the anomalous resistivity

Let us consider first the plasma microinstabilities caused by the plasma density nonuniformity and by the current across the magnetic field lines. It is well known that even the weak plasma nonuniformity favour the excitation of a wide class of drift oscillations (MIKHAILOVSKII, 1971). However in the case of a high β plasma, such as the magnetotail plasma sheet, most of the drift instabilities are stabilized. This stabilization is caused by the drift wave damping due to its resonance interaction with the group of particles whose gradient drift velocity is equal to the wave phase velocity. This effect is weak for the cold electron drift waves (i.e. for $T_e \ll T_i$), which are still growing in the plasma with sufficiently sharp gradients.

We consider here the lower-hybrid-drift instability whose development in the neutral sheet shown in Fig. 1 is expected under the following conditions (MIKHAILOVSKII, 1971):

$$1 \gg \frac{r_{\mathrm{Li}}}{2} \left| \frac{\mathrm{d} \ln n}{\mathrm{d}z} \right| \gtrsim 2(m_e/m_i)^{1/4} \tag{8}$$

where $r_{Lj} = v_j/\Omega_j$ is the local Larmor radius, $v_j = \sqrt{2T_j/m_j}$ is the thermal velocity and $\Omega_j = |e_j| B/m_j c$ is the Larmor frequency, all for the particles of the j-th species. The lower-hybrid-drift mode is characterized by strongly magnetized electrons and unmagnetized ions, with frequency and wavenumber satisfying

$$\Omega_e \gg \omega \gtrsim \Omega_i, \quad kr_{\mathrm{Li}} \gg 1 \gtrsim kr_{\mathrm{Le}}, \quad \boldsymbol{k} \cdot \boldsymbol{B} = 0 \tag{9}$$

where ω and $\boldsymbol{k}$ are the mode frequency and the wave-vector. The mode is electrostatic and its potential can be chosen in a form:

$$\varphi_1(\boldsymbol{r}, \, t) = \tilde{\varphi}_1 \mathrm{e}^{-i\omega t + i\boldsymbol{k} \cdot \boldsymbol{r}} \,. \tag{10}$$

Here we have used the geometric optics approximation which is valid for

$$k_z^2 \gg (\mathrm{d} \ln n/\mathrm{d}z)^2 \, , \qquad (\mathrm{d} \ln B/\mathrm{d}z)^2 \, . \qquad (11)$$

The linearized Boltzman equation for the perturbed distribution function

$$\left\{ \frac{\partial}{\partial t} + \boldsymbol{v} \cdot \nabla + \frac{e_j}{m_j c}[\boldsymbol{v} \times \boldsymbol{B}_0] \cdot \frac{\partial}{\partial \boldsymbol{v}} \right\} f_{1j} = \frac{i e_j}{m_j} \left(\boldsymbol{k} \cdot \frac{\partial f_{0j}}{\partial \boldsymbol{v}} \right) \varphi_1 \qquad (12)$$

can be represented in the form of an integral along particle trajectories (see Eq. (A.8) of the Appendix but note that $k_y \neq 0$ here):

$$f_{1j} = -\frac{e_j f_{0j}}{T_j} \left\{ 1 + i(\omega - k_y u_j) \int_{-\infty}^{t} \exp\left\{ -i\omega(t'-t) + i\boldsymbol{k} \cdot (\boldsymbol{r}(t') - \boldsymbol{r})] \mathrm{d}t' \right\} \varphi_1 \, .$$

$$(13)$$

To derive this expression we have assumed that the unperturbed distribution function, a function of constants of motion only, has the particular form given by the Eq. (1).

When calculating the perturbed ion distribution function we neglect influence of the magnetic field on the ion motion in the mode electric field with frequency $\omega \gg \Omega_i$; and wavenumber $k r_{Li} \gg 1$. Then the integration in Eq. (13) is performed along the straight ion trajectories $\boldsymbol{r} = \boldsymbol{v}t$ with the result:

$$f_{1i} = -\frac{e f_{0i}}{T_i} \left\{ 1 - \frac{\omega - k_y u_i}{\omega - \boldsymbol{k} \cdot \boldsymbol{v} + i0} \right\} \varphi_1 \, . \qquad (14)$$

Electrons, on the contrary, are assumed to be magnetized. The wave phase along the electron trajectories oscillates with the electron Larmor frequency. Since the latter is much larger than the mode frequency then during the integration along the electron trajectory we first perform averaging over these fast oscillations:

$$\int_{-\infty}^{t} \varphi_1[\boldsymbol{r}(t'), \, t'] \mathrm{d}t' \simeq \int_{-\infty}^{t} \varphi_1[\boldsymbol{r}_c, \, t'] \mathrm{d}t' \cdot \left[\frac{1}{2\pi} \int_0^{2\pi} \mathrm{d}\theta \, \exp\left(-\frac{i k v_\perp}{\Omega_e} \sin \theta \right) \mathrm{d}\theta \right]$$

$$(15)$$

where $\boldsymbol{r}_c = \boldsymbol{r} - [\boldsymbol{v} \times \boldsymbol{B}]/\Omega_e B$ is the guiding center coordinate for the electron with velocity $\boldsymbol{v}$ and position $\boldsymbol{r}$ at the moment t, θ is the azimuthal angle in the polar coordinate system $(v_\perp, \, \theta, \, v_\parallel)$ with the polar axis along the magnetic field. With the help of Eq. (15) we obtain

$$f_{1e} = \frac{e f_{0e}}{T_e} \left\{ 1 - \left(1 - \frac{k_y u_e}{\omega} \right) J_0 \left(\frac{k v_\perp}{\Omega_e} \right) e^{-(i[\boldsymbol{k} \times \boldsymbol{v}] \cdot \boldsymbol{B}/\Omega_e B)} \right\} \varphi_1 \, . \qquad (16)$$

The Poisson equation

148 A. A. GALEEV

$$k^2\varphi_1 = \sum_j 4\pi e_j \int f_{1j}\, d\mathbf{v} \tag{17}$$

is transformed into the dispersion relation

$$\varepsilon(\omega,\,\mathbf{k}) \equiv 1 + \frac{2\omega_{pe}^2}{k^2 v_e^2}\left[1 - \left(1 - \frac{k_y \mathbf{u}_e}{\omega}\right) I_0(\lambda) e^{-\lambda}\right] + \frac{2\omega_{pi}^2}{k^2 v_i^2}[1 + \zeta Z(\zeta)] = 0 \tag{18}$$

where

$$\zeta = (\omega - k_y u_i)/|\,k\,|v_i\,;\qquad Z(\zeta) = \pi^{-1/2}\int_{-\infty}^{+\infty} dt\, \exp\,(-t^2)/(t - \zeta)$$

is the plasma dispersion function; $I_0(\lambda)$ is the modified Bessell function, $\lambda = k^2 r_{Le}^2/2$.

We simplify this equation assuming here $|\,\omega - k_y u_i\,| \ll |\,k\,|v_i$, and $k^2 r_{Le}^2 \ll 1$ (MIKHAILOVSKII, 1971)

$$\varepsilon(\omega,\,\mathbf{k}) \equiv 1 + \frac{\omega_{pe}^2}{\Omega_e^2} + \frac{k_y u_e}{k^2 \lambda_{De}^2 \omega} + \frac{1}{k^2 \lambda_{Di}^2}\left[1 + i\pi^{1/2}\frac{\omega - k_y u_i}{|\,k\,|v_i}\right] = 0 \tag{19}$$

where $\lambda_D = \sqrt{k\,T_j/4\pi n_0 e^2}$ is the Debye length for particles of species j. This mode is known as the lower-hybrid-drift mode whose frequency and growth rate are (for $\omega_{pe}^2 \gg \Omega_e^2$):

$$\omega = k_y u_i/(1 + \lambda_*)\,;\qquad \gamma_k = \pi^{1/2}(k_y u_i)^2 \lambda_*/|\,k\,|v_i(1 + \lambda_*)^3 \tag{20}$$

where $\lambda_* = \lambda(T_i/T_e)$. The growth rate reaches its maximum at $\lambda_* = 1$ for $k_y^2 \gg k_z^2$

$$\gamma_M = \frac{\pi^{1/2}}{16\sqrt{2}}\,\omega_{LH}\left(r_{Li}\frac{d\ln n}{dz}\right)^2 \gtrsim \Omega_i\,. \tag{21}$$

Where $\omega_{LH} = \sqrt{\Omega_e \Omega_i}$ is the lower hybrid-frequency. If the above inequality is satisfied we can neglect the magnetic field influence on the Cherenkov interaction with ions. In fact it defines the threshold for the density gradient given by Eq. (8). In case of the smaller density gradients we should consider the ion cyclotron resonance instead of the Cherenkov resonance. The corresponding instability —the drift-cyclotron one— develops when $|\,r_{Li} d \ln n/dz\,| > \sqrt{m_e/m_i}$.

The main consequence of development of the lower-hybrid-drift instability is the particle scattering by electric field fluctuations and the anomalous resistivity related to it. We estimate the anomalous resistivity with the help of the weak plasma turbulence theory (SAGDEEV and GALEEV, 1969). In this approach the plasma turbulent motion is considered as a superposition of collective plasma oscillations weakly interacting among themselves and with plasma particles:

$$\varphi(r_1 t) = \sum_k \varphi_k(t) e^{-i\omega_k t + i k \cdot r} \tag{22}$$

where the frequencies ω_k obey the linear dispersion relation $\mathrm{Re}\,\varepsilon(\omega_k, k) = 0$ and the wave amplitudes φ_k are slowly changing in time due to their interaction with waves and particles.

Anomalous resistivity is described in this language as the electron-ion friction resulting from the collective oscillations exchanging momentum between them. We find the effective collision frequency v_{eff} of electrons by equating the macroscopic expression for the electron drag with the ion momentum loss due to the wave emission:

$$n_0 m_e v_{\mathrm{eff}}(u_i - u_e) = 2 \sum_k \gamma_k \frac{k_y}{\omega_k} W_k. \tag{23}$$

Here $W_k = (\partial/\partial\omega)[\omega\varepsilon(\omega, k)] k^2 |\varphi_k|^2 / 8\pi$ is the spectral energy density of waves. The ion momentum change here is assumed to be equal to the change of the product of the quantum number density $W_k/\hbar\omega_k$ and the momentum $\hbar k$ of one quantum. With the help of Eq. (20) we rewrite this as

$$v_{\mathrm{eff}} = (2\pi)^{1/2} \frac{m_i}{m_e} \omega_{\mathrm{LH}} \sum_k \frac{k_y^2}{k^2} \frac{\lambda_*^{3/2}}{(1+\lambda_*)^2} \frac{W_k}{n_0(T_e + T_i)}. \tag{24}$$

To find the wave growth saturation level we should consider the nonlinear wave-wave interaction. Instead of the kinetic approach we use for this the fluid type description based on the electron continuity equation in the guiding center approximation

$$\frac{\partial n_{1e}}{\partial t} + \mathrm{div} \left\{ c \frac{[E_1 \times B]}{B^2} - \frac{c}{\Omega_e B} \frac{dE_1}{dt} \right\} n_0 = 0. \tag{25}$$

Here we have taken into account only electric and inertial drifts assuming that the diamagnetic drift is small for $T_e \ll T_i$. This is sufficient for recovering the electron density perturbation in the linear approximation. In the nonlinear approximation we should include the finite electron Larmor radius correction to the electric drift:

$$V_{1E} = c \frac{[E_1 \times B]}{B^2} [1 - k^2 r_{\mathrm{Le}}^2 / 2]. \tag{26}$$

Without this correction the main nonlinear term in Eq. (25) disappears. To obtain it we have averaged the mode electric field first over the electron Larmor orbit with the help of Eq. (15) and then over the velocity distribution. Combining Eq. (25) and the Boltzman distribution of ions in the slowly varying mode electric field:

$$n_{1i} \simeq -\frac{e\varphi_1}{T_i} n_0 , \tag{27}$$

we obtain the dynamic equation for the wave amplitudes:

$$(1 + \lambda_*)\left(\frac{\partial \varphi_k}{\partial t} - \gamma_k \varphi_k\right) = \sum_{k' + k'' = k} \frac{c[\mathbf{k'} \times \mathbf{k''}]_x}{2B_x} (k''^2 - k'^2) r_{Le}^2 \varphi_{k'} \varphi_{k''} e^{-i(\omega_{k'} + \omega_{k''} - \omega_k)t} . \tag{28}$$

Here the wave amplitude is slowly varying due to its interaction with particles and other waves. Introducing the probability amplitudes

$$|C_k|^2 \equiv \frac{W_k}{|\omega_k|} = (1 + \lambda_*)n_0 e^2 |\varphi_k|^2 / 2|\omega_k| T_i , \tag{29}$$

we rewrite Eq. (28) in the canonical form

$$\frac{\partial C_k}{\partial t} = \sum_{k' + k'' = k} V_{k, k', k''}(t) C_{k'} C_{k''} + \gamma_k C_k \tag{30}$$

where

$$V_{k, k', k''}(t) = -\frac{\Omega_i}{|\omega_k|}[\mathbf{k'} \times \mathbf{k''}]_x (k'^2 - k''^2) r_{Li}^2 r_{Le}^2$$

$$\cdot \sqrt{\frac{|\omega_k \omega_{k'} \omega_{k''}|}{8 n_0 T_i (1 + \lambda_*)(1 + \lambda'_*)(1 + \lambda''_*)}} \exp\left[-i(\omega_{k'} + \omega_{k''} - \omega_k)t\right] .$$

With the help of this equation we obtain the wave kinetic equation in the random phase approximation (SAGDEEV and GALEEV, 1969);

$$\frac{\partial n_k}{\partial t} = 2\gamma_k n_k + 4\pi \sum_{k' + k'' = k} |V_{k, k', k''}|^2$$

$$\cdot \{n_k n_{k'} - n_k n_{k'} \, \text{sign} \, \omega_{k''} - n_k n_{k''} \, \text{sign} \, \omega_{k'}\} \delta(\omega_k - \omega_{k'} - \omega_{k''}) \tag{31}$$

where

$$n_k = |C_k|^2 .$$

Balancing here the linear growth and wave-wave coupling terms we obtain an order of magnitude estimate for the wave energy density

$$\sum_{k} W_k \simeq \gamma_k \omega_k / 2\pi |V_{k,\,k',\,k-k'}|^2$$

$$\simeq n_0 T_i \left(\frac{m_e T_i^2}{m_i T_e^2} \right) \left(\frac{r_{Li} \mathrm{d} \ln n}{2 \mathrm{d} z} \right)^3 . \tag{32}$$

In writing this we have assumed that the main contribution here comes from the waves with $k^2 r_{Le}^2 \lesssim 1$. Introducing this result into Eq. (23) we find the effective collision frequency

$$\nu_{\mathrm{eff}} \simeq \omega_{\mathrm{LH}} \left(\frac{r_{Li}}{2} \left| \frac{\mathrm{d} \ln n}{\mathrm{d} z} \right| \right)^3 \left(\frac{T_i}{T_e} \right)^2 . \tag{33}$$

Some other estimates for the wave energy density can be found in literature. HUBA *et al.* (1977) have assumed that an upper bound on the wave energy is given by the total directed kinetic energy, i.e. $W \lesssim n_0 m_e (u_e - u_i)^2 / 2$. But the electric current is maintained in fact by the plasma pressure gradient and thus the free energy reservoir is much larger (of the order of $\approx n_0 T_i$). SOTNIKOV *et al.* (1980) have pointed out that the lower-hybrid-drift mode collapse can be an effective instability saturation mechanism. Comparison with the weak turbulence estimate (32) shows that the collapse effects dominate for $|r_{Li} \mathrm{d} \ln n / \mathrm{d} z| \gtrsim (T_e / T_i)$.

The anomalous resistivity results in an enhanced magnetic field diffusion. The rate of this diffusion in plasma with $\beta = 1$ is equal to the plasma diffusion rate. Thus, the diffusion coefficient is

$$D_\perp \simeq \nu_{\mathrm{eff}} r_{\mathrm{Le}}^2 \left(1 + \frac{T_i}{T_e} \right) . \tag{34}$$

Here we used the fact that diffusion is caused by the plasma drift under the action of the electron-ion drag and therefore it is proportional to the total diamagnetic current.

For the typical plasma sheet parameters given in the introduction the magnetic field diffusion happens to be much slower than the magnetic field configuration change due to the tearing mode development considered below. Therefore the lower-hybrid-drift instability of the quiet plasma sheet cannot significantly influence its dynamics though these modes are observed in the plasma sheet (SCARF *et al.*, 1974; GURNETT *et al.*, 1976).

2.2 *Pitch angle scattering modulation by plasma convection*

In the case of one-dimensional neutral sheet model an equilibrium is maintained due to balance of plasma and magnetic field pressures across the magnetic field lines. In two-dimensional configuration it is also necessary to balance the magnetic field line tension. Earlier we have assumed that the plasma very quickly reaches the local thermodynamic equilibrium without pressure anisotropies and plasma flows. Then the tension can be balanced by the weak pressure gradient along the quasineutral sheet. But the equilibrium of an infinitely long plasma sheet can also be reached due to the plasma pressure anisotropy (RICH *et al.*, 1972):

$$p_\perp + \frac{B^2}{8\pi} = \text{const}, \qquad p_\parallel - p_\perp = \frac{B^2}{4\pi} \tag{35}$$

where $p_\parallel$ and $p_\perp$ are plasma pressures along and across magnetic field lines. The first condition here provides the balance of force across the field lines and the second is well known as the neutral equilibrium of the firehose modes in an anisotropic plasma. It is easy to see that for the finite thickness plasma sheet the second condition is violated at its boundary, where the pressure drops and the magnetic field increases. Thus it is impossible to reach equilibrium by the pressure anisotropy forces only.

Nevertheless the pressure anisotropy can arise either in the process of plasma convection or due to particle precipitation through the loss-cone. At the same time the excited plasma turbulence tends to quench the pressure anisotropy by the quasilinear relaxation. The quasistationary state is reached due to the balance of quasilinear relaxation with the anisotropy recovery processes. The residual anisotropy and pitch angle scattering rate can be estimated in the model of two-dimensional plasma sheet equilibrium given by Eq. (7). In this model the characteristic time of convection is

$$T_E = L_x \left(\frac{cE_y}{B_n} \right)^{-1} \tag{36}$$

where E_y is the steady state electric field across the magnetotail and $L_x = B_0 \Delta / B_n$ is the plasma sheet charcteristic length. We have assumed here that most of the particles is trapped between the magnetic mirrors and experiences the electric drift with velocity cE_y/B_n (see Subsection 3.3). On the other hand the rate of particle loss through the loss-cone should be slower than the convection rate in order to maintain the plasma sheet (KENNEL, 1969):

$$T_{\min} \simeq \frac{2L_x}{V_j} \frac{B_0}{B_n} > T_{\mathrm{E}} \tag{37}$$

where the factor (B_n/B_0) accounts for the phase volume of the loss-cone.

Thus we expect that plasma is in marginal state when the relaxation time is of the order of the convection time. In the case of the whistler type turbulence the relaxation time happens to be longer than the pitch angle scattering time of single particle (GALEEV, 1975). This is due to the quasiplateau formation in the velocity space that slows down the relaxation. To describe this effect we should consider the system of equations consisting of the quasilinear equation for the distribution function and the wave kinetic equation. Such an ambitious program is out of the scope of this review. Here we limit ourselves to an order of magnitude estimates of the dependence of the electron pitch angle scattering rate on the convection rate. Our estimate is based on the energy conservation law in the wave particle interaction

$$\sum_k \gamma_k \frac{|B_k|^2}{4\pi} \simeq (n_0 T_{\mathrm{e}})/T_{\mathrm{E}} \tag{38}$$

and the approximate form of the wave kinetic equation

$$\gamma_k |B_k|^2 \simeq \Omega_{\mathrm{e}} |B_k|^2 \sum_{k'} |B_{k'}|^2/B_0^2 . \tag{39}$$

Here we assume that during the time T_{E} the electron pressure anisotropy $(T_\perp/T_\parallel - 1)$ of the order of one is generated so that even the thermal electrons come into resonance with whistlers. The nonlinear term in the wave kinetic equation could be for example due to the mode-mode coupling conserving the wave energy (thus, it does not influence the energy balance Eq. (38)). Numerical factors of the order of unity are omitted at the right-hand sides of both these equations. From Eqs. (38) and (39) we obtain

$$\sum_k \frac{|B_k|^2}{B_0^2} \simeq (\Omega_{\mathrm{e}} T_{\mathrm{E}})^{-1/2} . \tag{40}$$

Using the estimate (36) for T_{E} in the case of electric voltage $\Delta\varphi_0 \simeq 10\,\mathrm{kV}$ across the magnetotail with width $40\,R_{\mathrm{E}}$ we find the fluctuating magnetic field of the order of $100\,\mathrm{m}\gamma$ for $B_0 \sim 10\gamma$ as it is observed (GURNETT et al., 1976). The corresponding pitch angle diffusion coefficient

$$D \simeq \Omega_{\mathrm{e}} \sum_k \frac{|B_k|^2}{B_0^2} \tag{41}$$

is so high that we should include it in the tearing mode stability analysis (see Subsection 3.5).

3. Tearing Instability of the Magnetospheric Tail as the Mechanism for Substorms

3.1 Energetics of substorms

Magnetospheric substorm is a complex of phenomena that reflects development of global instability inherent to the magnetosphere as a magnetohydrodynamic system. One of the most impressive manifestations of the substorm—aurora—has been known already since older times, but the generation of powerful ($\sim 10^9$ watt) sporadic radiobursts in the kilometric wave-length range was discovered only in the last decade. The consequences of the substrorm instability were discovered in different elements of the strongly coupled "magnetosphere/ionosphere" system and are now studied very intensively. Among those are strong magnetic disturbances resulting from the currents flowing into the ionosphere (magnetic substorms), fast plasma streams in the magnetospheric tail and bursts of high energy particles in the MeV-range, that are generated in the tail and escape from the magnetosphere into the interplanetary space.

Though a number of different substorm mechanisms and chains of events leading to a substorm are presently being discussed, we will describe here in detail only the concept of a substorm as a magnetospheric tail instability. This concept is supported by various pieces of evidences. The main argument in its favor is the fact that the long magnetospheric tail, which has a diameter of the order of thirty earth radii (Earth radius $R_E = 6{,}400$ km) and extends beyond the lunar orbit, i.e. farther than 60 R_E, plays a role of gigantic energy reservoir for substorms. Here the energy is stored in a form of the magnetic field whose strength is of order of $B \approx 20\,\gamma$ ($1\,\gamma = 10^{-5}$ gauss).

Satellite measurements of the magnetic field inside the magnetotail show an increase of its strength prior to the substorm break-up. Simultaneously the magnetotail increases its diameter. Such a behaviour of the magnetic field is thought to be a result of the magnetic flux transfer from the dayside magnetopause into the tail and corresponds to the growth phase of substorm. The piling-up of magnetic field lines in the magnetotail is accompanied by the plasma sheet thinning and the decrease of z-component of the magnetic field threading the plasma sheet.

The mechanisms proposed to be responsible for magnetic field line transfer from the dayside magnetopause to the tail are the reconnection of the interplanetary and earth magnetic fields followed by transport of open field lines to the tail (DUNGEY, 1961) and convection of closed field lines along the flanks of the magnetosphere (Axford and HINES, 1961). The magnetic field energy storage in the magnetotail and its subsequent dissipation during substorm are discussed here in the framework of the reconnection approach. For a quantitative discussion we will limit ourselves to the consideration of the magnetic field convection in the (x, z) plane, where x and z are solar magnetospheric coordinates, x being directed towards the sun along the earth-sun line and z—to the north perpendicular to the ecliptic plane. For simplicity we assume here that the earth's dipole is within the same plane and the interplanetary magnetic field has only B_z-component, positive or negative (see Chapter 1, Fig. 1). In the case of the northward interplanetary magnetic field we believe that the magnetosphere is in a state of minimum energy; and therefore there is no energy available for a substorm. Of course, to reach this lowest energy state, the interplanetary magnetic field should be directed northward for a prolonged period of time sufficient for the magnetosphere to relax to this state.

If after such a period the interplanetary magnetic field abruptly changes its direction to southward, then reconnection processes at the dayside magnetopause immediately react to this change and start to open previously closed magnetic field lines. The flux of open magnetic field lines increases with a rate $\dot{\Phi}_D$ depending on the plasma and field characteristics at the magnetopause. Open field lines frozen in the highly conducting solar wind plasma are blown away towards the magnetosphere tail and finally could be again reconnected in the vicinity of another neutral line, situated farther in the tail.* This leads to increasing the flux of closed field lines with the rate $\dot{\Phi}_N$. Stationary state is reached when $\dot{\Phi}_D = \dot{\Phi}_N$. However, if reconnection in the tail started simultaneously with the beginning of reconnection at the dayside magnetopause and exactly balanced it, then there would not be any storage of energy in the magnetospheric tail necessary for the substrom development (CORONITI and KENNEL, 1973). Therefore the global magnetospheric tail instability which we suggest here as a mechanism of substorm should possess some margin of stability against finite deviation of the system from the minimum energy state. As observations show, external perturbations of finite amplitudes are able to trigger the instability before the magnetosphere gains an amount of free energy sufficient to drive it internally unstable. For

---

* In the three dimensional picture dayside and nightside neutral lines are connected into one neutral line.

example, a positive pulse of B_z-component of the interplanetary magnetic field very often causes a substorm (KAWASAKI *et al.*, 1971; CAAN *et al.*, 1977). This happens in conditions when B_z-component has been negative and therefore energy storage in the magnetotail has been operative for some period of time. Thus, the global instability of the magnetospheric tail can be excited by finite amplitude disturbances while the tail is in a state which is stable against infinitely small perturbations. In this case one talks about nonlinear instability. Finally, in the absence of any significant external disturbances the evolution of the plasma and the field configuration in the magnetospheric tail during energy storage should lead to a state where the instability can develop spontaneously.

The amount of energy stored in the magnetospheric tail has been estimated using direct satellite measurements (CORONITI and KENNEL, 1972; TVERSKOY, 1968; SISCOE, 1969). According to MAEZAWA (1975) the magnetic flux in the magnetospheric tail increases prior to a substorm by about 10 to 30% of the initial flux. Auroral oval photographs obtained from ISIS-2 and DMS P-1, 2 polar satellites have given an independent estimate of both the free energy stored in the magnetotail and the energy dissipated in the ionosphere by particles precipitating from the magnetosphere (AKASOFU, 1976). The free energy can be expressed as the excess of the magnetic flux in the magnetotail

$$\varepsilon_F(\tau) = \frac{L_T}{8\pi^2 R_T^2} \left[\int_0^\tau (\dot{\Phi}_D - \dot{\Phi}_N)\mathrm{d}t \right]^2 \tag{42}$$

where L_T and R_T are the length and the radius of the magnetospheric tail. Magnetic flux in turn is estimated as

$$\int_0^\tau (\dot{\Phi}_D - \dot{\Phi}_N)\mathrm{d}t = B_p(S - S_{min}) \tag{43}$$

where B_p is the earth's magnetic field strength in the polar cap; S is the area of the polar cap inside the equatorial boundary of the auroral oval in the disturbed period; S_{min} is the same in the case of the minimal auroral oval corresponding to the state of the magnetosphere with the lowest energy.

Thus, if the substorm phenomenon is a consequence of magnetospheric tail instability, we must seek an instability satisfying the following requirements: a) it should be able to transform magnetic field energy into the energy of plasma streams and energetic particles, b) it should allow some margin of stability for the magnetospheric tail when amount of the stored free energy available for the instability is small, and c) the instability should develop either spontaneously after storage of a finite amount of free energy in the tail, or under the influence of external finite disturbances at some earlier stage.

The first of the above requirements means that magnetic field line reconnection or merging is required to be one of the consequences of this instability. Current-driven instabilities causing a violation of the frozen-in magnetic field condition could really favour the reconnection since they make the magnetic Reynolds number smaller, thus increasing the reconnection rate. However, for the characteristic plasma parameters in the magnetospheric tail the critical current velocity for Buneman instability could be reached only in the limit of very thin plasma sheet with the thickness of the order of 1,000 km (CORONITI and EVIATAR, 1977). Therefore we should not expect this instability to develop in the quiet plasma sheet whose typical thickness is of the order of 1 to 2 R_E; it therefore cannot be responsible for (at least) spontaneous reconnection in the tail. The lower-hybrid-drift instability of a thick plasma sheet has been actively studied recently as the reconnection mechanism (HUBA *et al.*, 1977; SOTNIKOV *et al.*, 1980). It has a very low threshold current velocity and therefore can easily develop in the plasma sheet. However, as we have shown in Subsection 2.1 the lower-hybrid-drift turbulence provides too slow plasma diffusion and reconnection. Moreover for the typical thickness of plasma sheet ($r_{\mathrm{Li}} \mathrm{d} \ln n / \mathrm{d}z \lesssim \sqrt[4]{m_e/m_i}$) there can develop only the drift-cyclotron mode which is stabilized in the neutral sheet by the high-β effects for the local $\beta \geq 2$ (MIKHAILOVSKII, 1971). Thus this mode is unable to violate the frozen-in magnetic field condition near the neutral sheet as one needs for reconnection.

The only mode naturally associated with magnetic field line reconnection is the tearing mode which as we will show, satisfies all three requirements mentioned above. First it was proposed as the reconnection mechanism for the magnetotail by COPPI *et al.* (1966).

3.2 Linear stability analysis of the plane neutral sheet

We will discuss here the tearing instability in detail from the point of view that it is the best candidate for the magnetospheric tail instability responsible for substorms. Let us consider first the plane neutral sheet shown in Fig. 2(a) as the simplest magnetic field model in the magnetotail. Such a configuration is supported by the plane current sheet which could be represented in a form of elementary current filaments uniformly distributed along the sheet. It is easy to see that this uniform distribution is unstable towards pairing of current filaments since the attraction between pairs of

A. A. GALEEV

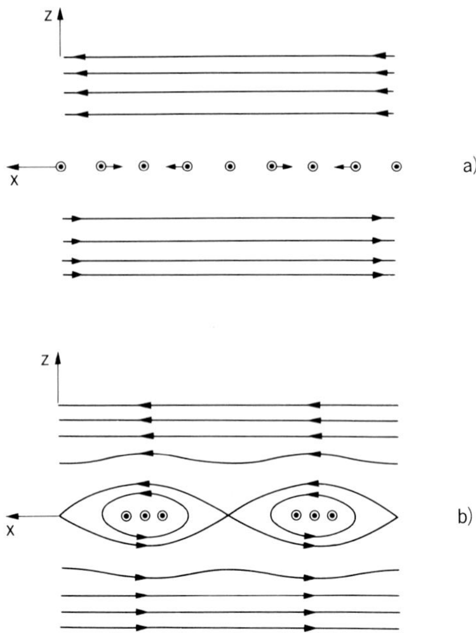

FIG. 2. Tearing instability as the current filaments pairing in a plane neutral sheet (a) and the magnetic field topology in the case of the weak tearing mode imposed on the unperturbed field (b).

filaments grows quickly as they draw together and the attraction of filaments from neighbouring pairs decreases as they move away. This redistribution of the currents changes the magnetic field topology: part of the open field lines in the middle of the sheet now reconnect and close around the pair of current filaments drawn together (Fig. 2(b)). Therefore it is clear that a necessary condition for the development of this tearing mode is the presence of the dissipative processes which could violate the magnetic field line frozen-in condition and result in relative motion of current filaments. As was shown by Laval *et al.* (1965), such dissipation in the collisionless plasma is provided by a Cherenkov interaction of particles with the standing electromagnetic wave created by the current concentration into the periodic structure of current filaments. This interaction can take place only in the narrow layer around the neutral sheet where the magnetic field is so weak that it cannot prevent Cherenkov interaction. Let us turn now to the quantitative consideration of the instability.

As is usual in stability analysis we impose on this equilibrium a small perturbation of the magnetic field corresponding to the pinching of the currents shown in Fig. 2(b).

$$A_{1y}(x, z, t) = \tilde{A}_1(z)e^{-i\omega t + ikx} . \tag{44}$$

It is most easy to do the stability analysis with the help of the energy balance in a form

$$\frac{\partial}{\partial t} \int dz \frac{|\boldsymbol{B}_1|^2}{8\pi} = -\int \boldsymbol{j}_1 \cdot \boldsymbol{E}_1^* dz \tag{45}$$

where $\boldsymbol{B}_1$, $\boldsymbol{E}_1$, and $\boldsymbol{j}_1$ are magnetic field, electric field and electric current perturbations, respectively. Here we see that the magnetic field energy growth is due to the work done by particles on the wave electric field. The latter consists of two parts. The first is the change of the elementary current filament interaction energy and the second is the energy dissipation due to Cherenkov interaction of particles with the wave. Adiabatic redistribution of the current is described by the particle distribution function (1) where we allow slow change of the vector potential by adding to the unperturbed value $A_{0y}(z)$ a small correction $A_{1y}(x, z, t)$ due to the presence of the tearing mode. Linearizing the expression for the adiabatically perturbed distribution function f_{1j}^{ad} we obtain (see Appendex) .

$$f_{1j}^{\mathrm{ad}} = (e_j u_j/cT_j)A_{1y}(x, z, t)f_{0j}(z_1 \boldsymbol{v}) . \tag{46}$$

Calculating the current with the help of this expression and using the result to obtain the adiabatic contribution to the right-hand side of Eq. (45) we find

$$\int j_{1y}^{\mathrm{ad}}E_{1y}^* dz = -\sum_j \frac{e_j^2 u_j^2}{2c^2 T_j} \frac{\partial}{\partial t} \int n(z)|\tilde{A}_1(z)|^2 dz \tag{47}$$

where u_j is the diamagnetic current velocity defined by Eq. (2). It is clear from (47) that this contribution represents an adiabatic change of current filament interaction energy and could be transferred to the left-hand side of Eq. (45). Then the energy balance takes a form

$$\frac{\partial}{\partial t}\left\{\frac{1}{8\pi}\int dz\left[\left|\frac{d\tilde{A}_1(z)}{dz}\right| + k^2|\tilde{A}_1|^2 - \frac{2|\tilde{A}_1|^2}{\Delta^2 \mathrm{ch}^2(z/\Delta)}\right]\right\} = -\mathrm{Re}\int j_{1y}^{\mathrm{res}}E_{1y}^* dz . \tag{48}$$

We see that for a sufficiently prolonged current filament concentration, whose size along the sheet is larger than the thickness of current layer, i.e. for $k^2\Delta^2 \ll 1$, the release of energy due to current concentration (pinching) more than compensates the consumption of energy by creation of the perturbed magnetic field. Thus, the total energy of the plasma with the tearing mode in it is lower than the energy of the unperturbed plasma and magnetic field. In other words the tearing mode energy is negative and therefore any dissipation of its energy results in its amplitude growth, i.e. instability. As we have already mentioned the tearing mode energy dissipation could be provided by Cherenkov interaction of particles with this mode. The magnetic field prevents Cherenkov interaction everywhere outside the narrow layer around the neutral plane $z=0$. The layer width is determined from the condition that the local cyclotron radius is larger than the distance from the plane $z=0$. In a magnetic field with the profile given by Eq. (4) this condition takes a form

$$|z| < d_{jz} = \sqrt{r_{Lj}^{(0)} \Delta} \tag{49}$$

where d_{jz} is the half-width of the layer where magnetic field does not influence the particle motion for species j; $r_{Lj}^{(0)} = v_j/\Omega_{j0}$ is Larmor radius, and $\Omega_{j0} = |e_j| B_0/m_j c$ is the cyclotron frequency far from the neutral sheet.

A crude estimate of the dissipation rate can be made by using the following expression (ARTZIMOVICH and SAGDEEV, 1979):

$$j_{1y}^{res} \simeq \begin{cases} \sigma_{eff} E_{1y}, & |z| \leq d_{ze} \\ 0, & |z| > d_{ze} \end{cases} \tag{50}$$

where $\sigma_{eff} = ne^2/m_e |k| v_e$. Here the collision time is assumed to be equal to the electron time of flight across the distance of the order of one wave length. When an electron moves across that distance the phase of the mode electric field E_{1y} is changed and thus its free acceleration is stopped. In Eq. (50) we have neglected the corresponding ion conductivity which is smaller by the factor $(m_e/m_i)^{1/2}$. With the help of Eqs. (48)–(50) one obtains a rough estimate for the long wave length tearing mode growth:

$$\gamma \sim |k| v_e \left(\frac{r_{Le}}{\Delta}\right)^{3/2} \left(1 + \frac{T_i}{T_e}\right). \tag{51}$$

To obtain this result we have assumed that the magnetic field perturbation is described by the smooth function $A_1(z)$ with the characteristic scale length $1/k > \Delta$ so the last term dominates in the integral on the right-hand side of Eq. (48). Besides that the relation

$$c^2/\omega_{pe}^2 = \left(1 + \frac{T_i}{T_e}\right) r_{Le}^{(0)2} \tag{52}$$

resulting from the pressure balance equation (5) has been used.

The formal solution of the stability problem requires determination of the eigenvalues of frequency ω_k as a function of the wave vector $\boldsymbol{k}$ corresponding to the finite eigenfunction $\tilde{A}_1(z) \to 0$ for $z \to \pm \infty$. To do this we analyze the Boltzman equation for the particle distribution f_{1j} and Maxwell equation for the vector potential A_{1y}. Integration of the Boltzman kinetic equation along the particle trajectories gives the result (see Appendix).

$$f_{1j} = \frac{e_j f_{0j}}{cT_j} \left\{ u_j A_{1y} + i\omega \int_{-\infty}^{t} v_y(t') A_{1y}[x(t'),\ z(t'),\ t'] \mathrm{d}t' \right\}. \tag{53}$$

The electrostatic potential contribution is neglected here under the assumption $\omega \ll kc$ (DOBROVOLNY, 1968). The first term in Eq. (53) corresponds to the adiabatic change of the distribution function. The main contribution to the second term comes from the unmagnetized particles moving within the neutral layer with width given by Eq. (49). Their motion there can be described as free motion between two ideally reflecting walls standing at $z = \pm d_{zj}$. Then the time integration in Eq. (53) can be performed along the straight trajectories $x = v_x t$. As the result of this we obtain:

$$f_{1j} = \frac{e_j f_{0j}}{cT_j} \left\{ u_j - \frac{\omega v_y}{\omega - kv_x + i0} \eta(d_{zj} - |z|) \right\} A_{1y} \tag{54}$$

where $\eta(z)$ is the step function equal to unity for $z > 0$ and to zero for $z < 0$. Besides that we have assumed that the mode amplitude $A_1(z)$ is almost constant for $|z| < d_{zj}$ and have taken it out of the integral in Eq. (53). Calculating the perturbed current with the help of this expression one can write the Maxwell equation (3) for the perturbed vector potential in a form:

$$\frac{\mathrm{d}^2 \tilde{A}_1}{\mathrm{d}z^2} - \left[k^2 + V_0(z) + \sum_j V_{1j}(z;\ \omega,\ \boldsymbol{k}) \right] \tilde{A}_1 = 0 \tag{55}$$

where

$$V_0(z) = -2/\Delta^2 \mathrm{ch}^2(z/\Delta)$$

$$V_{1j}(z;\ \omega,\ \boldsymbol{k}) = -i\pi^{1/2}(\omega_{pj}^2/c^2)(\omega/|k|v_j)\eta(d_{zj} - |z|)$$

Here $V_0(z)$ corresponds to the adiabatically perturbed current. The nonadiabatic contribution to the perturbed current $4\pi j_{1j}^{\mathrm{res}}/c = -\sum_j V_{1j} A_{1y}$ comes from the half residual of the velocity integral from the second term in Eq. (54). It describes the Cherenkov interaction of wave with the resonant particles ($v_x = \omega/k \ll v_j$). Equation (55) has a form of Schrödinger equation for the particle moving in the Teller potential well $V_0(z)$ with the narrow

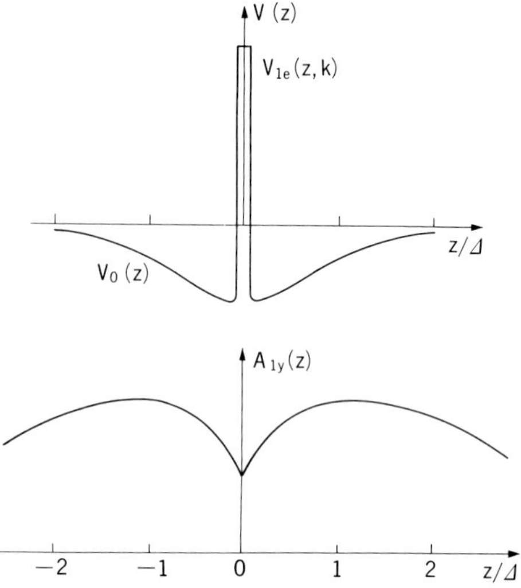

FIG. 3.   The effective potential $V(z)$ in the Schrödinger type equation for the tearing mode and its solution in the external region for $(k\Delta)=0.2$.

potential hump in its middle (Fig. 3). It is convenient to search for the solution of this equation in the outer $(|z|>d_{zj})$ and inner $(|z|<d_{zj})$ regions separately. The outer solution satisfying the boundary conditions $\tilde{A}_1(z)\to 0$ for $z\to\pm\infty$ is (WHITE *et al.*, 1977):

$$\tilde{A}_1(z)=\tilde{A}_1(0)\left[1+\frac{\text{th}(z/\Delta)}{k\Delta}\right]e^{-k|z|}. \tag{56}$$

The inner solution can be obtained only under special assumptions regarding the form of the potential hump $V_1(z)$. If it is, for example, rectangular then neglecting the adiabatic and ion contributions to Eq. (55) we find:

$$\tilde{A}_1(z)=A_1(0)\text{ch}(\sqrt{V_{1e}}z), \qquad |z|<d_{ze}. \tag{57}$$

Matching now the logarithmic derivatives of the inner and outer solutions one obtains the dispersion relation (DOBROVOLNY, 1968):

$$\Delta'(d_{ze})=\sqrt{V_{1e}}\,\text{th}(\sqrt{V_{1e}}d_{ze}) \tag{58}$$

where $\Delta'(d)=(1-k^2\Delta^2)/(k\Delta^2+d)$ is the outer solution's logarithmic derivative at the inner region boundaries. In the limit of not too long wave length $(k\Delta^2 \gtrsim d_{ze})$ we obtain from Eqs. (52) and (58) the tearing mode growth rate (compare with Eq. (51)):

$$\gamma_e = \pi^{-1/2}\Omega_e(r_{Le}/\Delta)^{5/2}\left(1+\frac{T_i}{T_e}\right)(1-k^2\Delta^2)\,. \tag{59}$$

In deriving Eq. (59) we have assumed $V_{1e}d_{ze}^2 \ll 1$. In this case of the transparent potential hump the condition for matching inner and outer solutions can be written in a general form without assumptions about the potential hump shape:

$$\Delta'(d_{ze}) = \frac{1}{2}\int_{-d_ze}^{+d_ze}\tilde{A}_1\frac{d^2\tilde{A}_1}{dz^2}\,dz \simeq \frac{1}{2}\int_{-\infty}^{+\infty}V_{1e}(z;\,\omega,\,k)dz\,. \tag{60}$$

Note that this case corresponds to the limit $k\Delta^2 \gg d_{ze}$, when $\Delta'(d_{ze})$ does not depend on the inner region boundary position.

Finally we should say that the energy balance equation in the form of Eq. (48) can be obtained from Eq. (55) if we note that

$$\frac{4\pi}{c}j_{1y}^{res} = -\sum_j V_{1y}(z;\,\omega,\,k)A_{1y}\,.$$

The procedure for that is to multiply Eq. (55) by A_{1y}^* and then to integrate over z. On the other hand if we use the particular form (56) of the outer solution to calculate the integral on the left-hand side of Eq. (48) then we obtain the equation very similar to Eq. (60):

$$\Delta'(z_0)|\tilde{A}_1(z_0)|^2 = \frac{1}{2}\sum_j\int_{-\infty}^{+\infty}dz V_{1j}(z;\,\omega,\,k)|\tilde{A}_1(z)|^2 \tag{61}$$

where z_0 is the inner region boundary position for the potential hump $V_1(z;\,\omega,\,k)$. Since $\tilde{A}_1(z)$ is slowly changing in the inner region and approximately equal to $\tilde{A}_1(z_0)$ then Eq. (60) and (61) for the particular case of $V_1 \simeq V_{1e}$ coincide. The left-hand side here is the tearing mode energy release (with minus sign) and the right-hand side is the energy loss due to dissipation or by the work done.

3.3 Marginal magnetotail stability

In the case of the one-dimensional plasma equailibria discussed in the previous section, the magnetic field configuration with the neutral sheet is known to be unstable against the tearing mode independent of how thick it is. On the other hand in reality plasma sheet equilibria are two-dimensional (Fig. 4) and involve a balance of forces both along z- and x-axes. For this plasma sheet configuration the magnetic field never passes through zero,

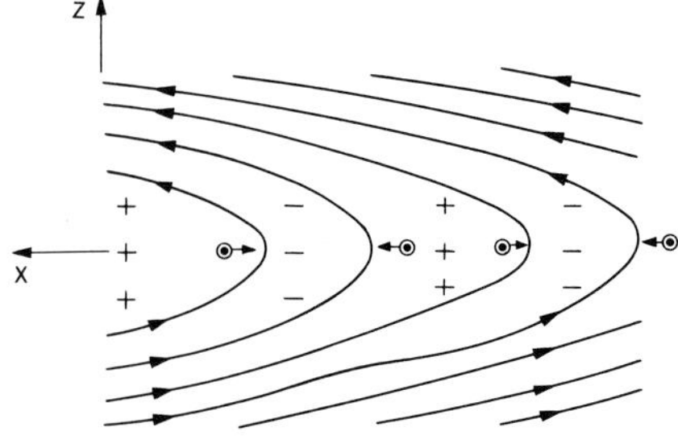

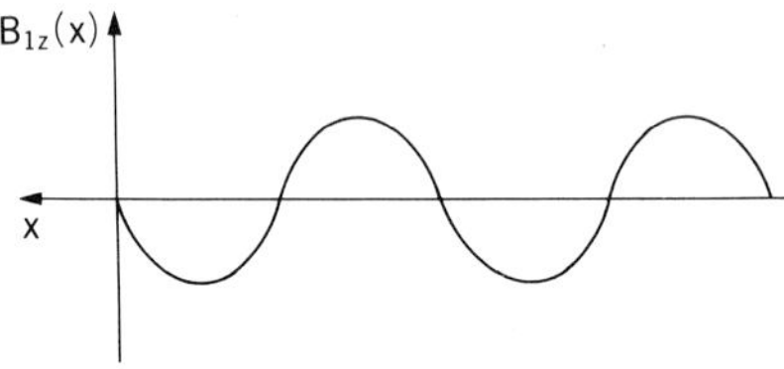

FIG. 4.   Two-dimensional model of the plasma sheet in the presence of a weak tearing mode. The plasma compression takes place in the field enhancement regions. $(B_{1z}>0)$ Space charge is shown by signs $+$ and $-$.

though it is very small near the neutral sheet. We assume here that the magnetic field in the neutral sheet is strong enough to magnetize electrons, but too weak to magnetize ions.

One of the effects of finite B_z-component in the neutral sheet is the supression of the electron Cherenkov interaction with the mode. But the tearing mode will still interact with ions moving within the neutral layer with width given by Eq. (49). This results in the so-called ion tearing mode instability with the growth rate γ_i (see Eq. (59)) (SCHINDLER, 1974).

Analysis shows that the real situation is much more complicated. As was first shown by GALEEV and ZELENY (1975, 1976) this real configuration is strongly stabilized by the contribution of electrons trapped in the vicinity of the neutral sheet. This helps to explain the stability of the magnetotail during the storage of its energy.

Let us turn now to the qualitative consideration of the plasma stability in the two-dimensional magnetic field configuration given by Eq. (7):

$$\boldsymbol{B}=B_0\mathrm{th}(z/\varDelta)\hat{\boldsymbol{e}}_x+B_n\hat{\boldsymbol{e}}_z \tag{62}$$

where B_n is assumed to be constant.

We have shown in the Introduction that though a finite B_z-component of the magnetic field changes the character of the particle motion, the equilibrium distribution function is still approximately the same. What is really important here is the change of the contribution of the magnetized electrons into the tearing mode energy. For the sake of simplicity we consider in detail the case when the magnetic field near the neutral sheet is frozen only in the electron component. This happens if the local Larmor radius of electrons is much smaller than the characteristic gradient scale $\sqrt{r_{\mathrm{Le}}\Delta}$

$$\frac{B_n}{B_0} > \sqrt{r_{\mathrm{Le}}/\Delta} \ . \tag{63}$$

Considering now the tearing instability in terms of the current filament pinching we should take into account the energy spent for plasma compression which occurs as field lines are drawn together (see Fig. 4). To estimate the density perturbation we use the continuity equation for electrons

$$-i\omega\langle n_{1\mathrm{e}}\rangle + ik\langle v_{1x}\rangle n_0 + \left\langle \frac{\partial}{\partial z} v_{1z} n_0(z) \right\rangle = 0 \ . \tag{64}$$

Since most of the electrons are trapped in the vicinity of the neutral sheet and oscillate along field lines between the magnetic mirrors much faster than the mode changes its amplitude, we have performed averaging over these oscillations. The y-component of the equation of the electron motion with inertia term neglected takes a form:

$$\langle E_{1y}\rangle - \langle v_{1x}\rangle B_n/c + \langle v_{1z}B_x(z)\rangle/c = 0 \ . \tag{65}$$

As a result of averaging over the Larmor rotation and bouncing between the mirrors this equation describes the motion of the drift orbits similarly to the "Larmor circles" motion in the conventional drift approximation. Mirroring impedes the bulk motion in z-direction. Therefore the last terms in Eqs. (64) and (65) disappear after averaging over the oscillating velocity v_{1z}. As a result of this the drift speed in x-direction is the same as without a B_x-component*

$$\langle v_{1x}\rangle = c\langle E_{1y}\rangle/B_n \ . \tag{66}$$

Substituting this result to the continuity equation we find the electron density perturbation (GALEEV and ZELENY, 1976; CORONITI, 1980):

---

* This situation is analogous to the "banana" drift in toroidal confinement devices.

$$\frac{\langle n_{1e}\rangle}{n_0}=\frac{ik\langle A_{1y}\rangle}{B_n}. \tag{67}$$

Again because of averaging this relation looks exactly like the frozen-in magnetic field condition for the case when only a B_n-component of the magnetic field is present. To the density perturbation at a distance z from the neutral plane contribute only those electrons whose orbit extends beyond this distance. The plasma compression (or rarefaction) is proportional to the magnetic field strength averaged over the electron orbit (see Fig. 4). Along the very extended orbits the phase of the perturbed magnetic field changes more than π, thus, resulting in the reduction of the perturbed density at large distances z. To find the extent of the density perturbation in z-direction we consider the phase variation of the perturbed field along the magnetic field lines. Near the neutral sheet the magnetic field line equation

$$\frac{\mathrm{d}x}{B_0(z/\varDelta)}=\frac{\mathrm{d}z}{B_n} \tag{68}$$

has a simple solution corresponding to the parabolic field line shape in the plane (x, z):

$$x(z)=(B_0/B_n \quad z^2/2\varDelta)+x(0) \tag{69}$$

where $x(0)$ is the coordinate of the field line in a neutral plane. The phase variation along the field line is represented now in a form:

$$kx(z)=q[b^2(z)-1]+kx(0) \tag{70}$$

where

$$q=(k\varDelta B_n/2B_0)\ll 1, \qquad b^2(z)=B^2(z)/B_n^2 .$$

The averaging over the electron orbit in Eq. (67) is in fact the averaging along the field lines between the mirror points. With the help of Eq. (70) we rewrite Eq. (67) in a form:

$$\langle n_{1e}\rangle=\frac{ikn_0}{B_n}\langle \tilde{A}_1(z')e^{iqb^2(z')}\rangle e^{-i\omega t+ikx-iqb^2(z)} . \tag{71}$$

Here the averaging of the perturbed magnetic field is performed over the drift orbits of the individual electrons passing through the point (x, z), with the subsequent summation over all of them. This procedure is simple but cumbersome (CORONITI, 1980).

To obtain an order of magnitude estimate it is sufficient to note that

$\tilde{A}_1(z)$ is approximately constant in the internal region and matches with the outer solution given by Eq. (56), i.e. $\tilde{A}_1(z) \approx \tilde{A}_1(z_0)$ at $|z| = z_0$ (see Fig. 3). Thus we have

$$\langle A_1(z')e^{iqb^2(z')}\rangle \approx \begin{cases} A_1(z_0) & \text{for} \quad |z| < z_0 \\ 0 & \text{for} \quad |z| > z_0 \end{cases} \tag{72}$$

where the size of internal region $z_0 = \Delta(\pi B_n/k\Delta B_0)^{1/2}$ is found from the condition that the phase variation is smaller than $\pi/2$.

We estimate the energy spent for a plasma compression by calculating the work done by the electron current generated by the perturbed pressure gradient along the quasineutral sheet

$$\langle j_{1y}\rangle B_n/c = ik(T_e + T_i)\langle n_{1e}\rangle . \tag{73}$$

As a result we obtain

$$\frac{1}{2c}\int \langle j_{1y}\rangle\langle A_{1y}^*\rangle dz = -\int dz \frac{|\langle n_{1e}\rangle|^2}{2n_0}(T_e + T_i) \simeq -\frac{k^2 B_0^2 |\tilde{A}_1(z_0)|^2}{2B_n^2} \frac{1}{8\pi} \cdot z_0 . \tag{74}$$

Stabilization of the tearing instability occurs when this energy exceeds the free energy available due to the current pinching in the neutral sheet. Estimating the latter with the help of Eq. (61) and using the expression (56) for the outer solution we write the stability condition in a form:

$$(1 - k^2\Delta^2)/(k\Delta^2 + z_0) < \frac{k^2 B_0^2}{2B_n^2} \cdot z_0 . \tag{75}$$

GALEEV and ZELENY (1975, 1976), who first considered this stabilization effect, have assumed that the spatial extent of the density perturbation is equal to the size of the region where $B_n \geq B_x(z)$ i.e. $z_0 \simeq \Delta(B_n/B_0)$. Later on LAMBERGE (1976) has obtained a correct estimate for z_0, but the final comprehensive analysis was done by CORONITI (1980). An estimate of $\frac{1}{2}c\int\langle j_{1j}\rangle\langle A_{1y}^*\rangle dz$ not based on electron drift orbit approximation is given in Appendix.

In a real configuration with a finite plasma sheet length L_x the magnetic field component normal to the neutral sheet is defined by the condition that the whole magnetic field flux is reconnected through the neutral sheet at the distance L_x, i.e. $B_n \simeq B_0\Delta/L_x$. On the other hand the maximum wave length is bounded by the condition $kL_x \gtrsim \pi$. Taking all of these into account we find from Eq. (75) that the equilibrium plasma sheet with magnetized electrons is stable against the tearing mode (CORONITI, 1980). This conclusion does not depend on numbers, since for $(k\Delta) \gg (B_n/B_0)^{1/3}$. Equation (75) is auto-matically satisfied and for $(k\Delta) < (B_n/B_0)^{1/3}$ it takes a form

$$\frac{k^2 B_0^2}{2 B_n^2} z_0^2 \sim \frac{\pi}{2} k\Delta \frac{B_0}{B_n} > 1 .$$

The situation changes when

$$\frac{r_{Le}}{\Delta} < \frac{B_n}{B_0} < \sqrt{\frac{r_{Le}}{\Delta}} . \tag{76}$$

Then most of the electrons can drift in the x-direction in the neutral layer with a size $d_{ze} = \sqrt{r_{Le}\Delta}$. At the same time they are bouncing between the magnetic walls at $z = \pm d_{ze}$ and participate in the Larmor rotation in the filed B_n. The magnetic field frozen in condition in a form of Eq. (67) still holds, but the size of the internal region is smaller ($z_0 \simeq d_{ze}$). Thus the plasma compression energy does not exceed the free energy available for instability if

$$\frac{1 - k^2 \Delta^2}{k\Delta^2} > \frac{k^2 B_0^2}{2 B_n^2} \sqrt{r_{Le}\Delta} \tag{77}$$

where we have assumed for simplicity that $k\Delta > \sqrt{r_{Le}/\Delta}$. Then the ion tearing instability with the growth rate γ_i given by Eq. (59) can develop under condition:

$$\gamma_i = \pi^{-1/2} \Omega_i (r_{Li}/\Delta)^{5/2} \left(1 + \frac{T_e}{T_i} \right) (1 - k^2 \Delta^2) > \Omega_i \frac{B_n}{B_0} . \tag{78}$$

Inequalities (77) and (78) define the instability region in the plane $(B_n/B_0, r_{Li}/\Delta)$ under the assumption (76) (GALEEV and ZELENY, 1975).

3.4   *Effect of the electrical coupling with the ionosphere*

The conducting ionosphere can affect the instability by shortcircuiting the electrostatic part of the tearing-mode electric field thus reducing the energy spent for the ion compression (GOLDSTEIN and SHINDLER, 1978). The mode electrostatic field results from the charge separation produced when the moving magnetic field lines carry electrons. Though ions are quickly redistributed under the influence of this space charge and the quasineutrality on an average is reached, a residual space charge is always left (Fig. 4). The electrostatic field of the tearing mode is found from the quasineutrality condition, assuming the Boltzman distribution for ions and using Eq. (67):

$$n_0 ik \langle A_{1y} \rangle / B_n = - n_0 e \varphi_1 / T_i \tag{79}$$

where φ_1 is the electrostatic field potential. (In the previous section the electrostatic field $E_y = -\nabla_y \varphi_1$ was zero since we assumed $k_y = 0$.)

Coupling to the conducting ionosphere where the magnetic field lines threading the plasma sheet are rooted results in uncompensated charges flowing down to the ionosphere. This effect can be taken into account through the continuity equation for electrons:

$$\frac{\partial}{\partial t}\langle n_{1e}\rangle + \frac{\partial}{\partial x}(c\langle E_{1y}\rangle n_0/B_n) - \frac{\partial}{\partial s}j_{\parallel}/e = 0 \tag{80}$$

where $j_{\parallel}$ is the field-aligned current from the plasma sheet to the ionosphere which is carried by untrapped electrons (cf. with Eq. (64)). Assuming the quasineutrality and the Boltzman distribution of ions we integrate Eq. (80) over time and represent the potential φ_1 in terms of electromotive force V_1 and capacity C of the inner part of the electrical circuit (Fig. 5).

$$\varphi_1 = V_1 + C^{-1}\int I_1 dt \tag{81}$$

where

$$V_1 = -ik\langle A_{1y}\rangle T_i/eB_n; \qquad C^{-1} = \frac{kT_i}{n_0 e^2 \pi L_y z_0}.$$

$I_1 \simeq j_{\parallel}\pi L_y/k$ is the perturbed current to the ionosphere, L_y is the width of the current sheet, and the characteristic depth, over which the charge is collected, has been taken equal to $z_0 = \Delta(B_n/k\Delta B_0)^{1/2}$. The external part of the circuit consists of the current sheet inductance L_M and the ionospheric resistance R_I (SATO and HOLTZER, 1973)

$$\varphi_1 = -L_M \frac{dI_1}{dt} - R_I I_1 \tag{82}$$

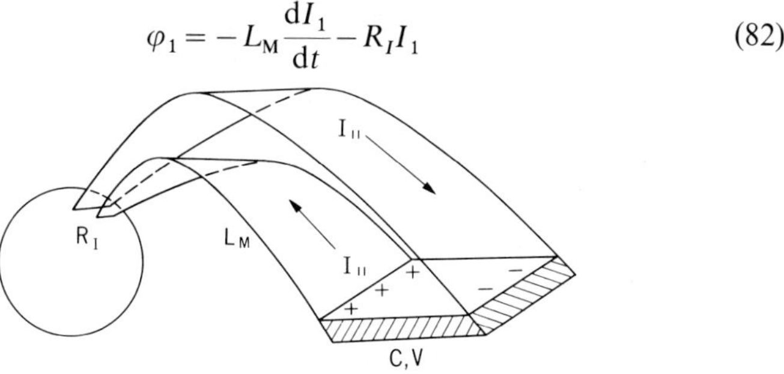

FIG. 5.   Schematic drawing of the longitudinal current sheets when the currents are generated due to space charge of the tearing mode (GOLDSTEIN and SCHINDLER, 1978). The elements of the equivalent electrical circuit are marked by symbols.

where $L_M = 8\pi(s/3)\langle L_x/L_y\rangle$, and s is length of the current sheet. Equations (81) and (82) should be solved together with the energy balance equation defining the growth of the e.m.f. in time. Simultaneously one should include in the energy equation the energy spent for the magnetic field generated by field-aligned currents (inductive losses) and for Joule dissipation in the ionosphere. This Joule dissipation can play a role similar to Landau damping.

It is known that the time constant of the LR circuit is of the order of ten minutes. The time constant of the LC-circuit has the same order of magnitude (GOLDSTEIN and SCHINDLER, 1978). Therefore if the characteristic time of the tearing mode growth is longer than 10 minutes we can expect that the inductive losses to be low and, hence, we succeed in shortcircuiting the mode electric field. But the simple estimate of capacitive charging shows that meanwhile we spend the same amount of energy to charge the capacitor so as to compress the ion component of the plasma ($CV_1^2/2 \simeq n_1^2 T_i/2n_0$). This means that we are not able to lower the threshold. However if the plasma sheet gets tearing unstable due to some other reason (see Section 3.5) then its coupling to the ionosphere causes the longitudinal current flow of the order of the observed one (GOLDSTEIN and SHINDLER, 1978).

Similarly IRBY et al. (1979) have tried to lower the threshold taking into account the reduction of the untrapped electron compression by the sound wave propagation along field lines. Since most of the electrons are trapped their compression does not relax through the sound wave and the net stabilizing effect is negligible.

3.5 Pitch-angle scattering destabilizing effect

An important destabilizing effect has recently been found by CORONITI (1980). He pointed out that the plasma sheet is noisy enough to give very efficient pitch-angle scattering of trapped electrons, thus reducing their compression by the leakage through the loss cone. The quantitative estimate of the density perturbation is obtained by including into the electron continuity equation (64) the term describing the trapped particle loss due to its pitch angle scattering:

$$-i\omega\langle n_{1e}\rangle - ikc\langle E_{1y}\rangle n_0/B_n = -v_{\rm eff}\langle n_{1e}\rangle \tag{83}$$

where $v_{\rm eff}$ is the bounce averaged effective collision frequency. The electron density variation in the tearing mode is estimated now as

$$\langle n_{1e}\rangle = \frac{\omega}{\omega + iv_{\rm eff}}\frac{ik\langle A_{1y}\rangle n_0}{B_n}. \tag{84}$$

We see that even very rare collisions are able to reduce significantly the density variation. Moreover the energy is spent not for the reversible plasma compression but it is irreversibly dissipated in plasma. The dissipation rate is defined as the work done by the mode electric field on current carrying electrons (compare with Eq. (74)):

$$\frac{1}{2c}\int \langle j^e_{1y}\rangle\langle A^*_{1y}\rangle \mathrm{d}z = \frac{n_0(T_e+T_i)}{2B_n^2}\int \frac{i\omega}{v_{\mathrm{eff}}}k^2|\langle A_{1y}\rangle|^2\mathrm{d}z. \tag{85}$$

Here we should take into account the fact that the effective collision frequency is larger for particles having mirror point at higher B. This is due to the differential character of the collisions where diffusion rate toward the loss cone ($\alpha \doteq 0$) increases for small pitch angles α as $v_{\mathrm{eff}} = v_0/\alpha^2$, where v_0 is rate of the $90°$ scattering. The electrons contributing to the density variation at $z > \Delta(B_n/B_0)$ have pitch angles of the order of $\alpha \simeq (B_n\Delta/B_0|z|)^{1/2} \ll 1$. Therefore the effective collision frequency can be estimated for them as $v_{\mathrm{eff}} = v_0(B_0|z|/B_n\Delta)$. As the result, the density variations at large distances $z \sim z_0$ give logarithmically larger contribution to the integral in Eq. (85).

The tearing mode growth rate is estimated from the energy balance equation (61) where we include both dissipation effects: the Cherenkov interaction with unmagnetized ions and the pitch angle scattering of electrons:

$$\bar{v}_{\mathrm{eff}} > \gamma = \frac{\gamma_i}{1 + \dfrac{(k\Delta)^{5/2}\gamma_i \ln(v_{\mathrm{eff}}/\gamma)}{2(B_n/B_0)^{3/2}(1-k^2\Delta^2)v_{\mathrm{eff}}}} > \frac{B_n}{B_0}\Omega_i \tag{86}$$

where $\bar{v}_{\mathrm{eff}} = v_0(\pi B_0/k\Delta B_n)^{1/2}$ is the effective collision frequency for the electrons with mirror point at $z=z_0$; γ_i is the linear ion tearing mode growth rate (see Eq. (78)). The γ has been derived as follows. Equation (61) in the present case become

$$\Delta'(z_0)|A_1(z_0)|^2 = \frac{1}{2}\int \mathrm{d}z\left\{V_{1i}|A_{1y}|^2 - \frac{4\pi}{c}\langle j^e_{1y}\rangle\langle A^*_{1y}\rangle\right\}$$

$$= \int_0^{d_zi}\mathrm{d}z\frac{(-i\omega)}{|k|v_i}\frac{\pi^{1/2}\omega_{\mathrm{pi}}}{c^2}|A_{1y}(0)|^2$$

$$+ \int_0^{z_0}\mathrm{d}_z\frac{(-i\omega)}{v_{\mathrm{eff}}(z)+\gamma}\cdot\frac{k^2B_0^2}{2B_n^2}|\langle A_{1y}(z)\rangle|^2.$$

Because of $|A_{1j}(0)| \simeq |A_{1j}(z_0)|$ (see Fig. 3) we find

$$\gamma_{\mathrm{i}} = -i\omega\left\{1 + \frac{\gamma_{\mathrm{i}}(B_n\varDelta/v_0 B_0)(k^2 B_0^2/2B_n^2)\ln \bar{v}_{\mathrm{eff}}/\gamma}{\varDelta'(z_0)}\right\}$$

where $\qquad \gamma_{\mathrm{i}} = \varDelta'(z_0)\dfrac{|\boldsymbol{k}|v_{\mathrm{i}}c^2}{\pi^{1/2}\omega_{\mathrm{pi}}^2 d_{zi}} \qquad$ with $\quad \varDelta'(z_0) \simeq \dfrac{1-k^2\varDelta^2}{k\varDelta^2}$

for $z_0/k\varDelta^2 \ll 1$, i.e. for $k\varDelta \sim 1$. The above inequalities define the instability regions in the plane $(B_n/B_0,\ r_{\mathrm{Li}}/\varDelta)$ for different ratios $(v_{\mathrm{eff}}/\Omega_{\mathrm{i}})$ (Fig. 6). Pitch angle scattering by lower-hybrid-drift instability (33) can lead to ion tearing instability when the effective collision frequency due to the lower-hybrid-drift instability (33) is higher than the growth rate. If it can be shown that the high-β effects do not suppress this instability near the neutral sheet, then the long wave length ion tearing mode will be unstable according to Eq. (86). The pitch angle scattering by the whistler type turbulence also seems to be sufficient for the tearing mode destabilization in the case of the strong enough plasma convection.

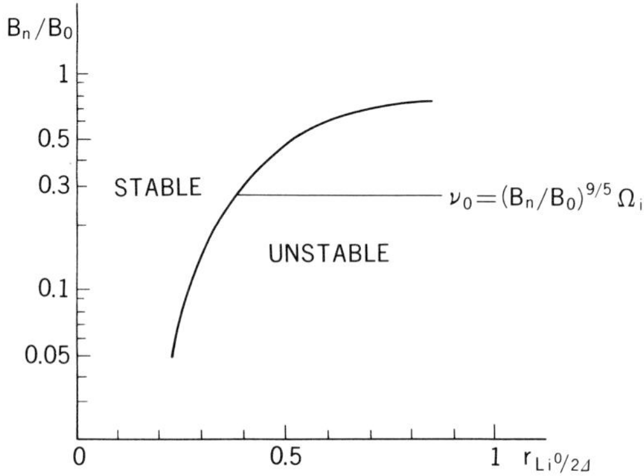

FIG. 6.   The linear tearing instability diagramm in the plane of parameters $(B_n/B_0,\ r_{\mathrm{Li}}/2\varDelta)$. The mode is destabilized by the pitch angle scattering if $v_{\mathrm{eff}} > \Omega_{\mathrm{i}}(B_n/B_0)^{9/5}$.

It is important to note that other modes of the electric and magnetic field fluctuations can be generated in the course of the finite tearing mode development (CORONITI, 1980). This is due to the fact that the electron perpendicular temperature is increasing with the perturbed magnetic field as well as the density:

$$\frac{T_{1\perp}}{T_{\mathrm{e}}} \simeq \frac{ik\langle A_{1y}\rangle}{B_n}. \tag{87}$$

This temperature perturbation is confined to the vicinity of a neutral sheet with the size $|z| < \Delta(B_n/B_0)$ much smaller than the extent of the density perturbation. Thus, we can expect the development of the firehose instability in the rarefaction regions and the mirror instability in the compression regions. To drive these instabilities the tearing mode amplitude need exceed a very low critical value. For example the firehose instability condition can be written in a form

$$n_0(T_\parallel - T_\perp) \sim \frac{|k| \cdot |A_{1y}|}{B_n} n_0 T_e > \frac{B_n^2}{4\pi}. \tag{88}$$

The unstable mode growth results in strong pitch angle scattering of the trapped electrons and thus reduces both the temperature anisotropy and the density perturbation. As a consequence the stabilization effect also decreases. Therefore we expect that the external perturbation of the tearing mode type can be driven unstable if its amplitude is large enough to produce enough scattering. Moreover, in case both the electrons and the ions are strongly pitch angle scattered, even a very thick plasma sheet ($\Delta \gg r_{\mathrm{Li}}$) can be unstable.

4. Magnetospheric Bursts of Energetic Particles

4.1   Explosive growth of the tearing mode at the nonlinear stage

The essential feature of the tearing instability of the Harris plasma equilibria in a plane neutral layer is that though the current sheet pinching leads to the global reconstruction of the magnetic field configuration, the current dissipation can take place only in the narrow vicinity of the neutral sheet where the conditions for Cherenkov resonance are fulfilled. Therefore, the stationary magnetic field reconnection assumes the continuous inflow of the new portions of magnetic field lines into this dissipation layer followed by their subsequent dissipation there. In the absence of the inflow, the weak Cherenkov interaction is unable to provide sufficient magnetic field annihilation and results only in the field line reconnection in the vicinity of the periodically spaced neutral X-lines. The reconnected field lines form closed magnetic surfaces along which the plasma pressure is equalized and thus a new quasistationary state is established. Influx of new portions of field lines blows up further this magnetic island. The instability development is described by linear theory until the width of the magnetic island becomes larger than that of the dissipation layer. The cross section of the magnetic surfaces is described by the equation:

$$A_{0y}(x, z) + A_{1y}(x, z) = \text{const.} \tag{89}$$

174 A. A. GALEEV

where $A_{0y}(x, z) = -B_0\Delta \ln \mathrm{ch}(z/\Delta) + B_n x$ is the zero-order vector potential; $A_{1y}(x, z)$ is the perturbed vector potential given by Eq. (56). The half-width of the magnetic island is defined as the distance from the 0-line to the above position of separatrix which passes through the X-line; for $w/\Delta \ll 1$, we find

$$\frac{w}{\Delta} = \frac{b_1}{(k\Delta)^2} + \left[\frac{b_1^2}{(k\Delta)^4} + \frac{4b_1}{k\Delta}\right]^{1/2} \tag{90}$$

where $b_1 = k\tilde{A}_1(0)/B_0$.

For the case of the ion tearing mode the nonlinear regime starts when

$$w > d_{zi} = \sqrt{r_{\mathrm{Li}}\Delta} \equiv \Delta\sqrt{\varepsilon_i} . \tag{91}$$

Now particle motions at different locations along the x-axis (i.e. near the X- and O-lines) are essentially different. Therefore the energy principle which we use to obtain an estimate of the growth rate is rewritten taking into account the plasma nonuniformity both across and along the neutral sheet

$$\iint dx\, dz \left[\left|\frac{\partial A_{1y}}{\partial z}\right|^2 + k^2|A_{1y}|^2 - \frac{2|A_{1y}|^2}{\Delta^2\mathrm{ch}^2(z/\Delta)}\right] = \frac{4\pi}{c}\iint dx\, dz j_{1y}A_{1y}^* . \tag{92}$$

Following the work of GALEEV et al. (1978) we restrict ourselves to the consideration of a diffuse neutral sheets and to not too long wave length ($\sqrt{\varepsilon_i} < k\Delta < 1$). The first integral here is evaluated using Eq. (56) and we obtain instead of Eq. (92):

$$\frac{\pi}{k}\frac{(1 - k^2\Delta^2)}{k\Delta^2}|A_1|^2 = -\frac{4\pi}{c}\iint dx\, dz j_{1y}A_{1y}^* . \tag{93}$$

To evaluate the second integral in Eq. (92) we must consider separately the contributions from the X-line and the magnetic island.

X-line contribution

The magnetic field near X-line forms a particle trap with the magnetic mirrors-"cusp," where the ions can be kept for a finite time (Fig. 7). The half height and width of the cusp region are

$$|z| < d_z = \Delta\sqrt{\varepsilon_i} ; \qquad |x| < d_x = \Delta[\varepsilon_i/k\Delta b_1]^{1/2} . \tag{94}$$

In the z-direction, the ions are confined by the unperturbed magnetic field $B_x(z)$ and in the x-direction—due to the B_z-component of the tearing mode having scale length of $\sim k^{-1}$. Inside the cusp the ions are unmagnetized. Since the cusp has 4 loss-cone holes at the edges the ions spend only a finite time τ before leaving the cusp. The confinment time is of

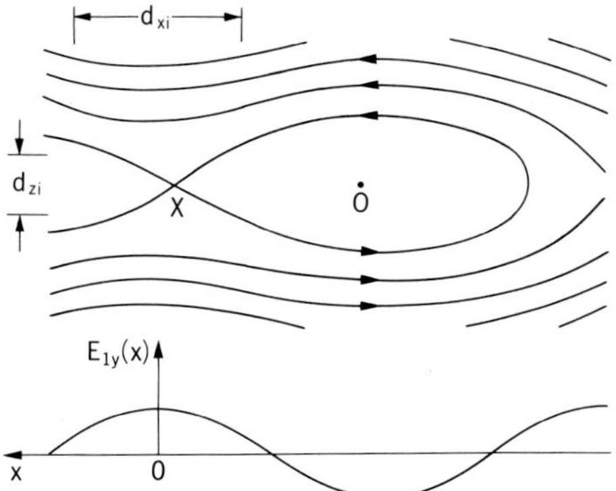

FIG. 7.   Magnetic field topology near the X-line and inside the magnetic insland formed during the tearing mode growth. Behaviour of the mode electric field is shown in the linear approximation.

the order of several flight times across the cusp (BISKAMP and SCHINDLER, 1971):

$$\tau = d_x/v_i .\tag{95}$$

The fact that the lifetime of ions in the trap is finite implies an irreversible character of the interaction between the ions and the tearing mode. The energy gained by ions from the mode electric field during their confinement in the cusp is lost later on when ions leave the cusp. The energy spent for such a heating of ions is described by the integral in Eq. (93). The ion electric current density in the cusp can be written in a form

$$j_{1y} = nev_{1y}\tag{96}$$

where $v_{1y} = -e\gamma\tau A_{1y}/m_i c$ is the ion velocity gained in the mode electric field during the lifetime τ. As a result the contribution of ions in the cusp to the integral of Eq. (93)

$$-\frac{4\pi}{c}\iint_{\text{CUSP}} dz\, dx\, j_{1y}A_{1y}^* = -\frac{4\omega_{\text{pi}}^2}{c^2}\gamma\tau d_x d_z |\tilde{A}_1|^2 .\tag{97}$$

The corresponding contribution of electrons to this integral proves to be smaller by the factor of $(m_e T_e/m_i T_i)^{1/2}$. Energy dissipation of the tearing mode because of its interaction with ions, having a finite lifetime in the cusp, is very similar to that of the electromagnetic wave due to the phenomenon of the so-called anomalous skin-layer, arising also due to the finiteness of the

lifetime of electrons in the skin-layer.
Magnetic island contribution

The contribution of the ions moving inside the magnetic island is a much more complex, essentially nonlinear, problem. We separate these ions into two groups: ions trapped within the minimum field region and ions circulating freely along closed field lines. If the tearing mode electric field penetrates into the island the electric drift of particles leads to increasing the plasma density and pressure in that part of the island where the magnetic field lines are getting denser. The latter is the consequence of the field line influx both due to the production of new flux from the O-line and due to the addition of the reconnected flux from the X-lines. However, in the absence of the field lines connection with the outer plasma the low pressure magnetic field is unable to sustain the plasma pressure excess resulting from the assumption of the electric field penetration into the island (we remind that here $\beta \gg 1$). Therefore, we expect that the plasma pressure will be adjusted to the local magnetic pressure in time of a sound wave propagation across the island, thus averaging the current denity over the island. Due to the finite inductance of the external electrical circuit connecting the plasma sheet to the ionosphere, the total electrical current cannot change during the growth of the tearing mode (see below). Estimates show (GALEEV *et al.*, 1978) that the magnetic island contribution is smaller than the destabilizing effect of the current pinching.

Explosive growth of the tearing mode

Thus, at the nonlinear stage of the tearing mode the energy balance is reached by the irreversible dissipation of energy released through current pinching in the region of the X-lines. By combining Eq. (93) and Eq. (97) we obtain the equation for the nonlinear tearing mode growth

$$\frac{db_1}{dt} = \frac{\pi v_i}{4\Delta} \varepsilon_i^{1/2} \left(1 + \frac{T_e}{T_i}\right) \frac{1 - k^2 \Delta^2}{k\Delta} b_1^2 . \tag{98}$$

The amplitude increases explosively in time as

$$b_1(t) = b_1(0)/(1 - t/\tau_R) \tag{99}$$

where $\tau_R^{-1} = (\pi/4)(v_i/\Delta)\varepsilon_i^{1/2}(1 + T_e/T_i)(1 - k^2\Delta^2)b_1(0)/k\Delta$. τ_R is the characteristic time of an explosive growth; $b_1(0)$ is the amplitude at the start of the nonlinear stage $b_1(0) = k\Delta\varepsilon_i$ (see Eq. (91)). Such an increase of the nonlinear growth rate with amplitude corresponding to its explosive growth is explained by the reduction of the dissipation region near the X-line along with the amplitude growth; the latter results in an electric field increase to dissipate the given amount of energy released due to the current pinching. The explosive mode growth should slow down or cease when the width of the island becomes of the order of (1 to 2) Δ since by then most of the available free energy will have been exhausted. Using Eq. (90) for a rough estimate we obtain the saturation amplitude $b_{1\ max} = (k\Delta)^2$.

4.2   Magnetospheric bursts of energetic particles

In the ion tearing instability concept of the magnetospheric substorms we could naturally expect the betatron acceleration of particles by the induced electric field (SCHINDLER, 1974). Analysis of the energetic particle bursts has really revealed generation of particles up to MeV-range at the substorm development phase (ROELOEF et al., 1976 TERASAWA and NISHIDA, 1976). At the same time the detailed data on their spectrum, anisotropy and temporal evolution (SARRIS et al., 1976a, b; KIRSCH et al., 1977) do not fit the simple generation scheme of such particles at the linear stage of the tearing mode. The first difficulty here is the low linear growth rate of the ion tearing mode and consequently its low electric field which is insufficient to accelerate particles up to the energies required even assuming the existence of a very long neutral line across the whole magnetosphere. We could assume of course that contrary to the observed tearing mode scale (see below) that its development takes place in the much narrower layer with the width of $\sim 1,000$ km (SCHINDLER, 1980). Then, particle energy and burst duration at a given energy would be reasonable in the case of a long neutral line. However, the time interval between the low and high energy particle generation would be much shorter than the observed one and the energy spectrum would be exponential (FRIEDMANN, 1969; BULANOV and SASOROV, 1975) instead of being the power law spectrum.

We will show now that the explosive tearing mode growth discussed in the previous section provides a natural explanation for a total compelx of the specific properties of the energetic particle bursts assuming the typical plasma and field parameters in the plasma sheet.

The problem of the particle acceleration is reduced to the solution of the particle motion equation for given electric and magnetic fields (GALEEV, 1979). Let us start with the analysis of the equation of ion motion near the X-line (Fig. 7):

$$\ddot{x} = \Omega_i b_1(t) k v_y(t) x \tag{100}$$

$$\ddot{z} = -\Omega_i v_y(t) z / \Delta \tag{101}$$

$$v_y(t) = (\Omega_i/k) b_1(t) - 0.5\Omega_i [b_1(t) k x^2 - z^2/\Delta] . \tag{102}$$

Here we have used the coordinate system with y-axis along the neutral X-line. We see that in the z-direction particles are bouncing between the magnetic walls formed by the unperturbed magnetic field $B_x(z) = B_0 z/\Delta$. Particle motion in x-direction is unbounded. The last equation is obtained from the conservation law of the generalized momentum along the neutral line. The general solution of these is difficult and we limit ourselves by the approximate solution for $1 - t/\tau_R \ll 1$:

$$x(t) = (1 - t/\tau_R)^{1/2}\{x_0 \mathrm{ch}[\sqrt{1/4 + (\Omega_i b_{10}\tau_R)^2}\ \ln\,(1 - t/\tau_R)$$
$$- [(\dot{x}_0\tau_R + x_0/2)/\sqrt{1/4 + (\Omega_i b_{10}\tau_R)^2}]$$
$$\cdot \mathrm{sh}[\sqrt{1/4 + (\Omega_i b_{10}\tau_R)^2}\ \ln\,(1 - t/\tau_R)]\} \tag{103}$$

where $x_0 = x(0)$, $x_0 = (\mathrm{d}x/\mathrm{d}t)(0)$, and $b_{10} = b_1(0)$.

We see that the particle starting its motion close to the neutral line is deflected gradually by the weak magnetic field and finally gets into the strong magnetic field region where the acceleration stops. The energy gained by the particle depends on the acceleration time and can be calculated from the generalized momentum conservation law (see Eq. (102)):

$$E_p = m_i v_y^2(t)/2 = E_0/(1 - t/\tau_R)^2 \tag{104}$$

where $E_0 = m_i \Omega_i^2 b_{10}^2/2k^2$.

The acceleration time is estimated from the condition that the magnetic field does not influence particle motion along the neutral line. Comparing the first and second terms on the right-hand side of Eq. (102) we write this condition in a form:

$$k^2 x^2(t) < 1\ . \tag{105}$$

Using the solution for $x(t)$ given by Eq. (103) we conclude that a particle can be accelerated up to the energy E_p or higher only if it starts its motion at the distance from the neutral line less than $x_0(E_p)$ and with the velocity less than $\dot{x}_0(E_p)$, i.e.

$$|x_0| < x_0(E_p) = (1/k)(E_0/E_p)^{\alpha_0/2 - 3/4}$$
$$|\dot{x}_0| < \dot{x}_0(E_p) = \alpha_0 x_0(E_p)/\tau_R \tag{106}$$

where

$$\alpha_0 = 1 + \sqrt{1/4 + (\Omega_i b_{10}\tau_R)^2}\ . \tag{107}$$

The total number of protons accelerated up to the energy $> E_p$ is proportional to the phase space volume:

$$\int_{E_p}^{\infty} N(E)\mathrm{d}E = n_0 |\, x_0(E_p)\dot{x}_0(E_p)\,|\, d_z L_y/v_i \tag{108}$$

where L_y is the neutral line length.

Differential flux of the protons with the energy E_p escaping from the acceleration region through the loss cones during the time interval τ_R can be estimated as

$$J(E_p) = -\frac{\partial}{\partial E_p}\, \frac{n_0 |\, x_0(E_p)\dot{x}_0(E_p)\,|\, d_z}{v_i \tau_R r_{Li}(E_p)}. \tag{109}$$

Here we have taken into account that the width of the loss cones is of the order of gyroradius $r_{Li}(E_p)$ of accelerated protons. Combining Eq. (106) and Eq. (109) we obtain

$$J(E_p) \simeq \frac{\alpha_0^2 n_0}{k^2 \tau_R^2 \sqrt{\varepsilon_i}\,(2E_0^3/m_i)^{3/2}} \left(\frac{E_0}{E_p}\right)^{\alpha_0}. \tag{110}$$

Thus we have shown that the explosive phase of the tearing instability in the magnetospheric tail is accompanied by generation of the energetic particle bursts with a power law spectrum. All the burst parameters can be estimated. The latter is particularly interesting since the energetic particle bursts provides us with the primary information about the processes taking place near the neutral lines. This justifies the comparison of the theory with the burst measurements made far from the source.

In our discussion here we assume that the typical plasma and field parameters in the plasma sheet are

$$B_{0x} = 20\gamma\,; \quad n_0 = 0.2 \text{ cm}^{-3}\,; \quad T_i = 5 \text{ keV}\,; \quad \Delta = 1 R_E\,. \tag{111}$$

Besides these we will use the estimates for the tearing mode amplitudes at the initial phase of its explosive growth $b_{10} = (k\Delta)\varepsilon_i$ and at the saturation stage $b_{1\,max} = (k\Delta)^2$ (see Section 4.1). Numerically $(k\Delta)$ is estimated to be 0.6. The calculation of the burst parameters under these conditions gives results which agree quite well with experiments: The characteristic rise and decay time for a source $\tau_R = 10^2$ sec (SARRIS et al., 1976b); the size of the source $d_x \simeq 3,000$ km, $d_z \simeq 1,500$ km (KRIMIGIS and SARRIS, 1980); the highest possible energy of the accelerated protons $E_{max} m = {}_i b_1^2{}_{\,max}\Omega_i^2/k^2 = 1$ MeV (SARRIS et al., 1976a); and the acceleration time for these highest energy protons (from $(1/2)E_{max}$ to E_{max}) $\Delta t = \tau_R\,(b_{10}/b_{1\,max}) = 10$ sec (SARRIS et al., 1976a; KRIMIGIS and SARRIS, 1980). The very poorly defined quantity b_{10} enters in the power law index of the particle energy spectrum and, therefore, we can not give a good and reliable estimate for it. Nevertheless its value $\alpha_0 \sim 5$ calculated for the plasma and field parameters chosen above is reasonable (SARRIS *et al.*, 1976a; KRIMIGIS and SARRIS, 1980). The situation is much worse for the estimate of the energetic particle flux which contains this poorly defined quantity in high power. Here we get reasonable agreement with the experiment if instead of $E_0 \simeq T_i$ for injection energy given in Eq. (104) we use $E_0 \simeq (3 \text{ to } 5)T_i$. Besides that in deriving Eq. (110) we have neglected the preacceleration at the linear stage of the tearing mode. The latter is probably responsible for the suprathermal plasma observed (ROELOF *et al.*, 1976).

It is important to note here that the acceleration model based on explosive tearing mode growth explains the specific feature of the bursts as well. The inverse velocity dispersion in the burst (SARRIS *et al.*, 1976b; KRIMIGIS and SARRIS, 1980) (Fig. 8) is a good example of this kind. The earlier arrival of the lower energy portion of the burst is explained by its earlier generation by the explosively growing mode electric field. Considering the magnetic field geometry in the case when the tearing mode develops in the magnetotail we find that the induction electric field is directed near the X-line from dawn to dusk. Therefore accelerated protons should be observed preferentially on the evening side of the plasma sheet and electrons—on the morning side. The experiment seems to agree with this statement (KRIMIGIS

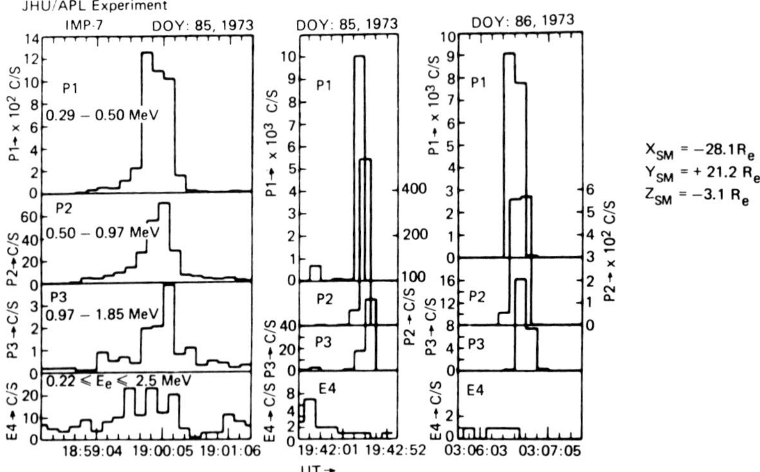

FIG. 8.   Time evolution of the counting rates in the different energy channels illustrating the inverse velocity dispersion in the energetic proton bursts (KRIMIGIS and SARRIS, 1979).

and SARRIS, 1978; CARBARY and KRIMIGIS, 1979). Electron acceleration takes place also, but because of the smaller size of their acceleration region (by the factor of $\sqrt{m_p E_p / m_e E_e}$) the electron burst intensity is usually much less intense. In the previous analysis we have neglected the particle acceleration near O-lines arguing that the particle motion there is unbounded both in the x- and z-directions and, therefore, the number of the accelerated particles is much smaller and the spectrum is softer. Nevertheless the acceleration near the 0-line by the tearing mode electric field directed from dusk to dawn provides an explanation for the detection of the lower electron (proton) peak in the evening (morning) sector (KRIMIGIS and SARRIS, 1980). In addition, after the tearing mode growth has ceased, the global magnetospheric electric field with an opposite direction (i.e. dawn to dusk) is able to continue the particle acceleration with a harder energy spectrum as a result of the bounded motion near the 0-line (STERN, 1979). Finally we should note that the above calculations are made assuming the infinite length of the neutral line for the sake of simplicity. In reality we are facing probably the tearing mode turbulence instead of a single neutral line. In this case a number of X- and O-lines are present in the magnetosphere (SCHINDLER and NESS, 1972) and the acceleration takes place near each of these lines. The length of the neutral lines then is of the order of the tearing mode wave length $2\pi/k \simeq (5$ to $10)R_E$. This length is still sufficient to provide an acceleration to higher energies (GALEEV et al., 1978). Numerical calculations of the particle acceleration in the localized induced electric fields performed independently by PELLINEN and HEIKKILA (1978) support this conclusion.

5. Macroscopic Consequences of the Ion Tearing Mode Instability in the Magnetotail

Let us consider first the case where the tearing mode growth saturates at a moderate amplitude of the order of, or a bit higher than, the B_z-component of the undisturbed magnetic field. Then the tearing mode growth results only in a small topology change: the small magnetic islands appear within the still regular closed magnetic field lines (Fig. 9(a)). The magnetic field topology change leads to the redistribution of the plasma pressure gradients earlier balancing the magnetic tension. Magnetic islands become decoupled from the external region. The plasma inside them reaches quickly their own equilibria. Outside the islands particles change their pitch-angle distribution and partially precipitate through the loss cone into the atmosphere. Then the balance between the plasma pressure and the magnetic tension is violated

                               A. A. GALEEV

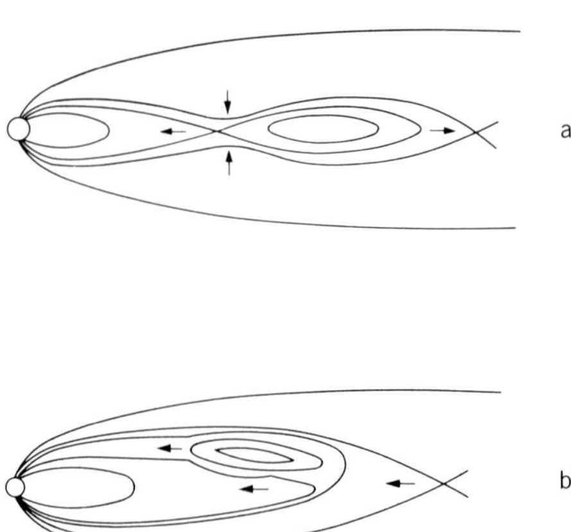

FIG. 9.   Topology of the magnetic field in the magnetotail during the explosive growth (a) and at the relaxation stage (b) in the case of a moderate tearing mode saturation level.

and magnetic field lines start to contract moving the frozen-in plasma earthwards. Since the plasma within the magnetic islands is decoupled from the outer region it moves independently of the outer plasma (Fig. 9(b)). The plasma flow speed can be estimated from the balance of the magnetic tension and the inertia force:

$$n_0 m_i v_x \partial v_x / \partial x \lesssim (B_n / 4\pi) \partial B_x / \partial z \, . \tag{112}$$

For typical plasma sheet parameters given by Eq. (111) we obtain $v_x \lesssim 200$ km/sec. This is much lower than the flow speed of $v_x \simeq 1{,}000$ km/sec sometimes observed in the tail (HONES, 1973; HONES *et al.*, 1973; FRANK *et al.*, 1976). Such high speed streams are observed usually at the boundary of the plasma sheet and in the localized regions which Frank *et al.* called "fireballs." We can get some insight into the fireball by considering Fig. 10, which shows results of the plasma and field measurements on board of IMP-7 spacecraft (CORONITI *et al.*, 1977). Part of the data corresponding to the high speed flow is blown up. We see that this flow is associated with magnetic field pulsations of the order of 5γ and its characteristic time scale is 10 sec. Coroniti *et al.* have assumed that the time variations result from the spatial nonuniformities passing by the spacecraft with the 1,000 km/sec speed. If it is so the 10 sec-variations correspond to a spatial scale of the order of 10,000 km which naturally appears as a result of the tearing mode instability

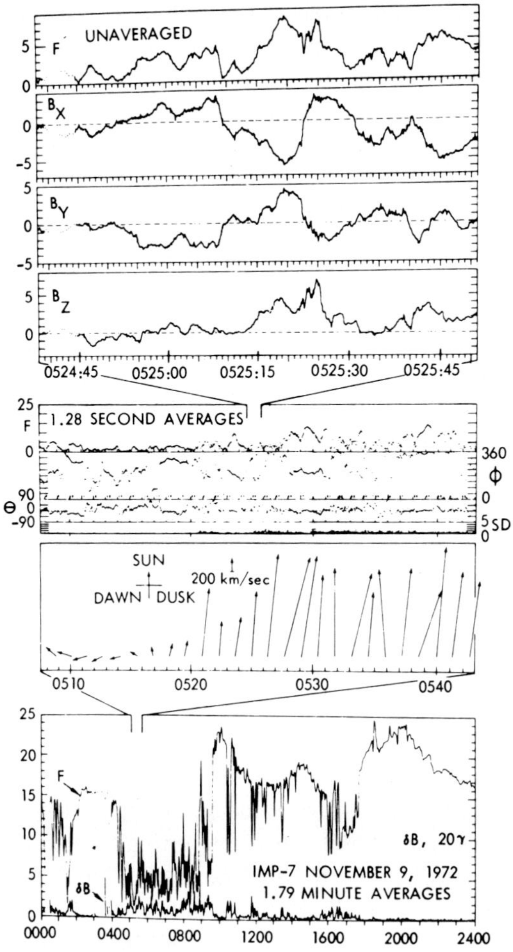

Fig. 10.   High time resolution magnetic field and plasma flow data. Here F, θ, and φ are strength, polar angle and azimuthal angle of the magnetic field vector. Plasma flow speed vectors are shown in the second panel from the bottom (Coroniti *et al.*, 1977).

of the plasma sheet with the thickness $\sim 1 R_E$. The recurrence of the high speed flows discovered later (Coroniti *et al.*, 1978) has the same time scale (15 to 30 min) as the linear tearing mode growth rate and the characteristic time of the magnetosphere-ionosphere electrical circuit. The above arguments show that we have very strong arguments to associate the fireball events with the tearing mode instability. What is then the localized region responsible for the plasma acceleration? We have already shown that the magnetic tension itself is unable to accelerate the plasma to such a speed. Therefore, we should consider the X-line region where the magnetic field energy has been already dissipated into heat. The heated plasma then flows through the loss-cones getting on its way some additional acceleration by the electrostatic electric field. The latter is the result of the electron density drop at X-line because of the magnetic field rarefaction and can be estimated as $e\varphi \sim T_i$. The combined effect of the free flow along field lines and the space charge acceleration is able to give the observed proton distribution (FRANK *et al.*, 1978). Computer simulation (SATO *et al.*, 1978) and laboratory experiments on the magnetic field line reconnection (STENZEL and GEKELMAN, 1979) strongly support this idea. Pressure gradient also contributes to the acceleration but its effect is comparable to the eectrostatic acceleration because $\nabla p \simeq -(T_i + T_e)\nabla n_1 \simeq -en_0 \nabla \varphi_1$.

Some data supporting the global configuration shown in Fig. 9 are also available from the measurements during the disturbed magnetosphere. So the preferential observation of the high speed plasma streams directed earthwards (FRANK *et al.*, 1976) agrees with the assumption that they could be generated even during the moderate tearing instability when there are no open field lines in the generation region. Simultaneous observations of the high speed plasma streams and the energetic particle bursts (Fig. 11, KRIMIGIS and SARRIS, 1980) support the idea that they are generated at the same source (X-line). The multiple neutral line configuration typical of the tearing mode corresponds to the case of the multiple energetic particle source discovered (SARRIS *et al.*, 1976b). It is interesting to note that the size of the source $L_y \sim 5$ to $10R_E$, $L_{x,z} \sim 1R_E$ (FRANK *et al.*, 1976; KRIMIGIS and SARRIS, 1980) corresponds to the auroral bright spots at the Earth with the size 300 to 600 km $\times$ 100 km. ($L_x \sim L_z$ follows Eq. (94) because of $k\Delta \sim 1$.)

In spite of these evidences in favor of the turbulent picture for the disturbed magnetotail we can not exclude the possibility of a coherent reconstruction of the global magnetotail configuration. CORONITI and KENNEL (1980) argue that the convection pattern characteristic for the substorm growth phase tends to drive the coherent tearing mode. In this case we expect the single neutral line to be established across the magnetotail (Fig. 12(a)). If the tearing mode grows to a large enough amplitude the growing magnetic island includes even the field lines which were connected earlier with the interplanetary plasma. Exactly this configuration has been advocated by a number of researchers as the phenomenological substorm

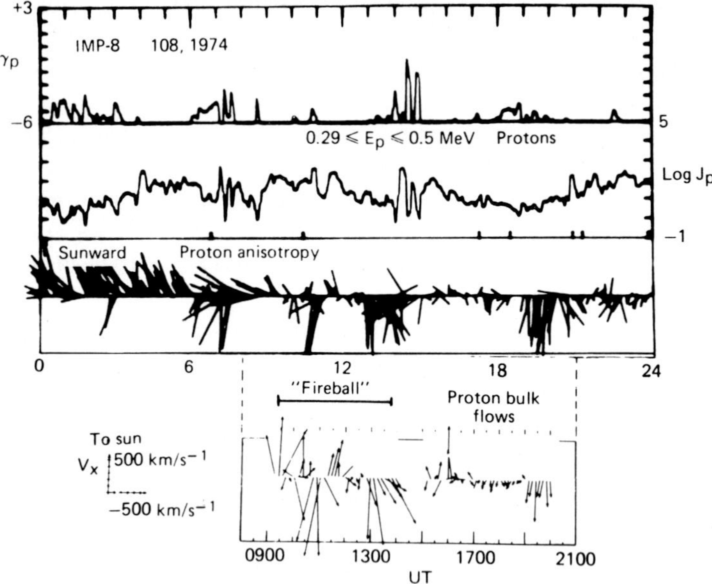

FIG. 11.   Simultaneous data on flux (J_p) anisotropy and spectral index (γ) of the energetic protons and on the plasma flow speeds inside "fireballs" (KRIMIGIS and SARRIS, 1980).

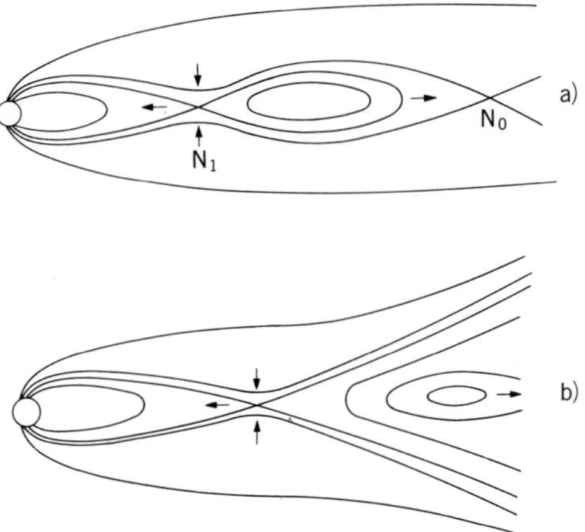

FIG. 12.   Phenomenological model of a substorm with the single neutral line (N_1) formation across the tail. The old neutral line position (N_0) is also shown. Here after the tearing-mode explosive growth (a) the disconnected plasma sheet is thrown through the tail into the interplanetary space (b).

model. The proposed time sequence of the events was the following (see e.g. review of HONES, 1973):

1)   During the substorm growth phase the dynamo processes at the dayside magnetopause provides an energy storage in the magnetotail and the latter evolves towards the unstable state (CORONITI and KENELL, 1972, 1973; RUSSELL and MCPHERRON, 1973).

2)   Substorm break-up is associated with the neutral line formation in the earthward portion of the plasma sheet at $x > -15R_E$ (HONES *et al.*, 1973; RUSSELL and MCPHERRON, 1973; NISHIDA and NAGAYAMA, 1973; NISHIDA and HONES, 1974).

3)   In between the new and old neutral lines the plasmoid is formed and then flows out along the tail under the action of the field line tension (HONES *et al.*, 1976, Fig. 12(b)).

4)   As the result of the plasmoid flow-out the plasma sheet becomes thin.

5)   Plasma sheet expands at the substorm recovery phase and the neutral line moves towards the tail to recover the equilibrium position.

As we have shown above the magnetotail field topology is usually so complex that it is not very easy to find the signature of the extremely simplified phenomenological substorm model based on the single neutral line formation across the tail. Because of that LUI *et al.* (1977) have come to the conclusion that the reconnection is not typical of substorms at all. However, recent statistical study by NISHIDA *et al.* (1981) has shown that at least fast tailward flows having speeds greater than about 400 km/sec are generated due to reconnection at the near earth neutral line. As the alternative model the plasma sheet current interruption model is considered (BOSTRÖM, 1974). One of the main arguments against this model is the absence of any satisfactory and appropriate mechanism of the current interruption. Physical mechanisms developed at least in some detail are the double layer and Buneman instability. Both of them can be generated only in the case when the current velocity exceed the electron thermal velocity. Such current velocities are not typical of the plasma sheet. Besides the double layer or Buneman instability that operates on field lines that connect magnetotail and ionosphere might be a result but is not the cause of interruption of the current that flows across the magnetotail. The high current density in the localized regions with the scale of the order of ~ 10 km could result from the tearing instability but apparently does not play an important role in the substorm energetics.

Author is indebted to Academician R. Z. Sagdeev for constant support and attention during this work. He is grateful to Professors F. V. Coroniti, C. F. Kennel, A. Nishida, and Dr. Zeleny for a number of interesting discussions.

REFERENCES

AKASOFU, S.-I., *Polar and Magnetospheric Substorm,* Reidel Pub. Co., Dordrecht, 1968.

AKASOFU, S.-I., Recent progress in studies of DMSP auroral photographs, *Space Sci. Rev.,* **19**, 169, 1976.

ARTZIMOVICH, L. A. and R. Z. SAGDEEV, *Plasma Physics for Physicists,* Atomizdat, Moscow, 1979 (in Russian).

AXFORD,W. I. and C. O. HINES, A unifying theory of high-latitude geophysical phenomena and geomagnetic storms, *Can. J. Phys.,* **39**, 1433, 1961.

BISKAMP, D. and K. SCHINDLER, Instability of two-dimensional collisionless plasmas with neutral points, *Plasma Phys.,* **13**, 1013, 1971.

BOSTRÖM, R., Ionosphere-magnetosphere coupling, in *Magnetospheric Physics,* edited by B. M. McCormac, pp. 45, Reidel Pub. Co., Dordrecht, 1974.

BULANOV, S. V. and P. V. SASOROV, Energy spectrum of particles accelerated in the vicinity of magnetic neutral line, *Astron. Zh.,* **52**, 763, 1975.

CAAN, M. N., D. H. FAIRFIELD, and E. W. HONES, Jr., Magnetic fields in flowing magnetotail plasmas and their significance for magnetic reconnection, *J. Geophys. Res.,* **84**, 1971, 1979.

CAAN, M. N., R. L. MCPHERRON, and C. T. RUSSELL, Characteristics of the association between the interplanetary magnetic field and substorms, *J. Geophys. Res.,* **82**, 4837, 1977.

CARBARY, J. F. and S. M. KRIMIGIS, Energetic particle activity at 5-min and 10-s time resolution in the magnetotail and its relation to auroral activity, *J. Geophys. Res.,* **84**, 7123, 1979.

COPPI, B., G. LAVAL, and R. PELLAT, Dynamics of the geomagnetic tail, *Phys. Rev. Lett.,* **16**, 1207 1966.

CORONITI, F. V., On the tearing modes in quasineutral sheets, *J. Geophys. Res.,* **85**, 6719, 1980.

CORONITI, F. V. and A. EVIATAR, Magnetic field reconnection in a collisionless plasma, *Astrophys. J. Suppl. Ser.,* **33**, 189, 1977.

CORONITI, F. V. and C. F. KENNEL, Changes in the magnetospheric configuration during substorm growth phase, *J. Geophys. Res.,* **77**, 3361, 1972.

CORONITI, F. V. and C. F. KENNEL, Can the ionosphere regulate magnetospheric convection?, *J. Geophys. Res.,* **78**, 2837, 1973.

CORONITI, F. V. and C. F. KENNEL, Magnetospheric reconnection, substorms and energetic particle acceleration, *J. Geophys. Res.* 1980 (to be published).

CORONITI, F. V., F. L. SCARF, L. A. FRANK, and R. P. LEPPING, Microstructure of a magnetotail fireball, *Geophys. Res. Lett.*, **4**, 219, 1977.

CORONITI, F. V., L. A. FRANK, R. P. LEPPING, F. L. SCARF, and K. L. ACKERSON, Plasma flow pulsations in Earth's magnetic tail, *J. Geophys. Res.*, **83**, 2162, 1978.

DOBROVOLNY, M., Instability of neutral sheet, *Nuovo Cimento*, **55B**, 427, 1968.

DUNGEY, J. W., Interplanetary mangetic field and the auroral zones, *Phys. Rev. Lett.*, **6**, 47 1961.

FRANK, L. A., K. L. ACKERSON, and R. P. LEPPING, On hot tenuous plasmas, fireballs, and boundary layers in the Earth's magnetotail, *J. Geophys. Res.*, **81**, 5859, 1976.

FRANK, L. A., K. J. DeCOSTER, and K. L. ACKERSON, Reply, *J. Geophys. Res.*, **83**, 3365, 1978.

FRIEDMAN, M., Possible mechanism for the acceleration of ions in some astrophysical phenomena, *Phys. Rev., Ser. II.*, **182**, 1408, 1969.

GALEEV, A. A., Plasma turbulence in the magnetosphere and plasma heating, in *Physics of the Hot Plasma in the Magnetosphere*, edited by B. Hultqvist and L. Stenflo, p. 251 Plenum, London-New York, 1975.

GALEEV, A. A. and R. Z. SAGDEEV, Nonlinear plasma theory, in *Voprosy Teorii Plasmy*, p. 3 Atomizdat, Moscow, 1973 (in Russian).

GALEEV, A. A., Reconnection in the magnetotail, *Space Sci. Rev.*, **23**, 411, 1979.

GALEEV, A. A. and L. M. ZELENY, Metastable states of diffuse neutral sheet and the substorm explosive phase, *JETP Lett.*, **22**, 170, 1975.

GALEEV, A. A. and L. M. ZELENY, Tearing instabilities in plasma configurations, *Sov. Phys. JETP*, **43**, 1113 (1976).

GALEEV, A. A., F. V. CORONITI, and M. ASHOUR-ABDALLA, Explosive tearing mode reconnection in the magnetospheric tail, *Geophys. Res. Lett.*, **5**, 707, 1978.

GOLDSTEIN, H. and K. SCHINDLER, On the role of the ionosphere in substorms: Generation of field-aligned currents, *J. Geophys. Res.*, **83**, 2574, 1978.

GURNETT, D. A., L. A. FRANK, and R. P. LEPPING, Plasma waves in the distant magnetotail, *J. Geophys. Res.*, **81**, 6059, 1976.

HAERENDEL, G., Microscopic processes related to reconnection, *J. Atmos. Terr. Phys.*, **40**, 343, 1978.

HARRIS, E. G., On a plasma sheath separating region of oppositely directed magnetic fields, *Nuovo Cimento*, **23**, 115, 1962.

HONES, E. W., Jr., Plasma flow in the plasma sheet and its relation to substorms, *Radio Sci.*, **8**, 979, 1973.

HONES, E. W., Jr., S. J. BAME, and J. R. ASBRIDGE, Proton flow measurements in the magnetotail plasma sheet made with Imp 6, *J. Geophys. Res.*, **81**, 227, 1976.

HONES, E. W., Jr., J. R. ASBRIDGE, S. J. BAME, and S. SINGER, Substorm variations of the magnetotail plasma sheet from $X_{sm} \approx -6\,R_E$ to $X_{sm} \approx -60\,R_E$, *J. Geophys. Res.*, **78**, 109, 1973.

HUBA, J. D., N. T. GLADD, and K. PAPADOPOULOS, The lower hybrid-drift instability as a source of anomalous resistivity for magnetic field line reconnection, *Geophys.*

Res. Lett., **4**, 125, 1977.

IRBY, J. H., J. F. DRAKE, and H. R. GRIEM, Observation and interpretation of magnetic-field-line reconnection and tearing in a theta pinch, *Phys. Rev. Lett.*, **42**, 228, 1979.

KAWASAKI, K., S.-I. AKASOFU, F. YASUHARA, and C.-I. MENG, Storm sudden commencements and polar magnetic substorms, *J. Geophys. Res.*, **76**, 6781, 1971.

KENNEL, C. F., Consequences of a magnetospheric plasma, *Rev. Geophys.*, **7**, 379, 1969.

KIRSCH, E., S. M. KRIMIGIS, E. T. SARRIS, R. P. LEPPING, and T. P. ARMSTRONG, Possible evidence for large transient electric fields in the magnetotail from oppositely directed anisotropies of energetic protons and electrons, *Geophys. Res. Lett.*, **4**, 137, 1977.

KRIMIGIS, S. M. and E. T. SARRIS, Energetic particle bursts in the Earth's magnetotail, in *Dynamics of the Magnetosphere,* edited by S.-I. Akasofu p. 599, D. Reidel Pub. Co., Dordrecht, 1980.

LAMBERDGE, B., These du doctorat 3ᵉ cycle, al 'Universite' de Paris, XI, 1976.

LAVAL, G., R. PELLAT, and M. VULLEMIN, *Plasma Physics and Controlled Fusion Research,* IAEA, Vienna, 1965.

LIN, R. P., K. A. ANDERSON, J. E. McCOY, and C. T. RUSSELL, Observations of magnetic merging and the formation of the plasma sheet in the Earth's magnetotail, *J. Geophys. Res.*, **82**, 2761, 1977.

LIPATOV, A. S. and L. M. ZELENY, Dynamics of the reconnection processes in neutral sheet during the passage of Alfven pulse, *Fizika Plasmy*, **5**, 936, 1979 (in Russian).

LUI, A. T. Y., C. I. MENG, and S.-I. AKASOFU, Search for the magnetic neutral line in the near-earth plasma sheet, *J. Geophys. Res.*, **82**, 1547, 1977a; **82**, 3603, 1977.

MAEZAWA, K., Magnetotail boundary motion associated with geomagnetic substorms, *J. Geophys. Res.*, **80**, 3543, 1975.

MIKHAILOVSKII, A. B., Instabilities of nonuniform plasma, in *Plasma Instabilities Theory*, Vol. 2, p. 40, Atomizdat, Moscow, 1971 (in Russian); Consultants Bureau, New York-London, 1974 (English translation).

NISHIDA, A. and E. W. HONES, Jr., Association of plasma sheet thinning with neutral line formation in the magnetotail, *J. Geophys. Res.*, **79**, 535, 1974.

NISHIDA, A. and N. NAGAYAMA, Synoptic survey for the neutral line in the magnetotail during the substorms expansion phase, *J. Geophys. Res.*, **78**, 3782, 1973.

NISHIDA, A. and C. T. RUSSELL, On the expected signatures of reconnection in the magnetotail, *J. Geophys. Res.*, **83**, 3890, 1978.

NISHIDA, A., H. HAYAKAWA, and E. W. HONES, Jr., Observed signature of reconnection in the magnetotail, submitted to *J. Geophys. Res.*, **86**, 1422, 1981.

PELLINEN, R. J. and W. J. HEIKKILA, Energization of charged particles to high energies by an induced substorm electric field within the magnetotail, *J. Geophys. Res.*, **83**, 1544, 1978.

RICH, F. J., V. M. VASYLIUNAS, and R. A. WOLF, On the balance of stresses in the plasma sheet, *J. Geophys. Res.*, **77**, 4670, 1972.

ROELOF, E. C., E. P. KEATH, C. O. BOSTRÖM, and D. J. WILLIAMS, Fluxes of $\gtrsim 59$-keV protons and $\gtrsim 30$-keV electrons at ~ 35 R_E, 1. Velocity anisotropies and plasma flow in the magnetotail, *J. Geophys. Res.*, **81**, 2304, 1976.

RUSSELL, C. T. and R. L. MCPHERRON, The magnetotail and substorms, *Space Sci. Rev.*, **15**, 205, 1973.

SAGDEEV, R. Z. and A. A. GALEEV, *Nonlinear Plasma Theory*, W. A. Benjamin Inc., New York-Amsterdam, 1969.

SARRIS, E. T., S. M. KRIMIGIS, and T. P. ARMSTRONG, Observations of magnetospheric bursts of high-energy protons and electrons at ~ 35 R_E with IMP 7, *J. Geophys. Res.*, **81**, 2341, 1976a.

SARRIS, E. T., S. M. KRIMIGIS, T. IIJIMA, C. O. BOSTRÖM, and T. P. ARMSTRONG, Location of the source of magnetospheric energetic particle by multispacecraft observations, *Geophys. Res. Lett.*, **3**, 437 1976b.

SATO, T. and T. E. HOLZER, Quiet auroral arcs and electrodynamic coupling between the ionosphere and the magnetosphere, *J. Geophys. Res.*, **78**, 7314, 1973.

SATO, T., T. HAYASHI, T. TAMAO, and A. HASEGAWA, Confinement and jetting of plasma by magnetic reconnection, *Phys. Rev. Lett.*, **41**, 1548, 1978.

SCARF, F. L., L. A. FRANK, K. L. ACKERSON, and R. P. LEPPING, Plasma wave turbulence at distant crossing of the plasma sheet boundaries and the neutral sheet, *Geophys. Res. Lett.*, **1**, 189, 1974.

SCHINDLER, K., A theory of the substorm mechanism, *J. Geophys. Res.*, **79**, 2803, 1974.

SCHINDLER, K., Microninstabilities of the magnetotail, in *Dynamics of the Magnetosphere*, edited by S.-I. Akasofu, p. 311, Reidel Pub. Co., 1980.

SCHINDLER, K. and N. F. NESS, Internal structure of the geomagnetic neutral sheet, *J. Geophys. Res.*, **77**, 91, 1972.

SISCOE, G. L. and W. E. CUMMINGS, On the cause of geomagnetic bays, *Planet. Space Sci.*, **17**, 1795, 1969.

SONNERUP, B. U. Ö., Magnetic field line reconnection, in *Solar-System Plasma Physics*, edited by C. F. Kennel, L. J. Lanzerotti and E. N. Parker, North-Holland Pub. Co., Amsterdam-New York-Oxford, 1979.

SOTNIKOV, V., V. D. SHAPIRO, and V. I. SHEVCHENKO, Nonlinear theory of the short wave length current instability of the drift waves, *Zh. Exper. Teor. Fiz.*, **78**, 512, 1980 (in Russian).

STENZEL, R. L. and W. GEKELMAN, Experiments on magnetic field line reconnection, *Phys. Rev. Lett.*, **42**, 1055, 1979.

STERN, D. P., The role of 0-type neutral lines in the magnetic merging during substorms and solar flares, *J. Geophys. Res.*, **84**, 63, 1979.

TERASAWA, T. and A. NISHIDA, Simultaneous observations of relativistic electron bursts and neutral line signatures in the magnetotail, *Planet. Space Sci.*, **24**, 855,

1976.

TVERSKOY, B. A., *Dynamics of the Earth's Radiation Belts*, Phys.-Math. Literature, Science Publ. House, Moscow, 1968.

WHITE, R. B., D. A. MONTICELLO, M. N. ROSENBLUTH, and B. V. WADDELL, Saturation of the tearing mode, *Phys. Fluids*, **20**, 800, 1977.

Appendix: Tearing Mode Stabilization in a Two-Dimensional Plasma Sheet Configuration with Magnetized Electrons

To describe a perturbed state of a magnetized plasma we will use the linearized Boltzman equation for the perturbed distribution function

$$\frac{\partial f_{1j}}{\partial t} + v \cdot \nabla f_{1j} + \frac{e_j}{m_j} \left\{ E_0 + \frac{1}{c} [v \times B_0] \right\} \cdot \frac{\partial f_{1j}}{\partial v}$$

$$= -\frac{e_j}{m_j} \left\{ E_1 + \frac{1}{c} [v \times B] \right\} \cdot \frac{\partial f_{0j}}{\partial v} \tag{A.1}$$

where

$$E_1 = -\frac{1}{c} \frac{\partial A_1}{\partial t} - \nabla \varphi_1 ; \qquad B_1 = [\nabla \times A_1]$$

are perturbed electric and magnetic fields. This equation must be supplemented by the Maxwell's equation for the perturbed vector potential

$$\nabla^2 A_1 = -\frac{4\pi}{c} \sum_j e_j \int v f_{1j} \, dv \tag{A.2}$$

and the plasma quasineutrality equation to find the perturbed electrostatic potential

$$\sum_j e_j \int f_{1j} \, dv = 0 . \tag{A.3}$$

We consider here the case of the generalized Harris equilibrium, when the magnetic field configuration slightly deviates from the neutral sheet model and provides the magnetization of electrons:

$$B = B_0 \text{th}(z/\Delta)\hat{e}_x + B_n \hat{e}_z$$

$$\sqrt{r_{\text{Li}}/\Delta} > B_n/B_0 > \sqrt{r_{\text{Le}}/\Delta} . \tag{A.4}$$

The unperturbed distribution function is a function of constants of motion:

$$f_{0j}(z, v) = n_0 \left(\frac{m_j}{2\pi T_j}\right)^{3/2} \exp\left\{-\frac{m_j v^2}{2T_j} - \frac{e_j \varphi_0}{T_j} + \frac{u_j}{T_j}\left(m_j v_y + \frac{e_j}{c} A_{0y}\right) - \frac{m_j u_j^2}{2T_j}\right\}.$$

$$(A.5)$$

The perturbations are chosen in a form:

$$A_{1y}(x, z, t) = \tilde{A}_1(z)e^{-i\omega t + ikx}$$

$$\varphi_1(x, z, t) = \tilde{\varphi}_1(z)e^{-i\omega t + ikx}.$$

$$(A.6)$$

Using the particular form of the equilibrium distribution function (A.5) we rewrite the Boltzman equation (A.1) as

$$\left[\frac{\partial}{\partial t} + v \cdot \nabla + \frac{e_j}{m_j}\left(E_0 + \frac{1}{c}[v \times B_0]\right) \cdot \frac{\partial}{\partial v}\right] f_{1j}$$

$$= \frac{e_j f_{0j}}{cT_j}\left\{\left(\frac{\partial}{\partial t} + v \cdot \nabla\right)[u_j A_{1y} - c\varphi_1] - \frac{\partial}{\partial t}[v_y A_{1y} - c\varphi_1]\right\}. \qquad (A.7)$$

The solution of this equation can be written in the form of the integral along particle trajectories (GALEEV and ZELENY, 1976):

$$f_{1j} = \frac{e_j f_{0j}}{cT_j}\left\{(u_j A_{1y} - c\varphi_1) + i\omega \int_{-\infty}^{t} dt'[v_y A_{1y} - c\varphi_1]\right\}. \qquad (A.8)$$

The first two terms here describe the adiabatic change of the distribution function (A.5) with the slow change of the potentials A_1 and φ_1. The integral along particle trajectories characterizes the nonadiabatic corrections to the distribution function and dominates in the small vicinity of the neutral layer. The latter we define as the internal region. The adiabatic change of the distribution function and the corresponding solution of the equation (A.2) in the outer region were already found in Section 3. Therefore we concentrate here on the nonadiabatic corrections defining the solution in the internal region.

Under assumptions of Eq. (A.4) about the magnetic field component B_n across the quasineutral sheet, most of the ions in the vicinity of neutral layer at $z = 0$ bounce between the magnetic walls at $z = \pm\sqrt{r_{Li}\varDelta}$ and move freely along the layer. In the internal region where the nonadiabatic corrections to the ion distribution are not small the perturbed vector potential practically does not depend on the z-coordinate. Thus while integrating along the ion trajectories we can easily average over the fast ion bouncing. Considering the ion motion along the neutral layer as free we integrate now along the straight ion trajectories:

$$f_{1i} = \frac{e f_{0i}}{c T_i} \left\{ (u_i A_{1y} - c\varphi_1) - \frac{\omega}{\omega - k v_x + i0} (v_y A_{1y} - c\varphi_1) \right\} . \tag{A.9}$$

Under assumptions (A.4) we use the drift approximation for electrons. The integral along their trajectories can be conveniently estimated by representing their motion as the superposition of the elementary motions with the essentially different time scales: a fast Larmor rotation, a bouncing along the magnetic field between the magnetic mirrors and a slow drift along y-axis:

$$v_y = v_\perp \sin (\theta + \Omega_e t) + v_D \tag{A.10}$$

$$x(t) = x - \frac{B_n v_\perp}{B \Omega_e} [\sin (\theta + \Omega_e t) - \sin \theta] + \frac{B_0}{2 B_n \Delta} [z_b^2(t) - z_b^2(0)] \tag{A.11}$$

$$z(t) = z + \frac{B_x v_\perp}{B \Omega_e} [\sin (\theta + \Omega_e t) - \sin \theta] + z_b(t) - z_b(0) . \tag{A.12}$$

Here the function $z_b(t)$ describes the electron bouncing between the magnetic mirrors. It obeys the equation

$$\frac{dz_b}{dt} = v_\parallel(z) \frac{B_n}{B(z)} \tag{A.13}$$

where $v_\parallel(z) = \pm \sqrt{2(\varepsilon - \mu B)/m_e}$ is the electron longitudinal velocity; ε and μ are the energy and the magnetic moment.

The curvature drift and the gradient-B drift are combined in Eq. (A.10) into one term

$$v_D = \frac{c}{eB^2} \{ \mu [\nabla B \times \boldsymbol{B}]_y + m v_\parallel^2 [(\boldsymbol{B} \cdot \nabla) \boldsymbol{B} / B^2 \times \boldsymbol{B}]_y \} .$$

In our model of the magnetic field with the parabolic field lines this can be rewritten as

$$v_D = - \frac{2c \sqrt{\varepsilon - \mu B}}{eB} \frac{\partial}{\partial z} (B_z \sqrt{\varepsilon - \mu B} / B) . \tag{A.14}$$

For the case of slow tearing mode the integration along the electron trajectories is then reduced to the averaging over the fast Larmor rotation and the fast bouncing.

The averaging over the Larmor rotation is equivalent to the averaging over the Larmor rotation phase. In the long wave length limit $(k v_\perp / \Omega_e \ll 1)$

we obtain:

$$\langle \varphi[x(t),\, z(t)]\rangle = \varphi(x_c,\, z_c)$$

$$\langle v_\perp \sin(\theta + \Omega_e t) A_{1y}[x(t),\, z(t),\, t]\rangle = \frac{v_\perp^2}{2\Omega_e B}\left[B_x \frac{\partial}{\partial z} - B_z \frac{\partial}{\partial x}\right] A_{1y}(x_c,\, z_c,\, t)$$

(A.15)

where

$$x_c = x + (v_\perp B_z/\Omega_e B)\sin\theta$$
$$z_c = z - (v_\perp B_x/\Omega_e B)\sin\theta\,.$$

The averaging over the bounce motion is performed with the help of relation:

$$\langle \varphi[z_b(t)]\rangle = \int \frac{B\mathrm{d}z}{v_\parallel B_z}\, \varphi(z)\,.$$

(A.16)

Using the expression (A.14) for the drift velocity we can show that the bounce average of it in the case of a parabolic field lines is equal to zero. Thus while averaging of the term $v_D A_{1y}$ one should take the A_{1y} dependence on z-coordinate into account. Besides the explicit dependence on z, which we assume to be weak, there is the dependence of the phase factor $\exp(ikx_b(z))$ on z. Using the integration by parts we obtain

$$\int \frac{B\mathrm{d}z}{v_\parallel B_n} v_D A_{1y} = \int \frac{B\mathrm{d}z}{v_\parallel B_n} \frac{v_\parallel^2 B_x}{\Omega_e B}\left(\frac{\partial}{\partial z} + \frac{B_x}{B_n}\frac{\partial}{\partial x}\right)\tilde{A}_1(z_c)\exp(-i\omega t + ikx_c)\,.$$

(A.17)

Combining the Eqs. (A.15) and (A.17) and introducing the new integration variable $b(z) = B(z)/B_n$ instead of z with the help of relation

$$\mathrm{d}z = \frac{\Delta B_n}{B_0}\frac{b\mathrm{d}b}{\sqrt{b^2 - 1}}$$

(A.18)

we find the perturbed distribution function in a form:

$$f_{1e} = -\frac{ef_{0e}}{cT_e} u_e A_{1y} + \frac{ef_{0e}}{T_e}\left[\varphi_1 - \langle \tilde{\varphi}_1 e^{iqb^2}\rangle e^{-i\omega t + ikx - iqb^2}\right]$$

$$-\frac{\mu f_{0e}}{T_e} e^{-i\omega t + ikx - iqb^2 - i(kv_\perp B_x/\Omega_e B)\sin\theta}\left[\int_1^{bm} \frac{b^2\mathrm{d}b}{\sqrt{(b^2 - 1)(b_m - b)}}\right]^{-1}$$

$$\cdot \int_1^{bm} \frac{b^2\mathrm{d}b\,\exp(iqb^2)}{\sqrt{(b^2 - 1)(b_m - b)}}\left\{ik\tilde{A}_1[b^{-1} + 2(b_m - b)(1 - b^{-2})]\right.$$

$$+\frac{\sqrt{b^2-1}}{b^2}(b_{\mathrm m}-2b)\frac{\partial\tilde A_1}{\partial z}\bigg\} \tag{A.19}$$

where $b_{\mathrm m}=\varepsilon/\mu B_{\mathrm n}$ is the magnetic field at the turning point of the particle with the energy ε and the magnetic moment μ.

To calculate the perturbed electric current it is convenient to use the variables $(\varepsilon,\ \mu,\ \theta)$ instead of $\boldsymbol v$

$$\mathrm d\boldsymbol v=\frac{\mathrm d\varepsilon B\mathrm d\mu\mathrm d\theta}{\sqrt{2m_{\mathrm e}^{\,3}[\varepsilon-\mu B]}}. \tag{A.20}$$

Then the integration on θ looks like the averaging over the Larmor rotation and the integration on z looks like the bounce averaging. Then the work done by the mode electric field can be written in a quadratic form:

$$\frac{1}{2c}\int j_{1y}A_{1y}^{*}\mathrm dz=\frac{15}{16}\,\varDelta\cdot\frac{n_0(T_{\mathrm e}+T_{\mathrm i})}{BB_{\mathrm n}}\int_1^{\infty}\frac{\mathrm db_{\mathrm m}}{b_{\mathrm m}^{7/2}}\bigg[\int_1^{b_{\mathrm m}}\frac{b^2\mathrm db}{\sqrt{(b^2-1)(b_{\mathrm m}-b)}}\bigg]^{-1}$$

$$\cdot\bigg|\int_1^{b_{\mathrm m}}\frac{b^2\mathrm dbe^{iqb^2}}{\sqrt{(b^2-1)(b_{\mathrm m}-b)}}\{ik\tilde A_1[b^{-1}+2(b_{\mathrm m}-b)(1-b^{-2})]$$

$$+b^{-2}\sqrt{b^2-1}\,(b_{\mathrm m}-2b)\frac{\partial\tilde A_1}{\partial z}\bigg\}\bigg|^{\,2}. \tag{A.21}$$

Here we took into account that the Boltzman correction to the distribution function does not contribute to the electron current, and the contribution of the term $ef_{0\mathrm e}\langle\varphi_1\rangle/T_{\mathrm e}$ is found with the help of the quasineutrality equation:

$$-n_0e\varphi_1/T_{\mathrm i}+en_0u_{\mathrm i}A_{1y}/cT_{\mathrm i}=\int f_{1\mathrm e}\mathrm d\boldsymbol v. \tag{A.22}$$

The last term in Eq. (A.21) contributes to the integral only in the neutral layer vicinity. The first term, in contrary, diverges if we do not take into account the oscillating phase factor and therefore dominates in the integral. With the phase factor taken into account the main contribution to the integral comes from the large $b\sim q^{-1/2}$. This permits to simplify the expression (A.21)

$$\frac{1}{2c}\int j_{1y}A_{1y}^{*}\mathrm dz=-I_0\left(\frac{\varDelta B_0}{B_n\sqrt q}\right)\frac{k^2|\tilde A_1(z_0)|^2}{8\pi} \tag{A.23}$$

where

$$I_0 = \frac{45}{16} \int_0^\infty d\alpha \left| \int_0^1 b\sqrt{1-b^2}\, e^{i\alpha b^2} db \right|^2 ; \qquad z_0 = \frac{B_n \Delta}{B_0 q^{1/2}} .$$

Here we have assumed that the function $\tilde{A}_1(z)$ is almost constant in the internal region $|z| < z_0$ and is equal to the its value $A_1(z_0)$ in the outer region at $z \to z_0$.

We see that the trapped electron contribution to the tearing mode energy agrees within a factor of the order of unity with its estimate on the basis of the avaraged drift of the electron "drift orbits." Therefore the stabilization criterion (75) is now rigorously justified.

Chapter 4

AURORAL PHYSICS

T. Sato

Institute for Fusion Theory, Hiroshima University, Hiroshima, Japan

1. Introduction

The solar wind carries away part of the solar energy and fills up the interplanetary space with plasmas of solar origin. On the way of expansion, part of the solar wind energy is deposited in the earth's magnetosphere and also in other magnetized planets like Jupiter and Saturn, whereby numerous attractive dramas are played therein. "Aurora" is the epilogue of a series of dramas that are played on the stage of the solar-terrestrial plasma system.

In spite of its ancient discovery, aurora is yet far from being completely understood. Aurora can be defined as light emitted from excited atmospheric atoms and molecules due to bombardment of hot electrons and protons that precipitate down into the ionosphere from the magnetosphere along magnetic field lines. This seemingly simple phenomenon, notwithstanding, the research of auroral physics is not that easy because of the complexity and diversity of occurrence conditions of auroral arcs. Auroral activities and forms are very different, depending on local times, latitudes and magnetic activities. There is no doubt that magnetotail dynamics plays a leading role in generating auroral arcs. However, a fundamental question arises as to whether auroral particles are generated directly in the magnetotail, or accelerated at relatively low altitudes ($\sim 1\ R_{\rm E}$) along the auroral field lines. This difference in the acceleration region is crucial in the construction of auroral physics. Suppose that auroral particles are always produced directly in the magnetotail. Then, most of the auroral features, including auroral forms and activities, must be manifestations of the nature inherent to acceleration mechanism in the magnetotail. There are several observational evidences, however, that quiet auroral arcs and even active discrete auroral arcs result from localized acceleration processes along the magnetic field lines that connect the magnetotail and the ionosphere, rather than direct acceleration in the magnetotail. If this point of view is true, then physical processes of auroras may well be divided into the following four stages: (1) Primary supply of energy of auroras, (2) field-aligned acceleration of auroral particles, (3) formation of auroral arcs, and (4) associated phenomena such as the auroral kilometric radiation (AKR). Strictly speaking, in particular for discrete auroras, energy supply, acceleration mechanism and formation mechanism, may not be possible to separate them from one another. In the present stage, when auroral physics has just toed the mark, however, it may be an appropriate approach to start by dividing the tangled causal chain, leading to a brightening of auroras, into several parts, in which energy is thought to flow unidirectionally from one part of the chain to another. The whole part can then be synthesized to construct a unified explanation.

Solar wind energy is transported into the magnetosphere both in a steady and transient manner. Accordingly, steady and transient plasma convections, hence, convection electric fields, are induced in the magnetosphere. Such energy deposit from the solar wind must eventually be released into the atmosphere (ionosphere) and/or returned back to the solar wind from the tail of the magnetosphere. Magnetospheric substorms must be the principal mechanism releasing the energy. Here a serious question arises, namely, whether magnetospheric substorms are responsive to transient energy inflow from the solar wind or responsive to steady energy inflow. Should the steady inflow be the principal cause, substorms would occur in a rather periodic manner, because the energy release would take place when the deposit exceeds the capacity of the magnetosphere. Observations, however, do not seem to show any convincing evidence of periodic occurrence but the occurrence frequency seems to be rather random. It can be said, therefore, that magnetospheric substorms must be closely connected to transient energy input. In order to clarify the dynamical process leading to substorm-associated active auroras, therefore, transient response of the entire magnetosphere has to be solved. At present it is premature to describe a self-contained theory that can explain the grand story connecting the solar wind to auroras. Most part of this Chapter will therefore be devoted to auroral dynamics of more gentle nature which may largely be associated with steady energy source in the magnetosphere.

2. Primary Energy Sources of Auroras

Energy deposited in the magnetosphere is divided into convection energy, magnetic field energy and thermal energy:

$$U = \frac{1}{2}\rho v^2 + \frac{B^2}{8\pi} + \frac{1}{\gamma-1}p\,, \tag{1}$$

where U is the total energy density, ρ is the mass density, v is the bulk plasma velocity, B is the magnetic field, p is the pressure (isotropy is assumed), and γ is the ratio of the specific heats. As a matter of course, not every term in Eq. (1) can be dedicated to acceleration of auroral particles. Energy of the dipole field, for example, can never be converted into plasma kinetic energy. In order to seek primary energy sources convertible to the energy of auroral particles, we must rely on equations that can adequately describe the dynamics of the magnetosphere (VASYLIUNAS, 1970; SATO, 1974; SOUTHWOOD, 1977).

Considering that the greater part of auroral energy is dissipated by Joule heating in the ionosphere, it would be natural to expect that field-aligned current carries the substantial part of the auroral energy. Let us then begin with seeking sources of field-aligned currents in the magnetosphere. Divergence-free condition of the magnetospheric current yields

$$\nabla\cdot\boldsymbol{j}_{\parallel} = B\frac{\partial}{\partial s}\left(\frac{j_{\parallel}}{B}\right) = -\nabla\cdot\boldsymbol{j}_{\perp} \tag{2}$$

where s is a field-aligned coordinate and subscripts "$\perp$" and "$\parallel$" designate "perpendicular" and "parallel" to the magnetic field, respectively. From this relation we know that field-aligned current sources can be found by examining the nature of the current perpendicular to the magnetic field (cross-field current).

2.1 Sources of field-aligned current

In the evaluation of field-aligned current sources in the magnetosphere the following MHD equations would be adequate:

$$\rho\frac{d\boldsymbol{v}}{dt} = \frac{1}{c}\boldsymbol{j}\times\boldsymbol{B} - \nabla p + \boldsymbol{F} \tag{3}$$

$$\frac{1}{c}\frac{\partial\boldsymbol{B}}{\partial t} = -\nabla\times\boldsymbol{E} \tag{4}$$

$$\nabla \times \boldsymbol{B} = \frac{4\pi}{c}\boldsymbol{j} \tag{5}$$

$$\boldsymbol{E} + \frac{1}{c}\boldsymbol{v} \times \boldsymbol{B} = 0 , \tag{6}$$

where assumed is that the density ρ and the pressure p are given as a function of position; $\boldsymbol{E}$ is the electric field; $\boldsymbol{F}$ represents the momentum source from the solar wind due to the viscous drag and/or dayside reconnection. The momentum source term is important only at the solar wind-magnetosphere boundary layer and plays a role of generating the convection electric field in the magnetosphere. When we consider that the plasma convection in the magnetosphere is given, this term can be omitted from Eq. (3). The role of this term will be discussed later on.

From Eq. (3) we obtain ($\boldsymbol{F}$ is omitted)

$$\boldsymbol{j}_\perp = \boldsymbol{j}_D + \boldsymbol{j}_{in} , \tag{7}$$

with

$$\boldsymbol{j}_D = c\boldsymbol{B} \times \nabla p/B^2 \qquad \text{(diamagnetic current)} , \tag{8}$$

$$\boldsymbol{j}_{in} = c\boldsymbol{B} \times \left(\frac{\rho}{B^2}\frac{\mathrm{d}\boldsymbol{v}}{\mathrm{d}t}\right) \qquad \text{(inertia current)} . \tag{9}$$

After straightforward but complicated vectorial manipulation, we arrive at the following relation (HASEGAWA and SATO, 1979; SATO and IIJIMA, 1979)

$$B\frac{\partial}{\partial s}\left(\frac{j_\parallel}{B}\right) = c\rho\frac{\mathrm{d}}{\mathrm{d}t}\left(\frac{\Omega}{B}\right) + \frac{2}{B}\boldsymbol{j}_\perp \cdot \nabla B - \boldsymbol{j}_{in} \cdot \frac{\nabla n}{n} , \tag{10}$$

where

$$\Omega = (\boldsymbol{B} \cdot \nabla \times \boldsymbol{v})/B \qquad \text{(vorticity)}. \tag{11}$$

This equation indicates that there are three independent $j_\parallel$ current sources in the magnetosphere. The first source is closely related to the convection of the magnetospheric plasma, while the other two sources are related to the configuration of the magnetosphere.

Let us consider the physical meaning of the first source. For the sake of simplicity, we suppose that the magnetic field is uniform. Performing the divergence operator on Eq. (6) yields

$$c\nabla \cdot \boldsymbol{E} = -\boldsymbol{B} \cdot \nabla \times \boldsymbol{v} = -B\Omega . \tag{12}$$

Thus, the vorticity Ω is expressed by

$$\Omega = -\frac{4\pi cq}{B},\tag{13}$$

where q is the charge. This relation states that vorticity is equivalent to charge. Substituting Eq. (13) into the first term of Eq. (10) gives

$$\frac{\partial}{\partial s}j_{\parallel} = -4\pi C\frac{dq}{dt},\tag{14}$$

where $C=c^2\rho/B^2$. From this equation it is concluded that the field-aligned current can be a discharging current of charges accumulated in the magnetospheric condenser in association with convective cell motions. Let us see how space charges are produced in the magnetosphere. (As a matter of fact, any electrostatic waves in the magnetosphere and ionospheric irregularities can be accompanied by space charges. However, we omit such small-scale sources and pay attention to a more large-scale and persistent source.) A large-scale charge source can be produced at the boundary layer between the magnetosphere and the solar wind. For example, a viscous drag force can be a potential candidate. Viscosity v is connected to the generation of vorticity by the following relation.

$$\frac{\partial\Omega}{\partial t} = v\nabla^2\Omega.\tag{15}$$

Through the viscous term on the right-hand side of Eq. (15) vorticity (or momentum source F in Eq. (3)) is transported into the magnetosphere from the solar wind. The viscosity-induced convective motion in the magnetosphere is likely to be most prominent on the equatorial plane. Its direction is clockwise in the morning sector and anti-clockwise in the evening sector when looked down from the northern hemisphere. The vorticity Ω defined by Eq. (11) is, therefore, negative in the morning sector and positive in the evening sector, hence from Eq. (13) it turns out that positive charges accumulate at the center of the morning convective cell and negative charges at the center of the evening cell. Thus, a large-scale field-aligned current system is established in the magnetosphere-ionosphere system as shown in Fig. 1. The field-aligned current flows from the center of the morning convective cell down into the high latitude ionosphere and flows back to the evening cell via the ionospheric (Pedersen) current. In view of the polarity of the current system and of the latitudinal relationship with the Heppner type persistent distribution of the convection electric field (HEPPNER, 1977),

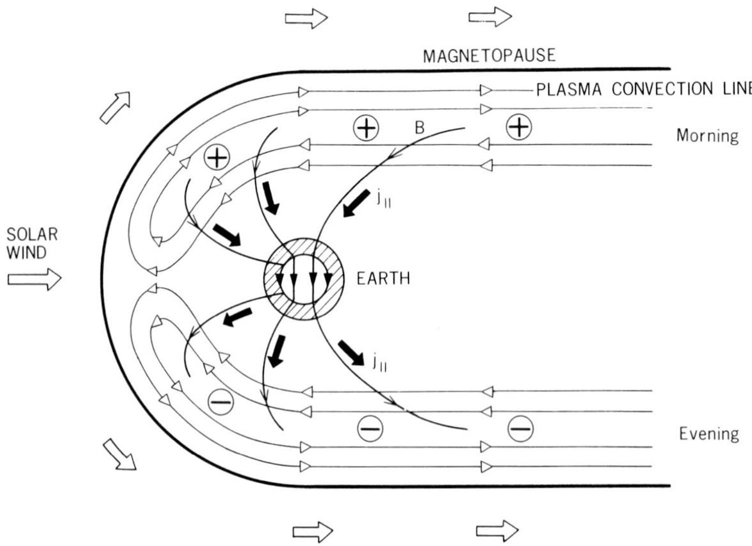

FIG. 1.   A schematic view showing a path of the field-aligned current originating from the twin convective cells at the magnetospheric equator.

"Region 1" field-aligned current (IIJIMA and PETEMRA, 1976) can well be explained by the large-scale twin convective cells in the magnetospheric equator.

In the above discussion, a viscous drag force is assumed to be a cause of the large-scale convective motion. We here remind the reader that the Dungey-type magnetic reconnection process (DUNGEY, 1961) can also cause a global circulation of the magnetospheric plasma. The plasma circulation caused by the reconnection process, however, is basically meridional rather than equatorial (namely, the convection cell is basically parallel to the noon-midnight meridian). Furthermore, the reconnection process cannot be of persistent type because it is vulnerable to the interplanetary magnetic field condition which is changeable from time to time. Rather than being the direct cause of a persistent vorticity-production, the reconnection process would play a different role, namely, it would play a role in bringing forth a violent change in the energy state of the magnetotail region—a magneto-spheric substorm. A possibility of the nightside reconnection process as producing the field-aligned current will be discussed briefly later on.

Thus, suppose that Region 1 field-aligned current is caused by a viscous drag at the magnetopause. Then, we must answer a question how the field-aligned current can persist as steadily as observed. The field-aligned current originating from vorticity is a discharging current through the load of ionospheric resistance. Therefore, it would decay with time. The Joule dissipation in the ionosphere is equivalent to the energy loss of the magnetospheric current system. In the present case where the current source is the plasma convection, we have, from Eq. (7) and (9),

$$ E_\perp \cdot j_\perp = \frac{1}{2}\,\rho\,\frac{\mathrm{d}}{\mathrm{d}t}\,v^2 \;. \tag{16} $$

This relation indicates that the ionospheric dissipation inevitably slows down the magnetospheric convection. Let us derive the slowing-down time τ. Integrating Eq. (14) from the top of the ionosphere $(s=h)$ to the magnetospheric equator $(s=l)$ we obtain

$$ j_\parallel(s=h) = -4\pi Cl\,\frac{\mathrm{d}q}{\mathrm{d}t}\;. $$

Here we have put $j_\parallel(s=l)=0$ because the northern and southern hemispheres can approximately be assumed symmetric around the equator. On the other hand, we have (c.f. Eq. 41))

$$ j_\parallel(s=h) = \Sigma_\mathrm{P}\nabla\cdot E = 4\pi\Sigma_\mathrm{p}q \;. $$

From these relations we have

$$ \frac{\mathrm{d}q}{\mathrm{d}t} = -\frac{\Sigma_\mathrm{p}}{Cl}\,q \;. $$

Thus, the slowing down time τ is given by

$$ \tau = Cl_\mathrm{eff}/\Sigma_\mathrm{P} \tag{17} $$

where $C=c^2\rho/B^2$ (magnetospheric capacitance per unit field-aligned length), Σ_P is the height-integrated ionospheric Pedersen conductivity and l_eff is an effective half field line length that can be evaluated by taking into account the geometry of the field line tube under consideration (see, DEWITT, 1968). The time constant τ is estimated to be of the order of 100 sec, this indicating that the magnetospheric convection cell would completely die out within an hour at longest, unless the lost momentum is supplied fast enough from the solar wind.

On the other hand, Eq. (15) means that the momentum (vorticity) supply rate γ, due to viscosity, is

$$\gamma = \nu L_{\text{eff}}^{-2} \qquad (18)$$

where L_{eff} is the effective thickness of the magnetospheric boundary layer where an instability, presumably the Kelvin-Helmholtz instability, gives rise to an anomalous viscosity. In order that the field-aligned current originating from the convective motion persists stably, it is required that $\gamma\tau > 1$, namely,

$$\nu > L_{\text{eff}}^2 \tau^{-1} . \qquad (19)$$

Put, for instance, $\tau = 100\,\text{sec}$ and $L_{\text{eff}} = 100\,\text{km}$. Then $\nu > 10^8\,\text{m}^2\,\text{sec}^{-1}$. Existing theories seem to support this inequality (AXFORD, 1964; TSUDA, 1967; EVIATOR and WOLF, 1968; MIURA and SATO, 1978; HASEGAWA and MIMA, 1978). From these considerations, therefore, it can be said that the slowing-down of the magnetospheric convection due to the ionospheric dissipation can be compensated by the viscous transport, so that a steady field-aligned current can be sustained.

The second field-aligned current source given by the second term on the right-hand side of Eq. (10) is related to the magnetic field configuration. It originates from the region where a magnetic field gradient is present in the direction of the magnetospheric cross-field current. Accordingly, the source region is restricted to the ring current region and/or the plasma sheet region where the tail current exists. These regions are considered to be located inside the viscous boundary layer discussed above, and hence, the field-aligned current system associated with this second source, if any, should be located at lower latitudes. Since the direction of the ring current is normally westward and the magnetic field gradient is considered to be increasing in the direction of the ring current in the evening sector and decreasing in the morning sector, it turns out that $j_\perp \cdot \nabla B > 0$ in the evening sector and $j_\perp \cdot \nabla B < 0$ in the morning sector; in other words, the polarity of the field-aligned current system is opposite to the previous case corresponding to Region 1 current system. These circumstantial evidences seem to support that this source can be a cause of "Region 2" current (see, for example, SATO and IIJIMA, 1979). Since the ring current $j_\perp (\approx B \times \nabla p / B^2)$ and the magnetic field configuration would not largely be modified by the ionospheric condition, this current source is considered to be a constant current generator. On the other hand, the source associated with the vorticity is a voltage generator because the current is dependent on the ionospheric resistance.

The third current source in Eq. (10) originates from the region where a density gradient is present in the direction of the inertia current. This source, however, is usually negligibly small. The proof is given in the following.

Comparing the intertia current j_{in} with the diamagnetic current j_D, we find that

$$\left|\frac{j_{in}}{j_D}\right| = \frac{\rho V/T}{p/L} = \frac{VL/T}{v_{ith}^2}$$

where V is the characteristic speed of the bulk plasma, v_{ith} is the ion thermal speed, L is the characteristic length of the pressure gradient, and T is the characteristic acceleration time. In a quiet plasma sheet, $v_{ith} \approx 700\,\text{km/s}$ (corresponding to the ion temperature of 5 keV), $V \approx 50\,\text{km/s}$ (corresponding to the electric field of 1 mV/m), $L \approx 20{,}000\,\text{km}$ (corresponding to 3 earth radii), and T cannot be less than 200 sec. Therefore, the inertia current is negligibly small compared with the neutral sheet diamagnetic current. During substorms it is reported that the proton speed becomes as high as 700 km/s (HONES, 1979; LUI, 1979). Reconnection simulation has also proved that proton jets of the order of Alfvén speed can be generated along the plasma sheet boundary as a result of slow shock formation (e.g., SATO, 1979). Build-up time of this proton jet is of the order of 100 sec and the shock width becomes as thin as one tenth of the normal plasma sheet, namely, $L \approx 1{,}000 \sim 2{,}000\,\text{km}$. Plugging these values in the above yields $|j_{in}/j_D| \sim 10^{-2}$. This shows that the intensity of the inertia current is negligibly small in either case. Provided, therefore, that the magnetic field gradient is of the same order as the density gradient, the third term in the right-hand side of Eq. (10) can be neglected compared with the second term.

Large-scale, persistent field-aligned currents like Region 1 and Region 2 current may well be attributed to the first and second sources as discussed above. Let us here consider a new field-aligned current source associated with substorms based on the above fundamental theory. We suppose that a magnetospheric substorm is triggered by an externally-driven magnetic reconnection (see, e.g., SATO, 1979). Based on the simulation result, we can deduce a three-dimensional field-aligned current system associated with reconnection in the tail. Figure 2 shows a pair of slow shock current sheets generated by an external driving force. This result indicates that as reconnection develops, the cross-tail current tends to concentrate in two thin shock layers. Besides, the magnetic field contours shown in Fig. 3 indicate that as reconnection develops, the field intensity at slow shocks increases. This is due to the pile-up of field lines which are driven toward the neutral sheet. Because of this pile-up of field lines at slow shocks, $j_\perp \cdot \nabla B$ becomes non-zero as shown in Fig. 4 a where $j_\perp$ is the cross-tail current. In detail, $j_\perp \cdot \nabla B > 0$ in the dawn sector and $j_\perp \cdot \nabla B < 0$ in the dusk sector, and hence field-aligned

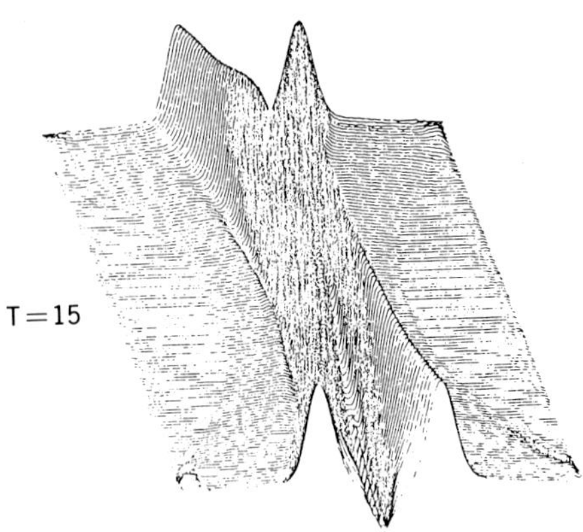

FIG. 2.   3-D display of fully-developed pair of slow shocks (current layer) resulting from an externally-driven magnetic reconnection. The X-type neutral line exists at the center of the picture.

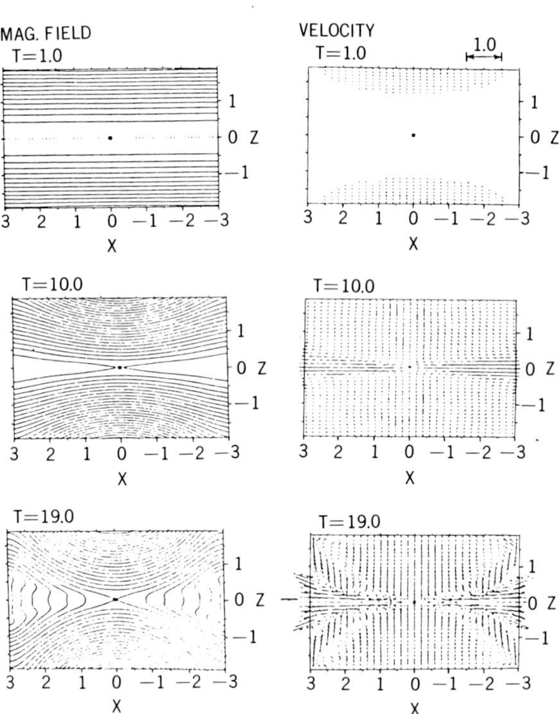

FIG. 3.   Development of reconnected magnetic fields (left) and plasma jets (right) as a result of an externally-driven magnetic reconnection.

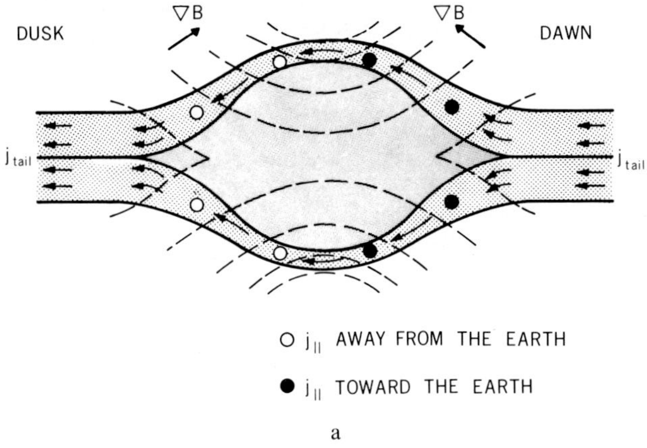

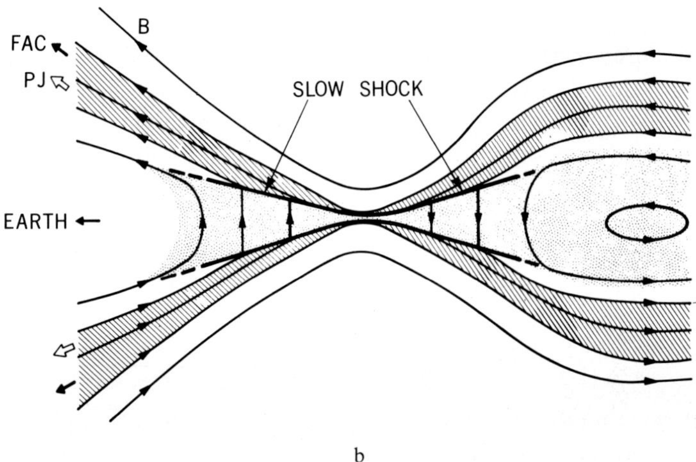

FIG. 4a.   Cross-sectional view on the earthside of the X-type neutral line of the magnetotail showing the tail current bifurcation and generation of nightside field-aligned current pairs.

FIG. 4b.   Meridional view showing the spatial relationship among field lines, slow shocks, field-aligned currents and plasma jets (slightly dawnward of local midnight).

currents flow toward the earth in the dawn sector and away from the earth in the dusk sector. A schematic view of the shock and magnetic field geometry in a meridian plane is shown in Fig. 4b and a three-dimensional view summarizing the present idea is drawn in Fig. 5. From this figure it can be inferred that the poleward edge of the field-aligned current sheet corresponds

                                   T. SATO

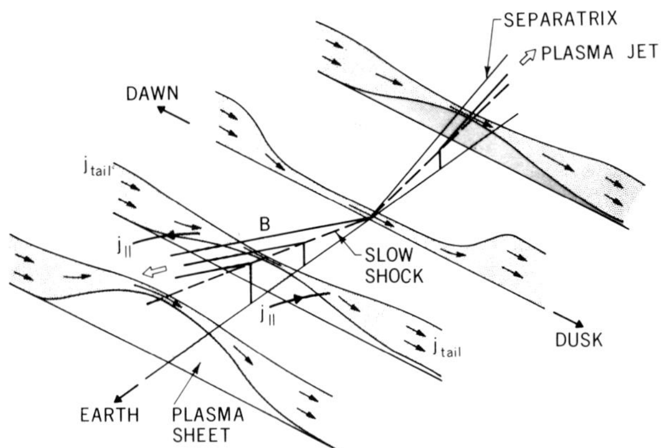

FIG. 5.   Magnetospheric substorm model based on externally driven reconnection.

to the reconnection line in the tail and the thickness of the current sheet in the north-south direction is determined by the extent of the shock wing. It is supposed that a current system localized in the midnight region such as discussed here would superpose on large-scale region 1 and 2 current systems during substorms.

To summarize there are two field-aligned current sources in the magnetosphere, regardless of whether it is a quiet time or a substorm time. They are:

$$c\rho\,\frac{\mathrm{d}}{\mathrm{d}t}\left(\frac{\Omega}{B}\right)\quad\text{and}\quad\frac{2\boldsymbol{j}_\perp\cdot\nabla B}{B}.$$

The first one represents a discharge current of charges accumulated in the magnetosphere capacitor due to convection motion. The other source is the diversion of the magnetospheric current due to the magnetic inhomogeneity. Needless to say, the field-aligned current thus driven, can be a potential candidate of auroral energy.

2.2   Sources of mechanical energy

In addition to the electrical (field-aligned current) sources, there may be mechanical sources convertible to auroral energy. As we have discussed in the previous subsection concerning the externally-driven reconnection, the stress that the magnetospheric plasma suffers as a result of plasma convection toward the neutral sheet would lead to local charge accumulation whereby a pair of field-aligned currents in the dawn-dusk sector are driven.

Besides this direct generation of field-aligned currents, non-current carrying, field-aligned plasma motion can also be generated by the reconnection process. The previous reconnection simulation (e.g., SATO, 1979) shows that a strong field-aligned plasma jet arises as a result of slow shock formation. This is because a steep pressure gradient $\nabla_{\|}p$ appears across the shock front and the slow shock front obliquely intersects the magnetic field. Were no mirror effect present between the magnetosphere and the ionosphere, the field-aligned accelerated plasma, the greater part of whose energy is carried by protons, would directly be lost in the ionosphere. In reality, because of the extremely large mirror effect, most particles are reflected back to the magnetosphere before reaching the lower ionosphere. Specifically, because the ratio of the field-aligned flow velocity to the thermal velocity is much smaller for electrons than for ions, electrons are generally reflected at higher altitudes, while protons penetrate deeper toward the earth. As long as a slow shock is maintained in the magnetosphere, therefore, an upward field-aligned electric field is expected to be generated as shown in Fig. 6. Since the extent of the shock wing is, as a matter of course, limited spatially, a V-shaped potential structure would naturally be formed by this mechanism. The simulation result shows that the speed of field-aligned plasma jets caused by slow shocks amounts to local Alfvén speed V_A. When we take $V_A = 700\,\mathrm{km/s}$, the energy of directed plasma flow becomes about $2.5\,\mathrm{keV}$, this implying that the field-aligned potential difference due to this mechanism would become large enough to produce auroral electrons. Thus, the field-aligned plasma jet, as

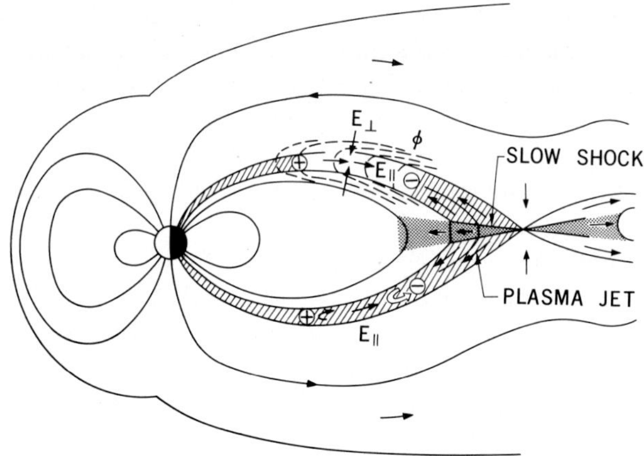

FIG. 6. A schematic picture showing generation of upward field-aligned electric field due to field-aligned plasma jets produced at slow shocks.

well as the field-aligned current, arising from a slow shock can be a powerful candidate of a direct energy source of auroral electrons.

3. Double Layers: Acceleration Mechanism

In the foregoing section we discussed available energy sources of auroral particles, i.e., a field-aligned current and a field-aligned plasma jet. Given the type and location of the direct energy source of auroral particles, we are able to go on to find the mechanism by which the source energy is converted into acceleration energy of auroral particles. Although the field-aligned plasma (non-current carrying) jet is a very likely and feasible energy source, we have not yet obtained much theoretical development based on this idea. In this section we pay attention to the case with a field-aligned current source.

It is possible that a field-aligned current is carried by hot electrons that are directly accelerated by reconnection electric field in the magnetotail (SPEISER, 1967; SONNERUP, 1971; MATSUMOTO et al., 1981). However, it is normally taken that the field-aligned current is primarily carried by background cold electrons. Then a fundamental question arises as to how the current carried by cold electrons yields auroral electrons.

Double layers are believed both observationally and theoretically to be a viable mechanism that can effectively convert the magnetospheric energy communicated in the form of field-aligned currents into auroral particles (see, a review paper by BLOCK, 1978 and papers cited therein). In this respect, the previous steady-state theories (BLOCK, 1972; KNORR and GOERTZ, 1974), numerical simulations (GOERTZ and JOYCE, 1975, DEGROOT et al., 1977), and laboratory experiments (TORVÉN and BABIC, 1975; QUON and WONG, 1976; COOKLEY and HERSHKOWITZ, 1979; IIZUKA et al., 1979), have all suggested that double layers must be a product of a Buneman two-stream instability. It is unlikely, however, that such an intense field-aligned current as bearing a Buneman instability can exist along auroral field lines in a normal state (KINDEL and KENNEL, 1971).

A steady state theory can give us conditions for the existence of a steady-state double layer; however, it does not guarantee that such conditions can be realized starting from realistic initial and boundary conditions. Since the final goal of space physics is to reveal the cause and effect relationship among the phenomena occurring in space, our effort must be directed toward finding mechanisms that can spontaneously generate double layers. Simultaneously, we must resolve a fundamental question as to whether or not an electron stream with a speed in excess of the electron thermal speed is indeed required for generation of double layers.

3.1 Ion acoustic double layers

The causal mechanism of double layers yet remains to be solved. Based on a recent numerical simulation (SATO and OKUDA, 1980, 1981), we here describe in some length one possible physical process, though it has not yet fully been developed. Previous simulations (BISKAMP and CHODURA, 1971; DEGROOT et al., 1977) have indicated that double layers are not generated in the velocity range of ion acoustic instability, $c_s < V_d < v_{eth}$, where $c_s = (kT_e/m_i)^{1/2}$, $v_{eth} = (kT_e/m_e)^{1/2}$ and V_d is the electron streaming velocity. If this is always true, then one must invoke anomalous resistivity or other mechanisms for the production of parallel electric fields in the presence of field-aligned currents. Observations (MOZER et al., 1977), however, suggest that mechanisms other than a simple anomalous resistivity are operative (see, SHAWHAN et al., 1978). In an attempt to reconcile double layers with anomalous resistivity, an extensive study by means of numerical simulation has recently been carried out. The initial motivation of performing this simulation was as follows: in space, the system length is almost unlimited. This indicates that however small the anomalous resistivity may be, an ion acoustic instability can be expected to produce a large (dc) potential drop across the unstable region. A test streaming electron that starts from the upstream boundary of the unstable region, gains the potential energy as it traverses. As they go down further, streaming electrons that are not trapped by unstable waves gain dc potential energy, while trapped electrons are decelerated at the expense of the wave growth. It can be expected therefore that the distribution function downstream becomes more unstable for ion acoustic waves. Further growth of ion acoustic waves would enhance the anomalous resistivity, which in turn would accelerate the build up of dc potential. The growth of the dc potential would then cause further change in distribution function leading to a positive feedback.

With this intuitive idea as a guide a numerical simulation has been carried out where the system size is taken to be much longer than that of the conventional simulation. Shown in Fig. 7 is a spatial potential distribution obtained by a one-dimensional simulation (SATO and OKUDA, 1981). The simulation parameters are as follows: system length $L = 1,024 \, \lambda_D (\lambda_D$ is Debye length), initial electron drift velocity $V_d = 0.6 \, v_{eth}$, temperature ratio $T_e/T_i = 20$ and mass ratio $m_i/m_e = 100$. The simulation model is shown in Fig. 8. As is shown in Fig. 8(a), the following situation is considered in this simulation: a field-aligned current is driven by twin convective cells in the

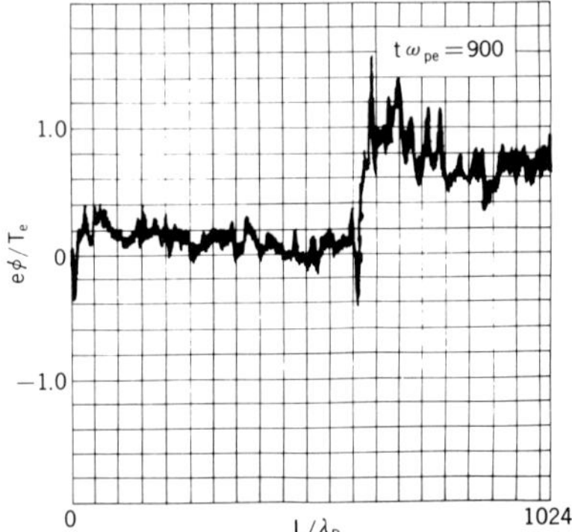

FIG. 7. Ion acoustic potential double layer obtained by a numerical simulation for L (system length) $= 1,024\ \lambda_D$ (SATO and OKUDA, 1981).

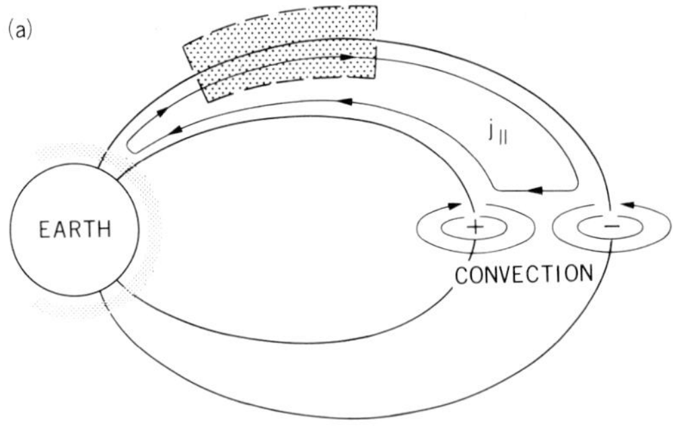

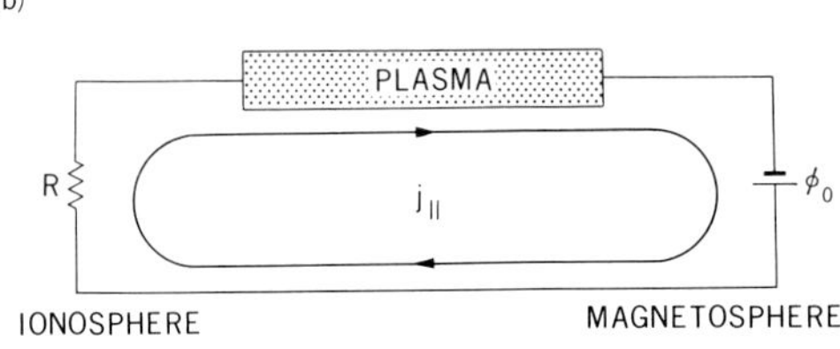

FIG. 8. Model used for the simulation of ion acoustic double layer (a) and its equivalent electric circuit (b).

magnetosphere and is closed by the ionospheric current through the Pedersen conductivity. From Fig. 7 it is evident that a double layer can indeed be generated starting from an initial condition with no potential drop across the system and with streaming electrons whose drift velocity is smaller than the electron thermal speed. The potential drop ϕ reached roughly $e\phi/kT_e \approx 1$. It should be noted here that double layer formation is closely associated with the drastic increase of anomalous resistivity reaching values roughly ten times as large as the normal anomalous resistivity (SATO and OKUDA, 1980). Note also that the double layer is transient and decays with an excitation of ion acoustic solitons, and that the anomalous resistivity completely disappears after the formation of double layer.

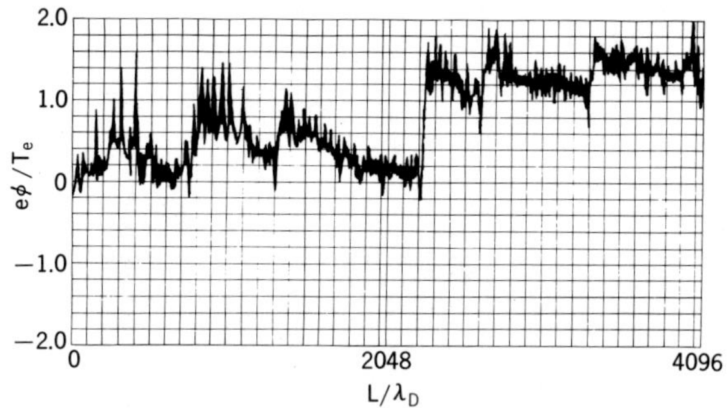

FIG. 9.   Ion acoustic potential double layers developed in a very long system with $L=4,096\ \lambda_D$ (SATO and OKUDA, 1981). Note the generation of multiple double layers and a larger potential drop compared with the case where $L=1,024\ \lambda_D$ (Fig. 7).

Figure 9 shows the potential distribution for another run where the system length is extended up to $L=4,096\ \lambda_D$ with other parameters remaining the same. It is seen from this result that a larger potential drop amounting roughly to $e\phi/kT_e \approx 2$ develops across the system in which formed are cascades of several double layers. We note that the double layers are rather randomly and independently formed both in time and space when the interspace between neighboring double layers is longer than a certain distance, say, $1,000\ \lambda_D$. Through those simulations, we can arrive at an important conclusion that double layers, though not steady, can be generated in a long system even when the initial electron streaming velocity is not so large as the thermal speed. Furthermore, the total potential drop is increased as the system length increases. One dimensional, notwithstanding, the simulation gives us a hope that low-intensity field-aligned currents existing along auroral field lines can yield double layers whereby auroral electrons are accelerated.

3.2   Steady-state theory

Thus, it is numerically proved that a potential drop exceeding the electron thermal energy can be created by a field-aligned current even when the streaming velocity is less than the electron thermal speed. With this proof, we can proceed to a steady-state theory that assumes *a priori* the existence of a potential drop between upstream and downstream boundaries.

The steady-state Vlasov equation and Poisson's equation are the starting equations (assuming one-dimensionality):

$$v_e \frac{\partial f_e}{\partial x} + \frac{e}{m} \frac{\partial \phi}{\partial x} \frac{\partial f_e}{\partial v_e} = 0 , \tag{20}$$

$$v_i \frac{\partial f_i}{\partial x} - \frac{e}{M} \frac{\partial \phi}{\partial x} \frac{\partial f_i}{\partial v_i} = 0 , \tag{21}$$

$$\frac{\partial^2 \phi}{\partial x^2} = 4\pi e(n_e - n_i) \tag{22}$$

with $n = \int f \, dv$.

Following BERNSTEIN-GREEN-KRUSKAL (1957), we can divide the electron distribution into free and trapped parts. The free and the trapped part are divided depending on whether $v_e^2 > 2e\phi/m$ or $v_e^2 < 2e\phi/m$. The free distribution can be defined differently depending on whether $v_e > (2e\phi/m)^{1/2}$ or $v_e < -(2e\phi/m)^{1/2}$, with a requirement that each be a function of the total energy $\varepsilon_e = (1/2)mv_e^2 - e\phi$. The trapped distribution must be an even function of v_e. Thus, the distribution function is written as

$$f_e = \begin{cases} f_+(\varepsilon_e) & \text{for} \quad v_e > \left(\frac{2e\phi}{m}\right)^{1/2} \\[2mm] f_{tr}(\varepsilon_e) & \text{for} \quad |v_e| < \left(\frac{2e\phi}{m}\right)^{1/2} \\[2mm] f_-(\varepsilon_e) & \text{for} \quad v_e < -\left(\frac{2e\phi}{m}\right)^{1/2} . \end{cases} \tag{23}$$

Likewise, the ion distribution function can be divided into three parts.

The task of finding a double layer (electrostatic shock) solution is to find the distribution functions that satisfy Poisson's equation along with physically reasonable constraints at boundaries (charge neutrality, potential values, current, etc.) (see MONTGOMERY and JOYCE, 1969; KNORR and GOERTZ, 1974; SWIFT, 1975; YAMAMOTO, 1976). Generally, the distribution function of trapped particles is assumed to be given arbitrarily. By noting that a step function in velocity space, $H(v - v_1)$ with $v_1 = [(2\varepsilon + e\phi)/m]^{1/2}$, is a

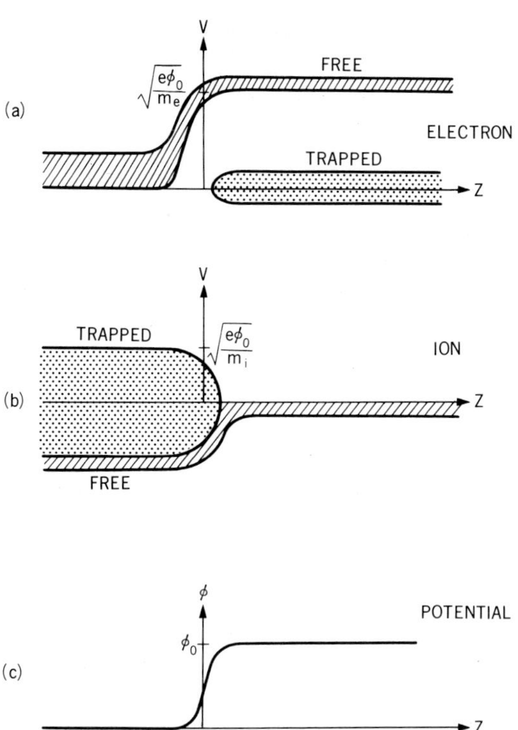

FIG. 10.   Stationary distributions of trapped and accelerated electrons and ions in phase space and the potential structure.

solution of the Vlasov equation (20), we are able to make some simplification in the analytical study of double layers if necessary, namely, a water-bag distribution may be useful. Electron and ion water-bag distributions in phase space and potential profile are schematically shown in Fig. 10.

Following the above-mentioned procedure we here show an example of how a BGK solution of an ion acoustic double layer can be constructed (HASEGAWA and SATO, 1981). The previous simulations have shown that an ion acoustic double layer is always accompanied by a stationary negative potential spike in front of it (see, Figs. 7 and 9). This fact leads us to an important clue to the causality of double layer formation. Suppose a stationary negative potential spike be formed in an electron stream by whatever mechanism. Then, those streaming electrons whose kinetic energies are less than the peak potential energy can be reflected by the potential spike, thus leaving an electron void in the velocity space on the downstream side of the spike. With this electron void, the potential would swing to a positive value on the downstream side forming a double layer structure.

Let us first attempt to prove the existence of a localized negative potential structure which is stationary in the ion frame. In the presence of a localized negative potential, ions would be evacuated from there so that an ion hole would be created in the phase space. At the foot of the localized potential structure where the potential barrier is low, however, the ion hole would be filled by the ions scattered by ion fluctuations or whatever. The ion distribution in the presence of a localized negative potential $\phi(x)(\leq 0)$ may therefore be assumed by

$$
n_i = \begin{cases} \dfrac{2n_0}{\sqrt{\pi}} e^{-e\phi/kT_i} \displaystyle\int_{\sqrt{-e\phi/kT_i}}^{\infty} e^{-t^2} dt & \text{for} \quad \phi \leqq -\phi_0 \\[2em] n_0 e^{-e\phi/kT_i} & \text{for} \quad \phi > -\phi_0 \end{cases} \tag{24}
$$

where ϕ_0 is the rms potential of ion fluctuations or something equivalent. The electron density, on the other hand, can be assumed by a Boltzman, i.e.,

$$
n_e = n_0 e^{e\phi/kT_e} \quad . \tag{25}
$$

Substitution of Eqs. (24) and (25) into Eq. (22) yields an integro-differential equation that can determine the potential structure $\phi(x)$. An example of a numerical solution with vanishing boundary conditions at $x = \pm\infty$ is shown in Fig. 11. This solution indicates that a stationary, negative potential solitary-wave structure can be a real existence, as was observed in the computer simulation of Sato and Okuda.

Let us then go on to find a potential structure for the case in which a stationary, negative wave-solitary structure stands in an electron stream. We designate the peak potential by $-\phi_m(\phi_m > 0)$. Only those electrons that have initial kinetic energies in excess of the peak potential energy $e\phi_m$ can pass the potential barrier and form an electron beam on the downstream side. Such electrons can be approximated by a water-bag distribution. Using the flux and energy conservations, the density of the beam can be given by $n(1 + \phi/\phi_s)^{1/2}$ with $\phi_s \equiv m_e v_0^2/2e$ where n_1 is the upstream density of the beam and v_0 is its average speed. On the downstream side, there exist thermal electrons along with the streaming electrons which passed the negative potential barrier. The density of the thermal electrons can be described by $n_2 \exp(e\phi/kT_e)$. Since the ion density must be continuous at $\phi = -\phi_m$, the first relation of Eq. (24) can be used for the downstream region as far as $\phi < 0$. If ϕ turns positive somewhere, then the ion density may be assumed to

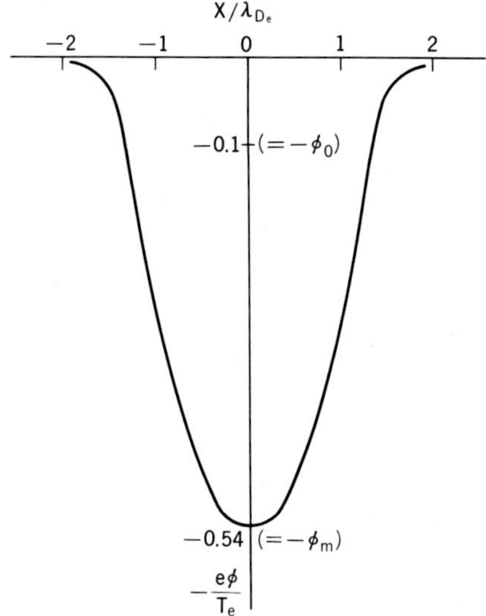

FIG. 11. A BGK solution of a negative potential solitary-wave structure: $T_i/T_e = 0.05$ and $e\phi_0/kT_i = 2$ are chosen.

be constant ($=n_0$). Substituting these relations into the right-hand side of Eq. (22) yields

$$\frac{d^2\phi}{dx^2} = 4\pi e \left[\frac{n_1}{\sqrt{1+\phi/\phi_s}} + n_2 e^{e\phi/kT_e} - n_i(\phi) \right] \tag{26}$$

where

$$n_i(\phi) = \begin{cases} \dfrac{2n_0}{\sqrt{\pi}} e^{-e\phi/kT_i} \displaystyle\int_{\sqrt{-e\phi/T_i}}^{\infty} e^{-t^2} dt & \text{for} \quad \phi \leq 0 \\[2ex] n_0 & \text{for} \quad \phi > 0. \end{cases} \tag{27}$$

From the continuity requirement of the electron density at $\phi = -\phi_m$ and the requirement $d\phi/dx = 0$ at $\phi = -\phi_m$, we obtain $\phi_s = \phi_m + kT_e/2e$ and $(1-n_2/n_0) \exp(-e\phi_m/kT_e) = (n_1/n_0)\sqrt{2e\phi_s/kT_e}$. Thus only one parameter, n_1/n_0 or n_2/n_0, remains to be determined so that the boundary conditions at $x = \infty$ can be satisfied. Impose now a boundary condition such that the potential approaches to a positive value at $x \to \infty$. Shown in Fig. 12 is a BGK solution that satisfies Eq. (26) and smoothly matches the previous negative solitary-wave structure at $\phi = -\phi_m$. It is interesting to remark that the theoretical solution is very similar to that observed in the numerical solution.

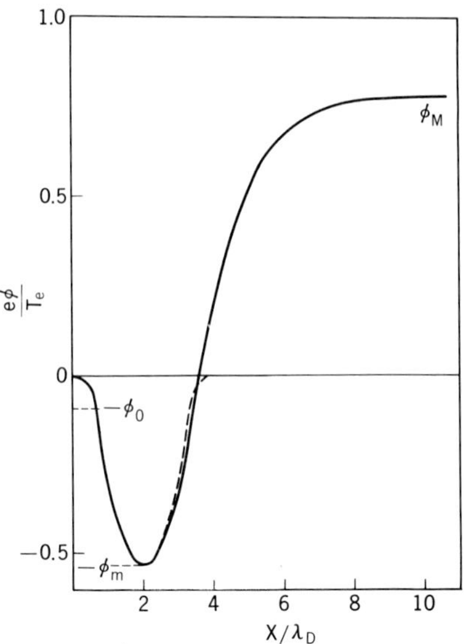

FIG 12.   A double layer solution following the solitary-wave structure of Fig. 11: $n_1/n_0 = 0.256$.

Note, however, that the above treatment may not be valid for long-time behavior since the stability of the solution is not guaranteed. In fact, the simulation indicated the collapse of the double-layer structure by emitting solitons downstream.

In actual auroral field lines, a strong mirror effect does exist, so a theory that includes the mirror effect is invoked. Furthermore, a two-dimensional theory that includes the finite Larmor effect must be developed. Steady-state theory, notwithstanding, it is very hard to seek a rigorous two-dimensional solution including both effects because of the appearance of many freedoms. An attempt has been made to find a two-dimensional solution for a uniform magnetic field geometry by SWIFT (1976, 1979). We shall see an essence of this procedure in the following.

We choose a right-handed Cartesian coordinate system where the z axis is parallel to a constant magnetic field and x is perpendicular to the field, as schematically shown in Fig. 13. We make a fundamental assumption that the spatial change in potential is small over a gyration motion. Then, the ion distribution function f_i can be specified in terms of four independent variables, namely, the parallel velocity v_{z0}, the gyration energy K_0, and the guiding center point x_{00} at the fiducial plane $z = z_0$ (see Fig. 13) through

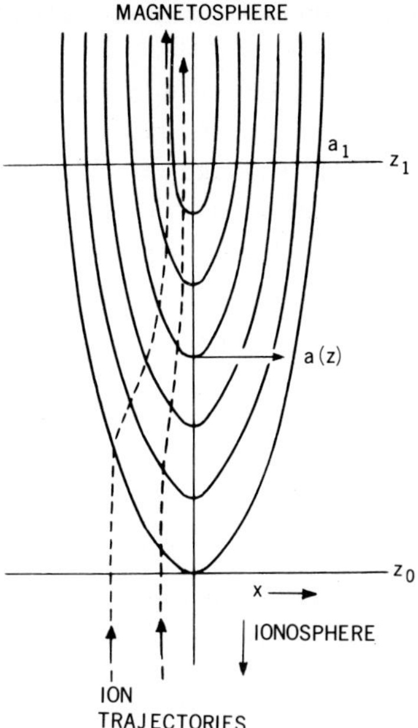

FIG. 13.   Coordinate system and schematic of two-dimensional equipotential model (SWIFT, 1979).

which all upgoing ions pass. In doing so, three constants of motion must be used. The first constant is the total energy H given by

$$H = \frac{\left(P - \dfrac{e}{c}A\right)^2}{2m} + e\phi \tag{28}$$

where $(e/c)A = m\Omega_0 x\hat{e}_y$; Ω_0 is the gyrofrequency and e_y is a unit vector along the y axis and P is the momentum. The second constant of motion is P_y which is given by

$$P_y = m(v_y + \Omega_0 x) \tag{29}$$

The third constant is an adiabatic invariant J_x with respect to the motion in the x direction defined by

$$J_x = \int v_x \, dx \tag{30}$$

This invariant comes from the assumption that the change in the potential is small over the gyration motion and it guarantees the validity of the guiding center approximation.

By using Eq. (28), (29) and (30) we first attempt to express v_x, v_y, v_z in terms of an instantaneous guiding center point x_0 at z, the gyrational energy K and the total energy H. Then we make a transformation from (x_0, K, H) to (x_{00}, K_0, v_{z0}) to obtain the final transformation expression.

The equation of motion in the x direction is given by

$$\dot{v}_x = \left(\frac{P_y}{m} - \Omega_0 x\right)\Omega_0 - \frac{e}{m}\frac{\partial \phi}{\partial x} \; . \tag{31}$$

When we define the instantaneous guiding center point x_0 as the point when $\dot{v}_x = 0$, we can obtain, by using Eq. (29),

$$v_y = \Omega_0(x_0 - x) + \frac{e}{m\Omega_0}\phi'(x_0, z) \tag{32}$$

where the prime is differentiation with respect to x. Furthermore, from Eq. (31) we can define a one-dimensional Hamiltonian K, by

$$K = v_x^2 + \frac{2e}{m}[\phi(x, z) - \phi(x_0, z)] - \frac{2e}{m}(x - x_0)\phi'(x_0, z) + \Omega_0^2(x - x_0)^2 \; . \tag{33}$$

Substitution of Eqs. (32) and (33) into (28) yields

$$\frac{2}{m}H = v_z^2 + K + \frac{2e}{m}\phi(x_0, z) + \frac{e^2}{m^2\Omega_0^2}\phi'^2(x_0, z) \; . \tag{34}$$

From Eqs. (32)–(34) we obtain

$$dv_x dv_y dv_z = V\,dx_{00}dk_0 dv_{z0} \tag{35}$$

where

$$V = \frac{\partial v_x}{\partial K}\frac{\partial K}{\partial K_0}\frac{\partial v_y}{\partial x_0}\frac{\partial x_0}{\partial x_{00}}\frac{\partial v_z}{\partial H}\frac{\partial H}{\partial v_{z0}}$$

$$= \frac{\Omega_0 v_{z0}[1 + e\phi''(x_{00})/m\Omega_0^2]\Omega(x_0)}{2v_x v_z \Omega(x_{00})}$$

where v_x and v_z are given by Eqs. (33) and (34). Thus, the ion density can be expressed by

$$n(x, z) = \int g(x_{00}, K_0, v_{z0})V\,dx_{00}dK_0 dv_{z0} \tag{36}$$

where g is the distribution function at z_0.

The electron density n_e, on the other hand, can be specified by the distribution function at the plane z_1 above the potential structure. Since the electrons are considered to be highly magnetized, we can assume that $e\phi''/m\Omega_0^2 \ll 1$ and the drift energy $e^2\phi'^2/m^2\Omega_0^2$ is negligible. Then the integration over x_{00} and K_0 in Eq. (36) can simply be replaced by the δ-function and we obtain

$$n_e(x, z) = \int \frac{g(v_{z1}, x)v_{z1}}{\left[v_{z1}^2 + \dfrac{2e}{m}(\phi - \phi_1)^2\right]^{1/2}}\, dv_{z1} \tag{37}$$

where g is the distribution specified at z_1; the subscript 1 denotes evaluation at z_1.

Formally speaking, therefore, combining Eqs. (36) and (37) with Poisson's equation can determine the two-dimensional potential structure if we specify the downgoing electron distribution at z_1 and the upgoing ion distribution at z_0 along with neutralizing background electrons and ions at either level. Notwithstanding this elegant formulation, it is a formidable task to obtain a numerical solution to satisfy a realistic boundary condition for a general distribution of ions and electrons. We show in Fig. 14 a numerical solution calculated by Swift for the case in which cold upward streaming ions from below and cold downward streaming electrons from above. The incident streaming energies of electrons and ions correspond, respectively, to

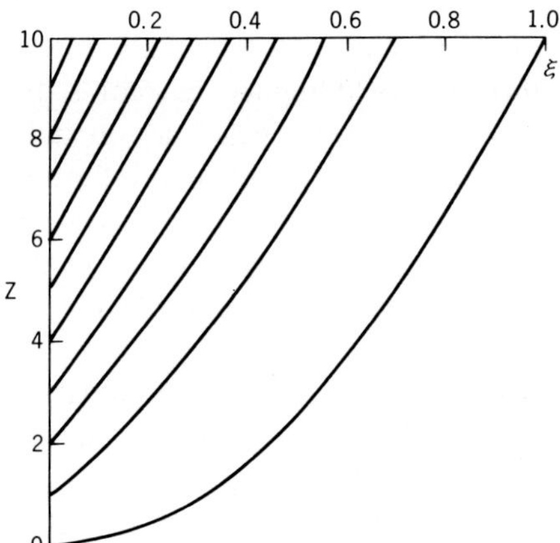

FIG. 14.   An equipotential structure obtained by SWIFT (1979). ξ is a perpendicular distance in units of the gyroradius of an ion of energy corresponding to the maximum potential drop.

100 eV and 10 eV when the maximum potential drop is assumed to be 2 kV. ξ is the distance in the x direction in units of the gyroradius of an ion of energy $e\phi_0$. Note that even in this simple example the solution cannot be a valid solution.

3.3   Some remarks on boundary conditons

A steady-state theory requires certain conditions on particle populations upstream and downsteam in order that a potential profile, which is rather arbitrarily conditioned, can be steadily maintained. In other words, it is required that the magnetotail, in which exists a primary source of field-aligned currents, can produce such an initial plasma distribution that can lead to the final distribution maintaining a double layer.

Suppose that a field-aligned plasma jet arising from a slow shock in the magnetotail is the primary source. Then, a field-aligned potential drop tends to distribute over the region between the mirror heights of electrons and protons constituting the forced jet. Since the plasma jet is continuously supplied, the charge separation can be maintained. The charge accumulation forces the magnetospheric plasma to carry a cross-field polarization current which inhibits further charge accumulation. Simultaneously, background cold electrons residing in the region between the mirror heights could be accelerated downward to produce auroral electrons, as is shown in Fig. 6.

In the case where the primary source generating a double layer is a field-aligned current, magnetically trapped (high pitch angle) particles above the height of the double layer are considered to be background particles. Those particles must redistribute so that the potential above the double layer is kept to a finite value. Since any field-aligned current is finite in thickness, a potential difference arises between the field lines on and off the field-aligned current sheet. Since such a cross-field potential difference develops with the development of double layer, a cross-field polarization current would be generated. Strictly, the effect of the polarization current connecting to the field-aligned current must be taken into account in the formulation of double layers.

These considerations imply that the boundary conditions on the magnetospheric side is complex. On the other hand, on the earth side they appear to be rather simple. This is partly because the earth side is downstream and passive for electrons, and because there are abundant cold ionospheric particles that can easily neutralize space charges. Furthermore, the fact that the magnetic mirror effect and the electron scattering at the ionosphere can provide sufficient number of trapped electrons on the downstream side makes the problem simpler.

Let us here make one comment on the inverted V- or S-potential structure. It may be true that the structure provides us with a key to elucidate the mechanism leading to auroral particle acceleration. However, we notice that there are two fundamentally different philosophies in this respect. One is that the structure is an internal structure which is constructed by the plasma parameters of the magnetosphere-ionosphere system such as the gyroradius represented by Swift's electrostatic shock model and the ionospheric conductivity (CHIU and CORNWALL, 1980). The essence of this philosophy lies in that the field-aligned current structure is nothing to do with the formation of the potential structure. Specifically, the structure of the field-aligned current is so modified that the resulting potential structure is sustained. Speaking without reserve, the potential structure results even if the initial current distribution is uniform.

There is another school of philosophy. Namely, the potential structure is essentially a product of the initial field-aligned current structure. Suppose that a thin sheet current is produced by some mechanism which is not directly related to production of potential structure, for instance by an ionosphere-magnetosphere feedback instability which will be discussed in Section 4.2.

We here show a procedure of how the potential structure is determined starting from a given field-aligned current distribution $j_{\parallel}(x)$. Given $j_{\parallel}(x)$, the potential distribution at the ionospheric height can be evaluated by the following equation (see, the next Section):

$$j_{\parallel}(x) = -\frac{\partial}{\partial x}\Sigma_{\mathrm{P}}\frac{\partial\phi}{\partial x}, \tag{38}$$

where Σ_{P} is the height-integrated Pedersen conductivity and the field-aligned current is positive for downward. The ionospheric potential distribution can readily be solved by properly imposing two boundary conditions, provided Σ_{P} is also given. Thus, the reference potential distribution perpendicular to the magnetic field is known. Along the field lines on which the field-aligned current is less than the instability threshold j_{c}, we can reasonably assume an equipotential. A potential drop $\Delta\phi(x)$ would arise along a field line on which $j_{\parallel}(x) > j_{\mathrm{c}}$ and a current-driven instability occurs. Furthermore, it would naturally be inferred that as the current increases, the instability activity increases and, hence, the resulting potential drop increases, as well. Without specifying a functional form of $\Delta\phi(x)$ in terms of $(j_{\parallel}(x) - j_{\mathrm{c}})$, the potential structure cannot be determined. However, there would be no doubt that the overall equipotential contours have a structure as shown in Fig. 15.

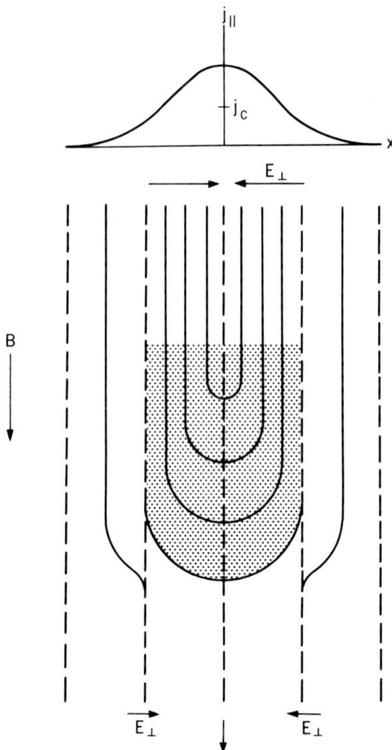

FIG. 15. Schematic equipotential contour lines associated with a field-aligned current with a finite thickness. Current-driven instability exists in a finite region denoted by dots. The vertical dashed lines represent the equipotential lines when the instability is not present.

4.   Magnetosphere-Ionosphere Coupling: Auroral Arcs

Can that spectacular auroral arcs with curtain-like structure be an art work of the magnetosphere alone? Or, can it be a cooperative work of the magnetosphere and the ionosphere? It is likely that active auroras associated with substorms are a primary work of the magnetosphere. Diffuse auroras, which usually appear in post breakup times, may be considered to be the work of the magnetosphere, since they are thought to result from strong pitch angle diffusion of energized particles in the magnetosphere disturbed by substorm activities (see, Chapter 5). Quiet auroral arcs, on the other hand, appear and move in a gentle manner keeping that naive shape. Judging from the gentle movement ascribable to the $E \times B$ drift, it seems not unlikely that they are able to be explained by some feedback mechanism between the ionosphere and the magnetosphere. Before going into this specific problem, we begin with the study of basic relations connecting the ionosphere to the magnetosphere.

4.1   Basic relations

Let us start with the ionospheric equations. For the purpose of attacking the ionosphere-magnetosphere coupling, it is convenient to use the height-integrated equations.

The height-integrated ionospheric current J_1 is given by

$$j_1 = N\Sigma_{PO}E_\perp - N\Sigma_{HO}\frac{E_\perp \times B_0}{B_0}, \tag{39}$$

where Σ_{PO} and Σ_{HO} are the height-integrated Pedersen and Hall conductivities, respectively, defined for a constant height-integrated electron density N_0, $E_\perp$ is the ionospheric electric field, B_0 is the magnetic field and N is the height-integrated electron density normalized by N_0.

The density N is described by the continuity equations of electrons and ions, which are combined to give (see, SATO, 1978).

$$\frac{\partial N}{\partial t} + c\frac{E_\perp \times B_0}{B_0^2} \cdot \nabla_\perp N = -\frac{j_\parallel}{eN_0} - \frac{\gamma j_h}{eN_0} - \alpha(N^2 - 1), \tag{40}$$

where $j_\parallel$ is the net (cold plus hot) field-aligned current (negative for upward), j_h represents the field-aligned current carried by hot (auroral) electrons, γ is the production rate of secondary electrons, and $\hat{\alpha}$ represents a quantity proportional to the recombination rate α defined by $\hat{\alpha} = \alpha N_0/h$ (h is the ionosphere thickness); the gradient operator $\nabla_\perp$ operates only on the plane perpendicular to B_0. Diffusion is neglected because it is usually negligible for the ionosphere-magnetosphere coupling problem that treats a spatial scale larger than 1 km.

Another important relation in dealing with the ionosphere-magnetosphere coupling is

$$j_\parallel = \nabla_\perp \cdot J_1. \tag{41}$$

This relation relates the field-aligned current at the ionospheric level with the ionospheric current. Note here that Eq. (38) can be derived when we put $E_\perp = -\nabla_\perp \phi$ and assume a uniform density ($N = 1$).

Equation (39)–(41) thus obtained constitute a set of fundamental equations which relate the field-aligned current at the ionospheric level to the

226   T. SATO

ionospheric parameters. We shall go on next to derive a relation that connects the field-aligned current to the magnetospheric parameters.

Equation (10), which is already derived from the MHD eqs., i.e., Eqs. (3)–(6), can provide a relation that connects the field-aligned current at the magnetosphere to the magnetospheric parameters. Since the last two terms on the right-hand side of Eq. (10) are of current-generator type, there would be no strong interrelation with the ionospheric condition, so that we may be able to disregard those terms for the present purpose. Neglecting the current sources and linearizing Eq. (10), we obtain

$$\frac{\partial}{\partial s}\left(\frac{j_\parallel}{B_0}\right) = -c^2 \frac{\rho}{B_0^3}\frac{d}{dt}\nabla_\perp \cdot \tilde{E}_\perp , \tag{42}$$

where the tilde implies the disturbance part; $\tilde{E}_\perp = -(1/c)\hat{v}\times B_0$ and $d/dt = \partial/\partial t + V_d \cdot \Delta_\perp$, with $V_d = cE_0 = B_0/B_0^2$ (convection velocity). Likewise, from Eqs. (4) and (5) we can easily obtain the following relation under the situation that V_d and B_0 are considered almost constant compared with the disturbance scale:

$$\frac{\partial}{\partial s}\nabla_\perp \cdot \tilde{E}_\perp = -\frac{4\pi}{c^2}B_0\frac{d}{dt}\left(\frac{j_\parallel}{B_0}\right). \tag{43}$$

This set of Eqs. (42) and (43) constitutes the transmission line equations whose inductance and capacitance per unit field line length are given by $4\pi/c^2$ and $c^2\rho/B_0^2$, respectively. Although we are here primarily concerned with a magnetospheric response to an ionospheric disturbance, they can of course be used for obtaining an ionospheric response to a magnetospheric disturbance. To get an explicit form of the response function, we must impose a terminating (boundary) condition at the other end of a transmission line, namely, at the magnetospheric equator or at the ionosphere.

Suppose a disturbance with frequency ω and wavenumber $k_\parallel$. Then, Eqs. (42) and (43) are solved to give

$$\nabla_\perp \cdot \tilde{E}_\perp = a \sin k_\parallel s + b \cos k_\parallel s$$
$$\tilde{j}_\parallel = iZ_0^{-1}(b \sin k_\parallel s - a \cos k_\parallel s) \tag{44}$$

where Z_0 is the characteristic impedance of the field (transmission) line defined by $Z_0 = (4\pi/c^2)V_A$, V_A being the Alfvén speed given by $V_A = B(4\pi\rho)^{-1/2}$; $k_\parallel = \omega/V_A$; a and b are constants determined by the terminating condition. In the above derivation a uniform medium was assumed for simplicity. From Eq. (44) the magnetospheric response seen from the ionosphere is solved as

$$\nabla_\perp \cdot \tilde{E}_\perp = i(bZ_0/a)\tilde{j}_{\parallel} \qquad\qquad \text{at} \quad s=0 . \tag{45}$$

Thus, we have arrived at a complete set of Eqs., (39), (40), (41) and (45), that can adequately describe the electrodynamic coupling between the ionosphere and magnetosphere.

4.2 Feedback instability: Formation of auroral arcs

Since we have derived the basic relations describing the ionosphere-magnetosphere coupling, we shall now examine the stability of the coupled system. Conjugate observations between the northern and southern hemispheres by airplane indicate that quiet auroral arcs appear with a high degree of conjugacy (BELON et al., 1969; STENBAEK-NIELSEN et al., 1972). Accordingly, we may well assume a symmetric conjugate model of the ionosphere-magnetosphere sysrem as shown in Fig. 16. Furthermore, an extremely elongated auroral form may allow us to assume uniformity along an arc. Thus, we can restrict our analysis on one direction which is perpendicular to an arc.

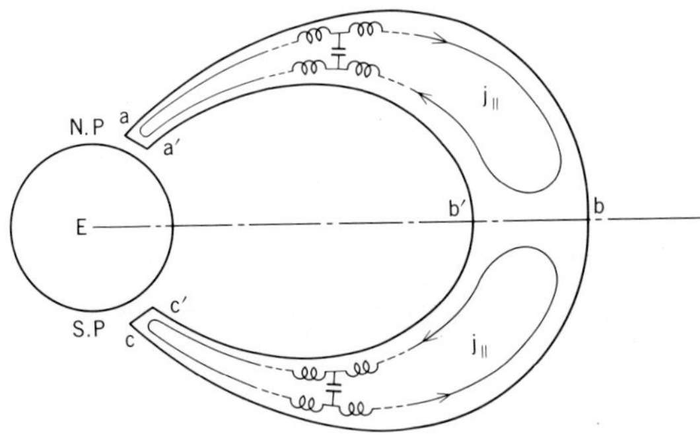

FIG. 16.   A symmetric ionosphere-magnetosphere coupling model.

First of all, we shall obtain an explicit magnetospheric response function Z. From the symmetricity assumption about the magnetic equator, the boundary condition at the magnetospheric equator $s=\imath$ is given by $j_{\parallel}(s=\imath)=0$. Thus, from Eq. (45) we obtain $b/a=\cot k_{\parallel}\imath$ (SATO, 1978), hence

$$\nabla \cdot \tilde{E}_\perp = \tilde{Z}\tilde{j}_{\parallel} , \tag{46}$$

with

$$\tilde{Z}=iZ_0 \cot k_{\parallel}\imath . \tag{47}$$

Let us introduce a right-handed coordinate system (x, y, z) in which for example, positive x is directed toward the equator from the pole, positive y toward the west from the east and positive $z(\equiv -s)$ downward along the field line. According to the above discussion, we assume $\partial/\partial y = 0$. To make the problem tractable, we rewrite the basic Eqs. (39), (40), (41) and (45) in the rest frame that moves with the $\boldsymbol{E} \times \boldsymbol{B}$ drift in the x direction:

$$\frac{\partial N}{\partial t} = -\frac{j_{\parallel}}{eN_0} - \frac{\gamma j_{\mathrm{h}}}{eN_0} - \hat{\alpha}(N^2 - 1), \tag{48}$$

$$j_{\parallel} = \frac{\partial}{\partial x}\{\Sigma_{\mathrm{PO}}N(E_{x0} + \tilde{E}_x) - \Sigma_{\mathrm{HO}}E_{y0}N\} \tag{49}$$

$$\frac{\partial \tilde{E}_x}{\partial x} = Z\tilde{j}_{\parallel}, \tag{50}$$

where $N = N_0 + \tilde{N}$ and $j_{\parallel} = j_{\parallel 0} + \tilde{j}_{\parallel}$; a quantity with subscript 0 denotes the background; Z is the magnetospheric impedance operator whose Fourier representation is given by Eq. (47).

Since we are interested in the stability of the quiet ionosphere-magnetosphere system, it is reasonable to neglect the hot electron precipitation, i.e., $j_{\mathrm{h}} = 0$ in Eq. (48). We assume the following perturbational form:

$$\tilde{N}, \tilde{j}_{\parallel}, \tilde{E}_x \propto \exp\left(-i\omega' t + ik_x x\right),$$

where ω' is a Doppler shifted frequency, $\omega' = \omega - k_x E_{y0}/B_0$, so that the magnetospheric impedance Z is given by Eq. (47). Substituting this form into Eqs. (48), (49) and (50), and linearizing them yield the following dispersion relation (SATO, 1978):

$$\omega' = \omega_{\mathrm{r}} + i(\omega_{\mathrm{r}}X - 2\hat{\alpha}), \tag{51}$$

with $\omega_{\mathrm{r}} = k_x(J_{x0}/eN_0)/(1 + X^2)$ and $X = \Sigma_{\mathrm{PO}}Z_0 \cot(\omega_{\mathrm{r}}'/V_{\mathrm{A}})$ (normalized magnetospheric reactance), where $J_{x0} = \Sigma_{\mathrm{PO}}E_{x0} - \Sigma_{\mathrm{HO}}E_{y0}$ (zeroth order ionospheric current).

From this dispersion relation an important conclusion is derived, namely, that the coupled ionosphere-magnetosphere system can spontaneously become unstable when the magnetosphere reacts inductively for a perturbation, i.e., $\omega_{\mathrm{r}}X > 0$ (OWAGA and SATO, 1971; SATO and HOLZER, 1973). It is interesting to note that the growth rate, $\Gamma = \omega_{\mathrm{r}}X - 2\hat{\alpha}$, becomes maximum when the magnetospheric inductance equals the ionospheric resistance ($X = 1$).

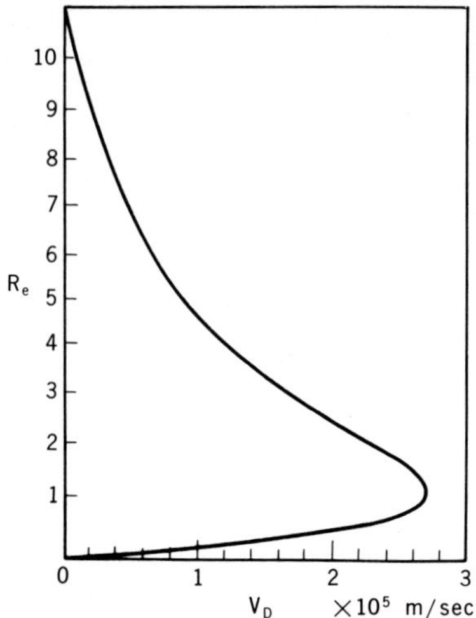

FIG. 17.   Eigenfunction of the field-aligned electron velocity of the most unstable mode of the ionosphere-magnetosphere feedback instability.

For example, we take the following typical parameter values at quiet times; $\Sigma_{PO}=0.5$ mho, $V_A=10^3$ km/s, $\iota=1.6\times10^5$ km ($\approx25\ R_E$), $J_{x0}=0.002$ Amp/m, $\alpha=3\times10^{-13}$ m^2/s, $h=30$ km and $N_0=1.2\times10^{15}$ m^{-2}. Then, the wavelength λ_x and the growth rate Γ of the most unstable fundamental mode turn out to be $\lambda_x=2\pi/k_x\approx20$ km and $\Gamma\approx4\times10^{-3}$ sec^{-1}, which indicate the feedback instability can be a viable mechanism that controls the auroral arc form. The field-aligned current distribution (eigenfunction) for an unstable mode is rigorously solved and shown in Fig. 17. This figure strongly suggests that a current-driven instability (KINDEL and KENNEL, 1971) could occur at rather lower altitudes ($\sim1\ R_E$) along the field lines as the ionosphere-magnetosphere feedback instability grows. As we have discussed in the previous section, double layers could be generated, accordingly, accelerated electrons could precipitate in coincidence with upward field-aligned currents.

Nonlinear coupled Eqs. (48)–(50) are numerically solved as an initial value problem including effects of accelerated electrons (SATO, 1978). Figure 18 shows a comparison between an observation and a theoretical solution. The upper panel shows a model current system of a quiet auroral arc constructed based on a rocket observation (PARK and CLOUTIER, 1971). The lower panel is a current system constructed based on a nonlinear solution of the feedback system. Agreements in field-aligned current intensities, their directions, the spatial relationship between the auroral arc and the field-

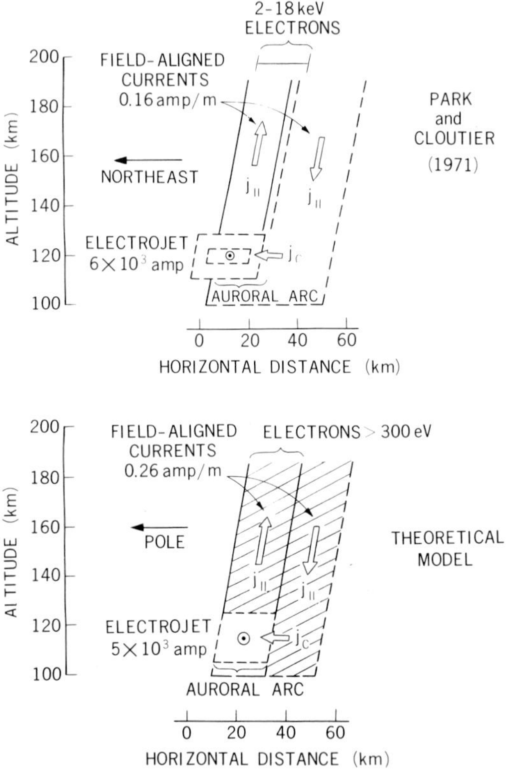

Fig. 18. Auroral arc models constructed based on a rocket experiment (Park and Cloutier, 1971) and on a nonlinear solution of the ionosphere-magnetosphere feedback instability theory.

aligned currents, and the electrojet intensities are surprisingly good.

Recently, a much more detailed global numerical simulation has been performed by Miura and Sato (1980). In this simulation the initial background configuration is so constructed that the large-scale Region 1 and 2 field-aligned currents (Iijima and Potemra, 1976), the dawn-dusk potential difference over the polar cap (Heppner, 1977), and the dayside-nightside electron density nonuniformity are taken into consideration. Under such a realistic background configuration, a global ionosphere-magnetosphere coupling is numerically simulated. Figure 19 shows a result of one such run. The upper three polar plots show temporal development of auroral arcs. It is clearly demonstrated in these plots that multiple auroral arcs are

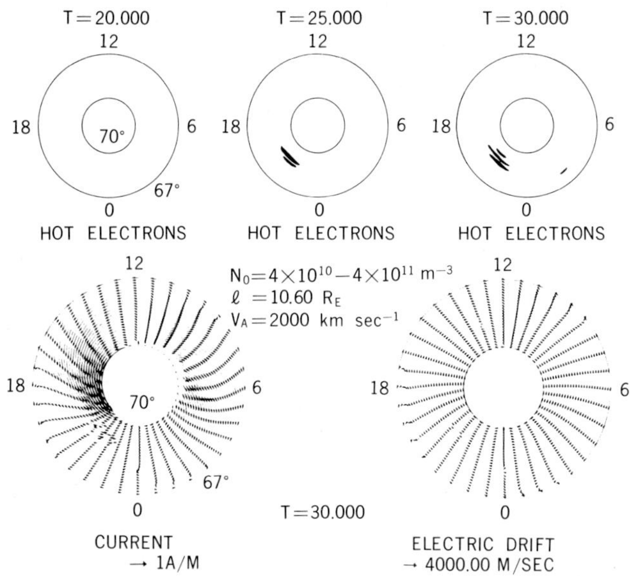

FIG. 19.   Results of a global numerical simulation of auroral arc formation. Note developments of auroral arcs in the evening-to-midnight sector (top) and of auroral electrojets (bottom left), and reduction of electric drifts inside auroral arcs (bottom right).

spontaneously generated preferentially in the evening-to-midnight local time sector of auroral oval. Their forms are very similar to actual auroral arcs. Their developing time is favorable and of the order of a couple of minutes, and fully-developed arcs persist very stably. The lower two polar plots show the electrojet development in association with auroral arcs (left) and the decrease of electric drift speed in auroral arcs (right). The latter fact is equivalent to the decrease of the north-sourth electric field within the arcs, this being consistent with observations (AGGSON, 1969; WESCOTT *et al.*, 1969). Cross-sectional distributions of the electron density and field-aligned current (all perturbed quantities) at 22 LT are shown in Fig. 20. It is to be noted that upward field-aligned currents always coincide with auroral arcs. Shown in Figs. 21 and 22 are results for another run in which the ionospheric density is reduced by a factor of 4 compared with the previous case with the other conditions remaining the same. It is seen that auroral activities such as the field-aligned current intensity and auroral electrojets are much more intensified, along with the increase of multiplicity of arcs. Parametrical dependence of fully-developed auroral arcs is summarized in Fig. 23.

232 T. SATO

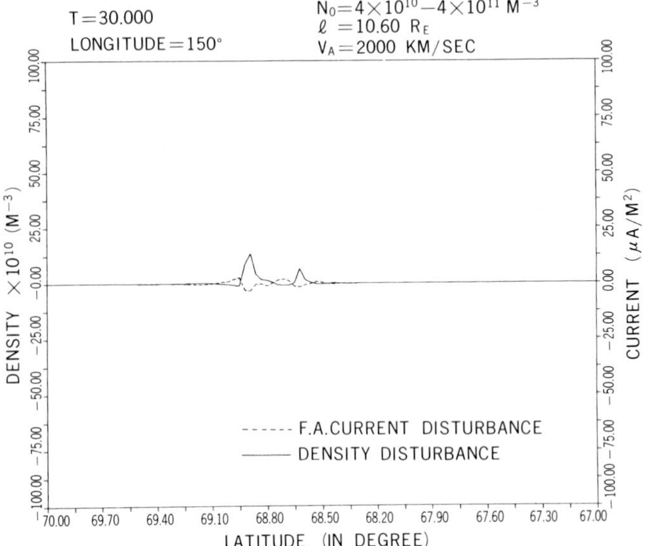

FIG. 20. Latitudinal profiles of developing auroral arcs at 22 local time at 110 km height. Then solid line represents the perturbed part of the electron density and the dotted line represents the perturbed part of the field-aligned current. Note one-to-one correspondence between the upward field-aligned current and the density peak.

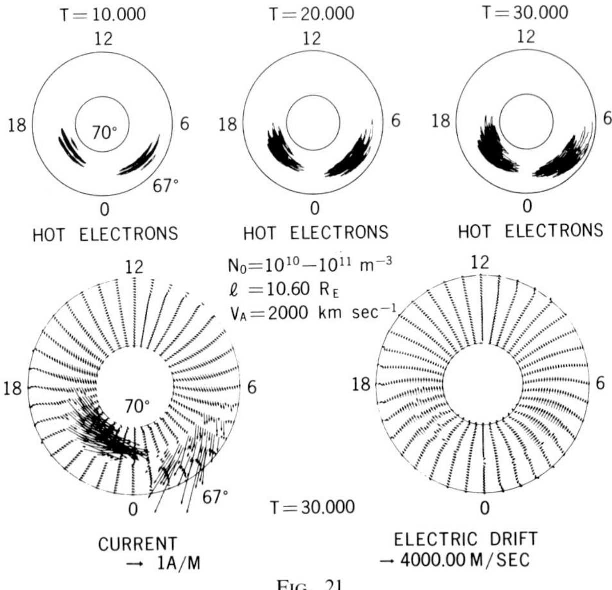

FIG. 21

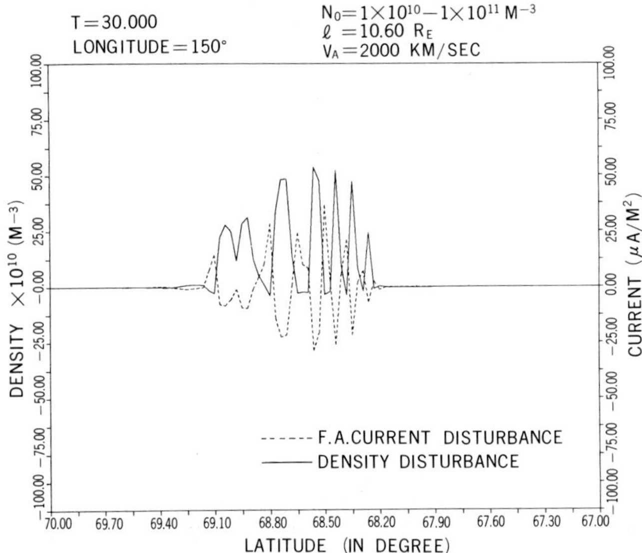

FIG. 22.   Same as Fig. 20 but for the case given in Fig. 21.

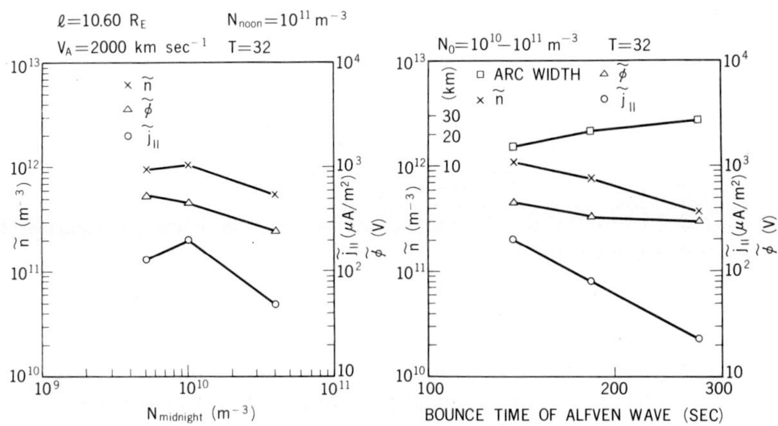

FIG. 23.   Parametric dependence of auroral arc characteristics (the arc thickness, the peak value of the induced field-aligned current, etc.). The field line length, the Alfvén speed in the magnetosphere, the ionospheric density distribution are changed.

FIG. 21.   Results for another run of the global numerical simulation of auroral arcs in the same format as Fig. 19. In this run, the overall electron density in the ionosphere was reduced by a factor 4 compared with Fig. 19.

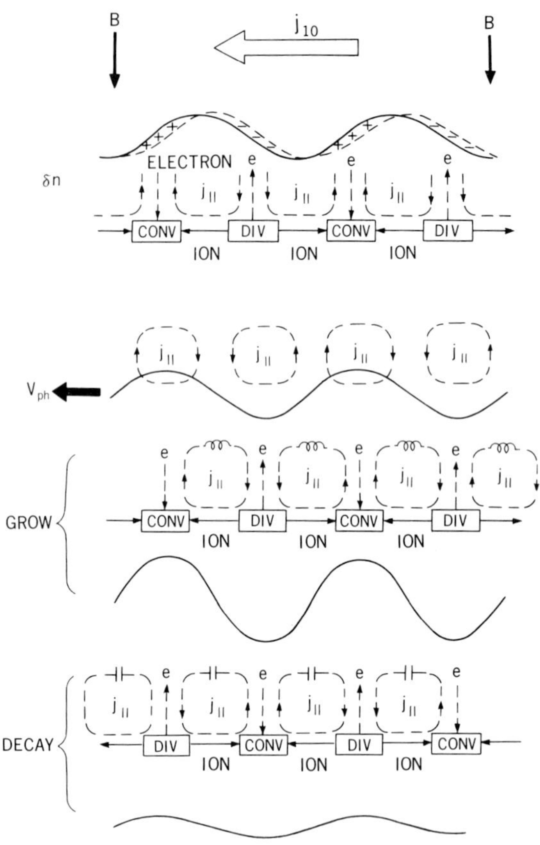

Fig. 24. Explanation of the physical mechanism of the ionosphere-magnetosphere feedback instability.

Before closing this section, we shall describe the physical mechanism of the instability in view of the importance of underlying basic concept in dealing with the ionosphere-magnetosphere coupling: As the zeroth order configuration, we assume an ionospheric current J_{x0} present in the north direction, which is a part of the ionospheric closure current of large-scale field-aligned currents. In this system that encompasses large-scale field-aligned currents, we suppose that a small-scale perturbation of ionospheric density δn, field-aligned current $\delta j_{\parallel}$, or whatever, appears. To make our explanation simpler, let us here begin with the density perturbation δn and take a rest frame that moves with the $\boldsymbol{E} \times \boldsymbol{B}$ drift in the x direction. Since the zeroth order current is flowing in the positive x direction, the density perturbation is accompanied by an electric field perturbation as shown in the top panel of Fig. 24. Associated field-aligned current perturbation $\delta j_\parallel$, therefore, should be such as shown in the second panel. Noting, however, that the perturbed ionospheric (Pedersen) current is mainly carried by ions, while the perturbed field-aligned current is mainly carried by high-mobility electrons, the density perturbation δn is forced to migrate in the positive x (zeroth order current) direction. Once such a spontaneous motion of the ionospheric density perturbation is induced, the magnetosphere feels the perturbation as an ac perturbation. Thus, an Alfvén wave is excited and propagates along the field lines to establish an ac field-aligned current in the magnetosphere. If the induced motion of the density perturbation is slow enough (i.e., $\omega_r l/V_A < \pi/2$), the magnetospheric response is capacitive. In other words, the ionospheric current is directly connected to the magneto-spheric polarization current. In this case, the perturbed field-aligned current leads to ionospheric perturbation by $\pi/2$ in phase. Consequently, as shown in the bottom panel of Fig. 24, the density perturbation suffers damping. On the other hand if the motion of the ionospheric perturbation becomes faster (i.e., $\pi/2 < \omega_r l/V_A < \pi$), the magnetosphere reacts as inductive. Thus, $\delta j_\parallel$ lags behind δn by $\pi/2$ in phase, indicating growth of the perturbation, as shown in the panel above the bottom of Fig. 24.

We further note that the global simulation model is now being developed so that the parallel resistivity can be included. The result shows that a strong concentration of parallel and perpendicular electric fields near $\sim 1\ R_E$ is observed.

5. Auroral Kilometric Radiation

The discovery of intense radiation from the earth (GURNETT, 1974; DUNKEL et al., 1970) has progressed one step forward the recognition of the importance of the earth's magnetospheric research. Since this discovery, a new concept that attempts to grasp the earth's magnetosphere as a model of magnetized planets has emerged. Subsequent new findings about Jovian magnetosphere by Pioneer 10 and 11 and Voyager 1 and 2 have stimulated the comparative magnetosphere study.

Much of the present knowledge on the terrestrial kilometric radiation was revealed by the first comprehensive study by GURNETT (1974). His findings about the terrestrial kilometric radiation, which is now preferred to be called auroral kilometric radiation AKR (KURTH et al., 1975), are

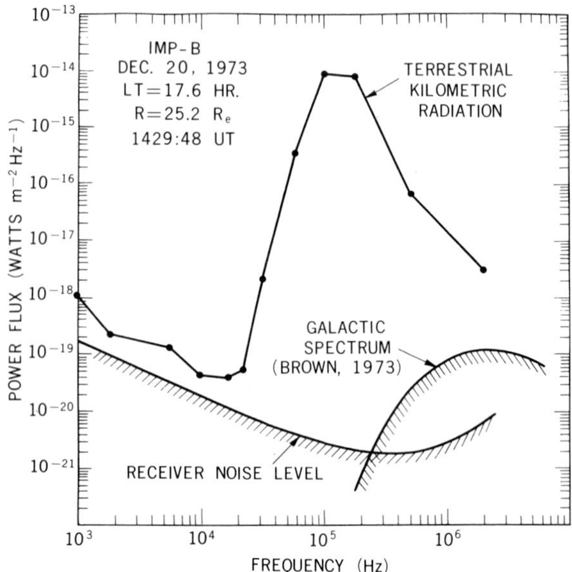

FIG. 25. Power flux versus frequency diagram of auroral kilometric radiation (AKR) observed by IMP-8 [GURNETT, 1974].

summarized as follows:

1) The frequency of AKR range from 50 kHz to 500 kHz, the peak frequency being around 200 kHz, see Fig. 25.

2) The total power amounts to 10^9 Watts.*

3) The radiation is likely to be the right-handed polarized (extraordinary) mode.

4) The radiation appears to originate at low altitudes less than 3 R_E and is closely correlated with discrete auroral arcs.

5) The radiation power is about 1% of the energy dissipation of auroral particles.

It is no exaggeration to say that the later observational works have contributed only to confirm and substantiate the findings of Gurnett's pioneering study. For example, it has become certain that the radiation is the extraordinary mode (GURNETT and GREEN, 1978; KAISER et al., 1978) and that the source locations are near 65° invariant latitude between 22[h] and 24[h] magnetic local time (GALLAGHER and GURNETT, 1979). Also noted is an indication of the role of auroral zone field-aligned currents in the generation of auroral kilometric radiation (GREEN et al., 1979). Recent near-source

* There is an observational argument that this value is $10^6 \sim 10^7$ Watts (Oya, private communication).

observations of AKR by the ISIS-1 satellite (Benson and Calvert, 1979) are also to be noted. They have observed that AKR is generated in the extraordinary mode just above the local cut-off frequency and emanates nearly perpendicular to the magnetic field. Furthermore, it is suggested that the source region is located within the region where substantial local depletion of electron density exists, the invariant latitudes ranging over $64°$–$75°$.

From these observations it may be a natural conclusion to connect the AKR generation mechanism to the generation mechanism of auroral particles. Several theoretical models have so far been proposed to explain the generation mechanism of AKR in conjunction with the Jovian (decametric) radiation. The proposed mechanisms may be classified into two classes: direct and indirect generation of an (x-mode) electromagnetic mode. Melrose (1976) proposed a direct generation mechanism of the x-mode as a result of a linear instability due to a large electron anisotropy. He supposed that a large anisotropy would result from large mirror effects of the converging geomagnetic field. Removing the requirement of Melrose's large anisotropy, Wu and Lee (1979) proposed a different direct mechanism which is based on a loss cone distribution of mirror reflected electrons.

In contrast, the indirect generation mechanism first requires the excitation of electrostatic waves of upper-hybrid mode or electron cyclotron mode due to accelerated auroral electrons, and then the conversion of electrostatic waves into electromagnetic wave. One possibility is a linear conversion mechanism of Benson (1975) based on Oya's conversion model (1971, 1974). The difficulty of this mechanism lies in polarization of the radiation. However, since in Oya's model the upper hybrid electrostatic mode is converted first into the x mode, the linear conversion theory may still reserve a possibility.

The other existing indirect mechanisms require the nonlinear conversion of electrostatic waves into an electromagnetic mode. Roux and Pellat (1979) showed that the electrostatic energy driven unstable by accelerated auroral electrons would accumulate at or near the upper hybrid frequency f_{UH} (and the lower hybrid frequency). They then concluded that an intense radiation should occur at $2 f_{UH}$ in the extraordinary mode as a result of nonlinear beatings of the unstable upper hybrid modes.

Palmadesso et al. (1976) proposed a mechanism that an electromagnetic wave is amplified via an electrostatic beat mode in the presence of an electron beam at the expense of ion wave turbulence. Recently, this principle is extended to a case in which coherent electrostatic ion cyclotron waves are assumed (GRABBE *et al.*, 1979). Assuming a density cavity as observed by the ISIS-1 detector (BENSON and CALVERT, 1979), they have shown that the *x*-mode can grow rapidly in a narrow frequency range that is accessible to free space.

As we have seen above, one crucial requirement that an indirect mechanism should satisfy is the existence of unstable electrostatic waves. As for electrostatic ion cyclotron waves required by GRABBE *et al.* (1979), we already have theoretical and observational support (KINDEL and KENNEL, 1971; MOZER *et al.*, 1977). The upper hybrid mode, however, is not yet observationally confirmed, although a theoretical possibility is discussed (ASHOUR-ABDALLA *et al.*, 1979).

As to whether the generation mechanism of AKR is direct or indirect, there is a point that can be checked observationally. The direct models (MELROSE, 1976; WU and LEE, 1979) require that the source region should be located above the region of localized parallel electric fields, whereas all indirect models require that it be at or below the region of parallel fields. Therefore, if the source region can be more precisely determined, then at least

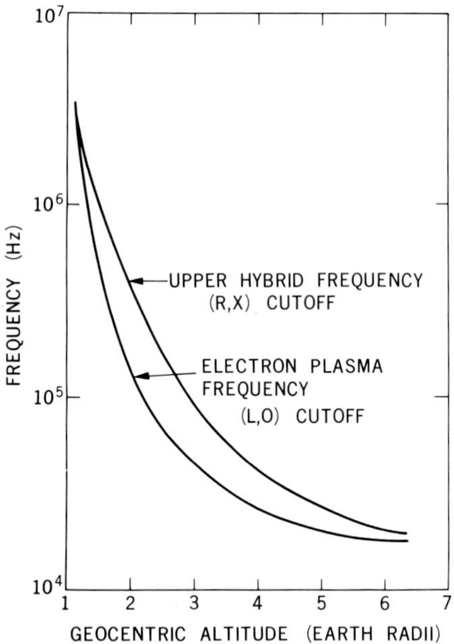

FIG. 26.   Diagram showing the characteristic plasma frequencies as a function of the distance from the earth [GURNETT, 1974]. This diagram is essential in constructing a theoretical model of AKR generation mechanism.

the issue of whether direct or indirect would be resolved. In any case, it is interesting to note that the frequency versus radial distance diagram, first introduced by GURNETT (1974), shows that the radiation must emanate near the cut-off frequency of the x-mode $f_x \approx f_H(1 + f_p^2/f_H^2)$ where f_H and f_p are the electron cyclotron and the plasma frequency, respectively; see Fig. 26.

6. Closing Remarks

Auroral physics, as part of magnetospheric physics, has just stepped toward quantitative understanding. Quantitative understanding of the whole causal relationship of auroral processes is as yet far from being completely understood. Nevertheless, some parts of the complicated picture puzzle of auroral physics are being filled in a quantitative manner, though fragmentarily and to different degrees. At this time therefore, it should be a formidable task to require description of well-established auroral physics. In this Chapter of auroral physics, an attempt is made to describe auroral physics with the author's strongly biased knowledge and philosophy. The prime object and hope of writing this Chapter lies in that young readers would do their best in completing this yet unaccomplished auroral picture puzzle with their own philosophy. In doing so, however, they should not be too urgent and sticky to explaining one particular observation, however important it appears; otherwise they would often be tempted to make an unsubstantiated assumption leading finally to a dead end, although at that time it might appear as if the observational fact were nicely resolved. It would be a royal road to attempt to construct a physically sound theoretical model without being overinfluenced by an individual observation. In this way, the grand auroral puzzle will steadily be resolved.

Acknowledgments
I wish to thank Dr. A. Hasegawa, Bell Laboratories. While I was temporalily visiting there, the first version of this manuscript was written.

REFERENCES

AGGSON, T. L., Probe measurements of electric fields in space, in *Atmospheric Emissions*, edited by B. M. McCormac and A. Omholt, D. Reidel, Dordrecht, Holland, and Hingham, Mass., 1969.

ASHOUR-ABDALLA, M., C. F. KENNEL, and W. LIVESEY, A parametric study of multiharmonic instabilities in the magnetosphere, *J. Geophys. Res.*, **84**, 6540, 1979.

AXFORD, W. I., Viscous interaction between the solar wind and the earth's magnetosphere, *Planet. Space Sci.*, **12**, 45, 1964.

BELON, A. E., J. E. MAGGS, T. N. DAVIS, K. B. MATHER, N.W. GLASS, and G. F. HUGHES, Conjugacy of visible auroras during magnetically quiet periods, *J. Geophys. Res.*, **74**, 1, 1969.

BENSON, R. F., Source mechanism for terrestrial kilometric radiation, *Geophys. Res. Lett.*, **2**, 52, 1975.

BENSON, R. F. and W. CALVERT, Isis I observations at the source of auroral kilometric radiation, *Geophys. Res. Lett.*, **6**, 479, 1979.

BERNSTEIN, I. B., J. M. GREENE, and M. D. KRUSKAL, Exact nonlinear plasma oscillations, *Phys. Rev.*, **108**, 546, 1957.

BISKAMP, D. and R. CHODURA, Computer simulation of anomalous resistivity, *Phys. Rev. Lett.*, **27**, 1553, 1971.

BLOCK, L. P., Potential double layers in the ionosphere, *Cosmic Electrodynamics*, **3**, 349, 1972.

BLOCK, L. P., A double layer review, *Astro. Space Sci.*, **55**, 59, 1978.

CHIU, Y. T. and J. M. CORNWALL, Electrostatic model of a quiet auroral arc, *J. Geophys. Res.*, **85**, 543, 1980.

COAKLEY, P. and N. HERSHKOWITZ, Laboratory double layers, *Phys. Fluids*, **22**, 1171, 1979.

DEGROOT, J. S., C. BARNES, A. E. WALSTEAD, and O. BUNEMAN, Localized structures and anomalous dc resistivity, *Phys. Rev. Lett.*, **38**, 1283, 1977.

DEWITT, R. N., Polarization of the auroral electrojet, *J. Geophys. Res.*, **73**, 6307, 1968.

DUNGEY, J. W., Interplanetary magnetic field and the auroral zones, *Phys. Rev. Lett.*, **6**, 47, 1961.

DUNKEL, N., B. FICKLIN, L. RORDEN, and R. A. HELLIWELL, Low frequency noise observed in the distant magnetosphere with Ogo 1, *J. Geophys. Res.*, **75**, 1854, 1970.

EVIATOR, A. and R. A. WOLF, Transfer processes in magnetopause, *J. Geophys. Res.*, **73**, 5561, 1968.

FRANK, L. A., K. L. ACKERSON, and R. P. LEPPING, On hot tenuous plasmas, fireballs and boundary layers in the earth's magnetotail, *J. Geophys. Res.*, **81**, 5859, 1976.

GALLAGHER, D. L. and D. A. GURNETT, Auroral kilometric radiation: Time-averaged source location, *J. Geophys. Res.*, **84**, 6501, 1979.

GOERTZ, C. K. and G. JOYCE, Numerical simulation of the plasma double layer, *Astrophys. Space Sci.*, **32**, 165, 1975.

GRABBE, C. L., P. J. PALMADESSO, and K. PAPADOPOULOS, A coherent nonlinear theory at auroral kilometric radiation, NRL Mem.Report 4087, 1979.

GREEN, J. L., N. A. SAFLEKO, D. A. GURNETT, and T. A. POTEMRA, A correlation between auroral kilometric radiation and field-aligned current, *J. Geophys. Res.*, **84**, 5216, 1979.

GURNETT, D. A., The earth as a radio source: Terrestrial kilometric radiation, *J. Geophys. Res.*, **79**, 4227, 1974.

GURNETT, D. A. and J. A. GREEN, On the polarization and origin of auroral kilometric radiation, *J. Geophys. Res.*, **83**, 689, 1978.

HASEGAWA, A. and K. MIMA, Anomalous transport produced by kinetic Alfvén wave turbulence, *J. Geophys. Res.*, **83**, 1117, 1978.

HASEGAWA, A. and T. SATO, Generation of field-aligned currents during substorm, in *Dynamics of the Magnetosphere*, edited by S.-I. Akasofu, D. Reidel Publ. Co., Boston, 529 p. 1979.

HASEGAWA, A. and T. SATO, Existence of negative potential solitary-wave structure and formation of double layer in a nonequilibrium plasma, *Phys. Fluids*, to be published, 1981.

HAYASHI, T. and T. SATO, Magnetic reconnection: Acceleration, heating, and shock formation, *J. Geophys. Res.*, **83**, 217, 1978.

HEPPNER, J. P., Empirical models of high-latitude electric fields, *J. Geophys. Res.*, **82**, 1115, 1977.

HONES, E. W., Jr., A. T. Y. LIN, S. J. BAME, and S. SINGER, Prolonged tailward flow of plasma in the thinned plasma sheet observed at $r \approx 18\ R_\mathrm{E}$ during substorms, *J. Geophys. Res.*, **79**, 1385, 1974.

HONES, E. W., Jr., Plasma flow in the magnetotail and its implication for substorm theories, in *Dynamics of the Magnetosphere*, edited by S.-I. Akasofu, D. Reidel, Boston, 1979.

IIJIMA, T. and T. A. POTEMURA, The amplitude distribution of field-aligned currents at northern high latitudes observed by Triad, *J. Geophys. Res.*, **81**, 2165, 1976.

IIZUKA, S., K. SAEKI, N. SATO, and Y. HATTA, Buneman instability, Pierce instability, and double-layer formation in a collisionless plasma, *Phys. Rev. Lett.*, **43**, 1404, 1979.

KAISER, M. L., J. K. ALEXANDER, A. C. RIDDLE, J. B. PEARCE, and J. W. WARWICK, Direct measurements by Voyagers 1 and 2 of the polarization of terrestrial kilometric radiation, *Geophys. Res. Lett.*, **5**, 857, 1978.

KINDEL, J. M. and C. F. KENNEL, Topside current instabilities, *J. Geophys. Res.*, **76**, 3055, 1971.

KNORR, G. and C. K. GOERTZ, Existence and stability of strong potential double layers, *Astrophys. Space. Sci.*, **31**, 209, 1974.

KURTH, W. S., M. M. BAUMBECK, and D. A. GURNETT, Direction finding measurements of auroral kilometric radiation, *J. Geophys. Res.*, **80**, 2764, 1975.

LUI, A. T. Y., Observations of palsma sheet dynamics during magnetospheric substorms, in *Dynamics of the Magnetosphere*, edited by S.-I. Akasofu, D. Reidel, Boston, 1979.

MATSUMOTO, H., K. NAGAI, and T. SATO, Particle acceleration in a developing magnetic reconnection process, submitted to *J. Geophys. Res.*, 1981.

MELROSE, D. B., An interpretation of Jupiter's decametric radiation and the terrestrial kilometric radiation as direct amplified gyromission, *Astrophys. J.*, **207**, 651, 1976.

MIURA, A. and T. SATO, Shear instability: Auroral arc deformation and anomalous momentum transport, *J. Geophys. Res.*, **83**, 2109, 1978.

MIURA, A. and T. SATO, Numerical simulation of global formation of auroral arcs, *J. Geophys. Res.*, **85**, 73, 1980.

MONTGOMERY, D. C. and G. JOYCE, Shock-Like solutions of the electrostatic Vlasov equation, *J. Plasma Phys.*, **3**, 1, 1969.

MOZER, F. S., C. W. CARLSON, M. K. HUDSON, R. B. TORBERT, B. PARADY, J. YATTEAU, and M. C. KELLEY, Observations of paired electrostatic shocks in the polar magnetosphere, *Phys. Rev. Lett.*, **38**, 292, 1977.

OGAWA, T. and T. SATO, New mechanism of auroral arcs, *Planet. Space Sci.*, **19**, 1393, 1971.

OYA, H., Conversion of electrostatic plasma waves into electromagnetic waves: Numerical calculation of the dispersion relation for all wavelengths, *Radio Sci.*, **6**, 1131, 1971.

OYA, H., Origin of Jovian decameter wave emission-conversion from the electron cyclotron plasma wave to the ordinary mode electromagnetic wave, *Planet. Space Sci.*, **22**, 687, 1974.

PALMADESSO, P., T. P. COFFEY, S. L. OSSAKOW, and K. PAPADOPOULOS, Generation of terrestrial kilometric radiation by a beam-driven electromagnetic instability, *J. Geophys. Res.*, **81**, 1762, 1976.

PARK, R. J. and P. A. CLOUTIER, Rocket-based measurements of Birkeland currents related to an auroral arc and electrojet, *J. Geophys. Res.*, **76**, 7714, 1971.

QUON, B. H. and . Y. WONG, Formation of potential double layers in plasmas, *Phys. Rev. Lett.*, **37**, 1393, 1976.

ROUX, A and R. PELLAT, Coherent generation of the auroral kilometric radiation by nonlinear beatings between electrostatic waves, *J. Geophys. Res.*, **84**, 5189, 1979.

SATO, T., Possible sources of field-aligned currents, *Rept. Iono. Space Res. Japan*, **28**, 179, 1974.

SATO, T., A. theory of quiet auroral arcs, *J. Geophys. Res.*, **83**, 1042, 1978.

SATO, T., Strong plasma acceleration by slow shocks resulting from magnetic reconnection, *J. Geophys. Res.*, **84**, 7177, 1979.

SATO, T. and T. E. HOLZER, Quiet auroral arcs and electrodynamic coupling between the ionosphere and the magnetosphere, 1, *J. Geophys. Res.*, **78**, 7314, 1973.

SATO, T. and T. HAYASHI, Externally driven magnetic reconnection and a powerful magnetic energy converter, *Phys. Fluids*, **22**, 1189, 1979.

SATO, T. and T. IIJIMA, Primary sources of large-scale Birkeland current, *Space Sci. Rev.*, **24**, 347, 1979.

SATO, T. and H. OKUDA, Ion acoustic double layers, *Phys. Rev. Lett.*, **44**, 740, 1980.

SATO, T. and H. OKUDA, Numerical simulation of ion acoustic double layers, *J. Geophys. Res.*, **86**, 3357, 1981.

SATO, T., T. HAYASHI, T. TAMAO, and A. HASEGAWA, Confinement and jetting of plasmas by magnetic reconnection, *Phys. Rev. Lett.*, **41**, 1548, 1978.

SHAWHAN, S. D., C.-G. FÄLTHAMMAR, and L. BLOCK, On the nature of large auroral zone electric fields at 1-R_E altitude, *J. Geophys. Res.*, **83**, 1049, 1978.

SONNERUP, B. U. Ö., Adiabatic particle orbits in a magnetic null sheet, *J. Geophys. Res.*, **76**, 8211, 1971.

SOUTHWOOD, D. J., The role of hot plasma in magnetospheric convection, *J. Geophys. Res.*, **82**, 5512, 1977.

SPEISER, T. W., Particle trajectories in model current sheets 2. Applications to auroras using a geomagnetic tail model, *J. Geophys. Res.*, **72**, 3919, 1967.

STENBAEK-NIELSEN, H. C., T. N. DAVIS, and N. W. GLASS, Relative motion of auroral conjugate points during substorms, *J. Geophys. Res.*, **77**, 1844, 1972.

SWIFT, D. W., On the formation of auroral arcs and acceleration of auroral electrons, *J. Geophys. Res.*, **80**, 2096, 1975.

SWIFT, D. W., An equipotential model of auroral arcs, 2, Numerical solutions, *J. Geophys. Res.*, **81**, 3935, 1976.

SWIFT, D. W., An equipotential model for auroral arcs: The theory of two-dimensional Laminar electrostatic shocks, *J. Geophys. Res.*, **84**, 6427, 1979.

TSUDA, T., Effective viscosity of a streaming collision-free plasma in a weakly turbulent magnetic field, *J. Geophys. Res.*, **72**, 6013, 1967.

TORVÉN, S. and M. BABIC, Current chopping space charge layers in a low pressure arc plasma, Proc. Int. Conf. on Phenomena in Ionized Gases, Netherlands, 1975.

VASYLIUNUS, V. M., Mathematical models of magnetospheric convection and its coupling to the ionosphere, in *Particles and Fields in the Magnetosphere*, edited by B. M. McCormac, D. Reidel, Dordrecht, 1970.

WESCOTT, E. M., J. D. STOLARIK, and J. P. HEPPNER, Electric fields in the vicinity of auroral forms from motions of barium vapor releases, *J. Geophys. Res.*, **74**, 3469, 1969.

WU, C. S. and L. C. LEE, A theory of the terrestrical kilometric radiation, *Astrophys. J.*, **230**, 621, 1979.

YAMAMOTO, T., On the formation of electrostatic shock waves associated with auroral electron precipitation, *Planet. Space Sci.*, **24**, 1073, 1976.

Chapter 5

ELECTROSTATIC WAVES AND THE STRONG DIFFUSION OF MAGNETOSPHERIC ELECTRONS

C. F. Kennel and M. Ashour-Abdalla

Department of Physics and Institute of Geophysics and Planetary Physics, University of California, Los Angeles, California, U.S.A.

Abstract   We present a comprehensive review of electron pitch angle scattering in the magnetosphere and the plasma waves responsible for it, emphasizing the strong diffusion of diffuse auroral electrons by electrostatic electron cyclotron harmonic waves. We review briefly the weak diffusion of energetic radiation belt electrons within the plasmasphere. Several new suggestions concerning the quasilinear diffusion from and saturation of electrostatic waves are included.

1. Introduction

Modern spacecraft detectors measure virtually the whole range of plasma waves of interest to theory, from below the ion cyclotron frequency to above the electron plasma frequency, both electromagnetic and electrostatic. They are sensitive enough to detect thermal fluctuations in many regions of space. Three dimensional ion and electron velocity distributions can be measured, with a resolution rarely matched in the laboratory, at the same time as the plasma wave spectra. Such exquisite diagnosis disturbs space plasmas much less than laboratory plasmas. Of course, space plasma physicists pay a price for these advantages. It is difficult to measure wavelengths. The magnetosphere's global structure, hard to deduce from a few point measurements, is always changing uncontrollably. But even the price contains a hidden benefit, for this variability creates a rich plasma wave phenomenology.

NON-LOCAL SHOCK
Beam, Heat flux
instabilities
COLLISIONLESS BOW SHOCK
STRONG MHD TURBULENCE
RECONNECTION
MHD turbulence
Tearing Mode
Hi-β anisotropy
LHR drift waves
MAGNETOPAUSE
LHR drift waves
B≠O tearing
Kelvin-Helmholtz
PLASMASPHERE
Scattering, diffusion
of energetic particles
AURORAL ARCS
Anomalous Resistance
EIC waves
Double layers
LHR beam instability
Cavitons
RF generation
DIFFUSE AURORA
Strong diffusion by
EM, ES cyclotron
waves

Fig. 1

COUPLING OF CROSS-FIELD TRANSPORT
TO MIRROR LOSSES
DIFFUSE AURORA
Coupling of flow
to strong pitch
angle diffusion
INNER EDGE
Variable
Electric
Field
CONVECTION CARRIES
HOT PLASMA TOWARDS
NIGHT SIDE
RADIATION BELTS
Turbulent convection
drives radial diffusion,
which couples to weak
pitch angle diffusion

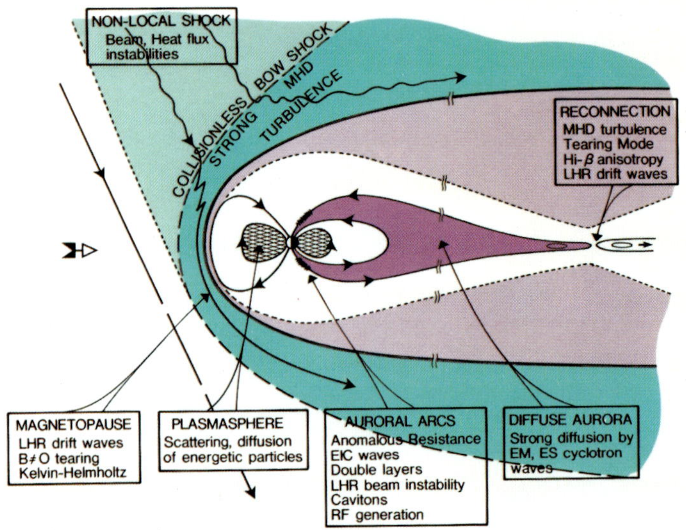

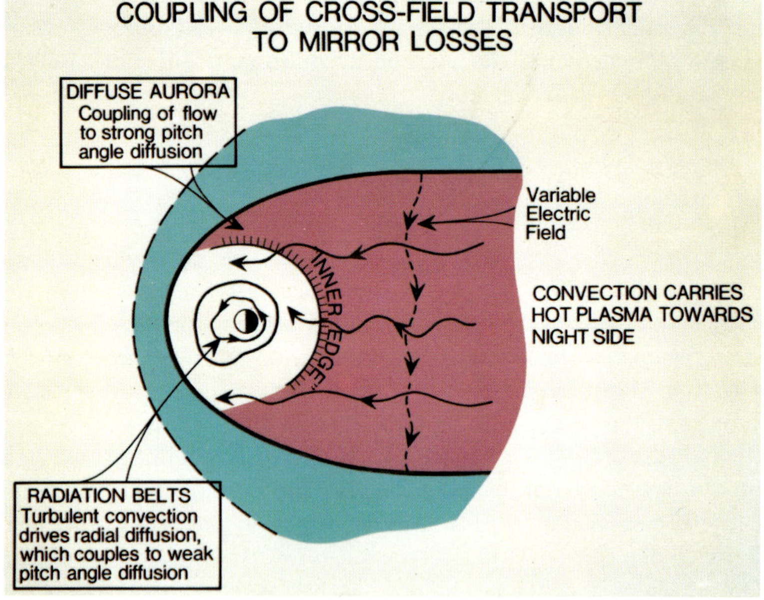

Fig. 4

FIG. 1.   Theoretician's magnetosphere. Shown in technicolor is a schematic of the earth's magnetosphere. The view is of a magnetic meridian plane from the local geomagnetic dusk direction. The solar wind streams from the left; its interaction with the geomagnetic field creates a standing collisionless fast shock. The magnetopause (dark line) separates shocked solar wind (dark blue) from magnetosphere (purple). The transport of momentum and energy across the magnetopause sets magnetospheric plasma and field lines in motion. The boxes list various plasma instabilities and the arrows indicate where they are thought to occur. For example, energetic particles and heat fluxes streaming from the bow shock lead to instabilities ahead of the earth in the solar wind. The magnetosheath contains highly nonlinear magnetohydrodynamic (MHD) turbulence. Some of the instabilities thought to contribute to magnetopause transport include lower hydrid resonance (LHR) drift waves, the Kelvin-Helmholtz instability, and the collisionless tearing mode in the guide field limit, which leads to reconnection. If reconnection at the nose of the magnetosphere occurs, there must also be reconnection in the tail to balance rates of flux transport. Tail reconnection and the associated plasma sheet turbulence again involve the tearing mode, and, possibly, LHR drift waves; since the plasma sheet has a high ratio β of plasma to magnetic pressure, it is subject to a variety of electromagnetic thermal anisotropy instabilities; it could also be strongly turbulent on hydromagnetic scales. Tail reconnection leads to a flow of hot plasma from the tail to the earth's night side. This flow is coupled to the auroral ionosphere by field aligned currents. Where these currents are most intense, instabilities leading to anomalous resistance and/or laminar potential jumps (double layers) may occur; electrostatic ion cyclotron waves (EIC) and lower hybrid instability of whistler waves are observed; and there is intense generation of radio (RF) emissions, possibly in nonlinear cavitons on the Debye length scale. This paper discusses two problems: the scattering into the atmosphere of plasma sheet electrons (dark purple) by electron cyclotron harmonic waves, and the scattering by whistler waves of energetic electrons in the plasmasphere (cross-hatched).

FIG. 4.   Coupling of radial transport to mirror losses. Shown here is a cut of the magnetosphere in the ecliptic plane looking down from the north. The dashed line symbolizes the time-variable convection electric field imposed upon the magnetosphere by its dissipative interaction with the solar wind. This electric field transports hot plasma in the plasma sheet from the tail towards the earth's night side. Cold ionospheric plasma also streams into these convecting flux tubes. Electron cyclotron harmonic waves scatter the plasma sheet electrons in pitch angle at the strong diffusion rate. When the minimum lifetime is less than the flow time, the flux tubes are emptied of their hot electrons, and there is a sharp inner edge to the electron plasmasheet. Energetic ($>40\,\mathrm{keV}$) electrons inside the plasmasphere gradient drift around the earth. Fluctuations in the convection electric field cause these to diffuse radially inward. They are scattered in pitch angle by whistler waves, which put them in weak diffusion. The combination of radial diffusion and mirror losses accounts for the observed pitch angle and radial profiles of the Van Allen Belt electrons (CORONITI and THORNE, 1973).

Figure 1, a theoretician's cartoon of the magnetosphere, lists some plasma waves discussed in the magnetospheric literature and indicates where they occur. It is an extraordinarily diverse list. The plasma waves range in scale from the strong hydrodynamic turbulence in the magnetosheath, with wavelengths a good fraction of an earth radius, to the Debye length scale cavitons that, it is suggested, form on auroral arc field lines and in the solar wind. Some instabilities, such as those associated with reconnection at the magnetopause and in the geomagnetic tail, may regulate the magneto-sphere's time-dependent behavior. Others are associated with the energetic electron beams responsible for auroral arcs, and may also populate the magnetosphere with ionospheric ions. Still others regulate the heat flux escaping upstream from the earth's bow shock along magnetic field lines.

Virtually all the waves and instabilities in the plasma literature, except those dependent upon machine geometry, can be found in magnetospheric research. Clearly an entire volume could be devoted profitably to magnetospheric plasma waves and their effects. In this one article, we limit ourselves to pitch angle scattering by plasma waves, the oldest microscopic turbulence problem in plasma physics. The first experiments in laboratory mirror devices and the earth's radiation belts took place at about the same time. And remarkably similar explanations for the anomalous losses of their ostensibly trapped particles emerged almost simultaneously (ROSENBLUTH and POST, 1965; KENNEL and PETSCHEK, 1966).

Figure 2 illustrates the most important difference between laboratory mirrors and the geomagnetic mirror. Laboratory devices have a small mirror ratio M, the ratio of maximum to minimum magnetic field on a flux tube, whereas the geomagnetic mirror ratio is huge. M is about 2 for laboratory machines, and $2L^3$ for the geomagnetic dipole field, where L is the geocentric distance in units of earth radii at which the field line crosses the geomagnetic equator. Particles are lost if they penetrate to the maximum magnetic field point, which is at the end plates in the laboratory and in the atmosphere at the earth.

In the absence of fluctuations near their cyclotron frequencies, particles conserve their energy E and their magnetic moment μ, where μ is the perpendicular energy $1/2\ mv_\perp^2$ divided by the magnetic field strength. In terms of the local pitch angle α, $\mu = E \sin^2\alpha/B$. Their orbits can be classified by their pitch angle α_0 at the minimum magnetic field point, an adiabatic constant of the motion equivalent to μ. Trapped particles, which satisfy $\sin \alpha_0 > M^{-1/2}$, mirror before reaching the end plates or atmosphere. Particles with $\sin \alpha_0 < M^{-1/2}$ are lost. $\bar{\alpha}_0 = \sin^{-1} M^{-1/2}$ defines the opening angle of the

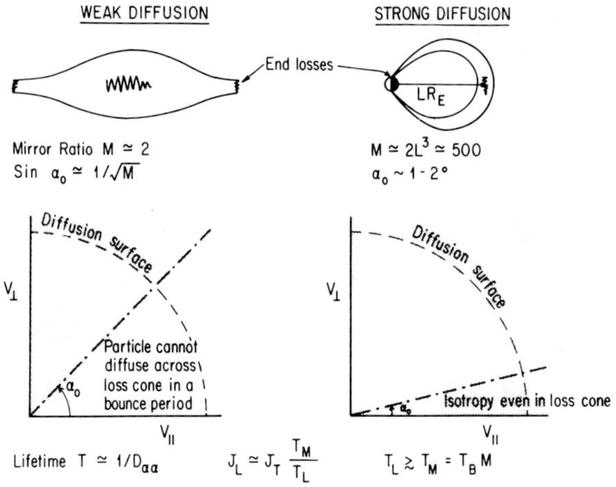

FIG. 2. Weak and strong velocity space diffusion. At the top left and right, respectively, are schematics of a typical laboratory mirror and the geomagnetic mirror. They have very different mirror ratios M. When waves (indicated by wavy lines) are present, this leads to significant differences in how the plasma diffuses in velocity space. The bottom left and right contain schematics of a $(v_\perp, v_\parallel)$ space, where $v_\perp$ and $v_\parallel$ are the particle velocities perpendicular and parallel to the magnetic field respectively. The loss cone, large in the laboratory and small in space, is indicated by the dot-dash line. Typical diffusion surfaces assuming particles scatter only in pitch angle for simplicity, are shown by dashed lines. In weak diffusion, particles random walk to the edge of the loss cone and are lost; velocity space is filled with particles (shading) only outside the loss cone; the distribution function gradient at the edge of the loss cone provides free energy for plasma instabilities. In strong diffusion, particles diffuse in and out of the loss cone before they reach the atmosphere; the pitch angle distribution is nearly isotropic; the particle lifetime approaches a minimum; and the free energy sources are subtle.

loss cone in velocity space.

$\bar{\alpha}_0$ is about $45°$ in laboratory devices, and about $3°$ at $L=6$ in the geomagnetic field. This leads to crucial differences in behavior when plasma turbulence that diffuses particles in pitch angle is present. Suppose the pitch angle diffusion coefficient is $D_{\alpha\alpha}$. Particles in mirror machines would diffuse to the loss cone and be lost on the next bounce. Their "precipitation" lifetime would be $D_{\alpha\alpha}^{-1}$, and the loss cone would be virtually empty. On the other hand, even a modest $D_{\alpha\alpha}$ can diffuse particles into, *and back out of*, the small geomagnetic loss cone in a quarter bounce time T_B, the time it takes them to reach the atmosphere. In this limit, the pitch angle distribution, even within the loss cone, is nearly isotropic. The distribution function becomes independent of position along the field line at all energies where the pitch angle distribution is kept isotropic. And the trapping lifetime T_L approaches a lower limit $T_M \simeq T_B M$. Space physicists have therefore introduced a distinction between the weak ($D_{\alpha\alpha} T_M < 1$) and strong ($D_{\alpha\alpha} T_M > 1$) pitch angle diffusion (KENNEL and PETSCHEK, 1966; KENNEL, 1969).

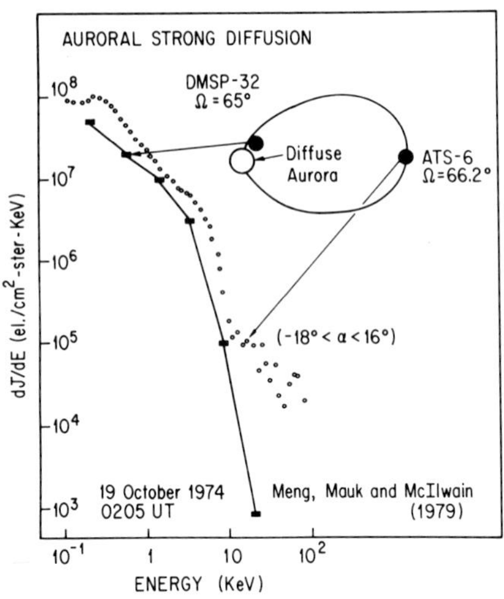

FIG. 3. Auroral strong diffusion. This figure has been adapted from MENG *et al.* (1979). Here are distribution functions measured near the geomagnetic equator by ATS-6 (open dots) and near the atmosphere *on the same* flux tube by DMSP-32 (solid line). The two differ by no more than a factor two between 100 eV and 10 keV energy. Thus the pitch angle distributions are maintained isotropic in this energy range.

Figure 3 demonstrates the occurrence of strong diffusion on auroral field lines (MENG *et al.*, 1979). DMSP-32 is a low-altitude polar orbiting spacecraft; and ATS-6, a geostationary orbiter ($L = 6.6$) in the geomagnetic equatorial plane. Detailed magnetic field models determine when DMSP-32 and ATS-6 are on the same field line. Figure 3 shows simultaneously measured energy distribution functions for precipitating electrons at DMSP-32 and those with $-18° < \alpha_0 < +16°$ at ATS-6. The two are virtually identical between 200 eV and 10 keV. The factor two difference between them may stem in part from the fact that the two detectors were not mutually intercalibrated before launch. MENG *et al.* (1979) could even detect inaccuracies in the magnetic field model. The most nearly identical distributions were encountered about one degree below the calculated geomagnetic latitude of the ATS-6 field line.

Strong pitch angle diffusion is a subtle plasma physical problem. The empty mirror machine loss cone is clearly a free energy source for instability. By contrast, the free energy sources of space plasmas in strong diffusion are not obvious and may be difficult to resolve experimentally. That there can be little structure in the energy as well as the pitch angle distribution only compounds the problem for events such as that of Fig. 3. Yet there the electrons between 200 eV and 10 keV were precipitating near their maximum possible rates. Evidently, the waves responsible for their pitch angle scattering must have been destabilized by small features in their distribution function.

Since magnetospheric particle precipitation is often quasi-steady, a complete understanding demands specification of particle sources as well as sinks. Although *in situ* acceleration of particles to high energies is often invoked, the greatest successes thus far have come from saying that the particles precipitated at one place come from somewhere else. We will therefore be concerned with the coupling of spatial transport to mirror losses.

Figure 4 (see p. 246, 247) sketches the two problems of coupled spatial transport and pitch angle diffusion that we will review. Auroral substorms and magnetic storms inject energetic (>40 keV) electrons into the inner magnetosphere. Following injection, they are diffused across the plasmapause by fluctuations in the geomagnetic convection electric field. Once inside the plasmasphere, they interact with a steady band of ELF whistler mode hiss which puts them in weak pitch angle diffusion. Radial and pitch angle diffusion combine to produce the quasi-steady structure of the decaying electron radiation belts (CORONITI and THORNE, 1973). Our second problem involves thermal (1–10 keV) plasma sheet electrons that convect from the geomagnetic tail towards the night-side under the action of the *steady* component of the convection electric field. When they reach auroral field lines ($L=6$–10) they are lost to the atmosphere by strong pitch angle scattering. Their precipitation creates the diffuse aurora.

2. Outline

Subsection 3.1 summarizes one theory of coupled spatial transport and pitch angle diffusion that agrees quantitatively with observation. Our understanding of radial diffusion and weak pitch angle scattering by ELF whistler hiss of energetic electrons within the plasmasphere sets a standard for our future comprehension of the strong pitch angle diffusion of plasma sheet diffuse auroral electrons by electrostatic cyclotron waves. The morphology of the diffuse aurora described in 3.2 has been understood for ten years in terms of the balance between convection of plasma sheet electrons into the inner magnetosphere and precipitation governed by the minimum lifetime. Our present more subtle difficulties relate to how convection maintains free energy for wave growth in the presence of strong diffusion; which electron cyclotron waves are unstable; whether and how they propagate and, consequently, what their spatial distribution is; how they saturate nonlinearly; and, surprisingly, what role cold electrons of ionospheric origin play in these processes. None of these difficulties plague the theory of the energetic electron radiation belts. Subsections 3.1 and 3.2 do demonstrate by implication one important methodological difference between theoretical laboratory and space plasma physics. In the laboratory, one studies the time evolution to nonlinear saturation of an instability, whereas in space one must address the delicate balances between sources and sinks of particles and waves that maintain the observed steady turbulent state.

Section 4 presents the basic linear theory of electrostatic cyclotron harmonic waves in a uniform magnetized plasma, beginning in 4.2 with the dispersion relation first derived by HARRIS (1959). In 4.2 we introduce separable distributions into this dispersion relation, and in 4.3 we solve in it the zero temperature limit. The stable electrostatic waves propagating perpendicular to the magnetic field in a warm Maxwellian plasma discovered by BERNSTEIN (1958) and discussed in 4.4 provide intuitive insight concerning oblique Harris instabilities. However, since the magnetosphere has both hot and cold electrons, we also discuss Bernstein modes in a two-component electron plasma, in 4.5. We outline general properties of Harris instabilities and of the distribution functions that cause them in 4.6 and 4.7. We then apply this information in 4.8 to derive general instability criteria for an anisotropic bi-Maxwellian plasma. Subsection 4.9 summarizes some insights useful in understanding Harris instabilities in space.

Magnetospheric electron cyclotron waves have been observed for ten years. What until recently has not been measured are the electron distributions responsible for them. In section 5, we classify various instability calculations pertinent to the observed waves by their assumed free energy sources; loss cones, bi-Maxwellian thermal anisotropy, antiloss cones, beams, and anisotropic power law distributions. This list alone indicates that several free energy sources can cause Harris instabilities in the observed frequency bands.

Space physicists must study the dependence of linear instabilities on a wide variety of plasma parameters to understand how they behave in the inhomogeneous, variable magnetosphere. Not only might the free energy source differ from case to case, but given a prevalent free energy source, Harris instabilities still depend crucially on the properties of hot and cold electrons. The only free energy source for which these parametric dependences have been studied is the loss cone. One such parametric study, reviewed in Section 6, emphasizes the control of the spatial growth rates by cold electrons.

We review the observations of magnetospheric electron cyclotron waves in Section 7. They occur not only on diffuse auroral field lines but throughout the magnetosphere beyond the plasmapause. A theoretically meaningful classification of the waves in both earth's and Jupiter's magnetospheres is now possible. First harmonic, multiharmonic, and intense upper hybrid emissions correspond to Harris instabilities in different cold electron density regimes, whereas diffuse low intensity multiharmonic bands may be enhanced thermal fluctuations.

Recent efforts to compare observations with linear theory are discussed in Section 8. In 8.2 we interpret the spatial morphology of terrestrial electron cyclotron waves in terms of what is generally known of the distribution of hot and cold electrons. In Subsection 8.3 we discuss attempts to identify the free energy source in electron distribution functions measured at the same time that intense upper hybrid emissions are observed.

In Section 9 we turn to the frontier of our subject—nonlinear theory. In 9.2 we outline the quasi-linear theory of the evolution of the electron distribution function. We show that only the first two cyclotron resonances with the first two harmonic wave bands are likely to drive the strong diffusion observed in the diffuse aurora, and we argue that nonlinear heating of cold electrons can modify how unstable waves propagate. In 9.3, rather than discuss what little is known about nonlinear saturation, we propose a strategy for future studies.

Section 10 comments on why our understanding of the diffuse aurora, despite ten years of intense effort, is less complete and quantitative than that of the inner electron radiation belts.

3. Two Problems of Coupled Spatial Transport and Pitch Angle Diffusion

3.1 Radiation belt diffusion and loss

A steady convection electric field only distorts energetic electrons' azimuthal guiding center drift orbits around the earth. However, electric field variations with time scales comparable to the drift period results in a net L-shell displacement. If these variations are random and conserve the first two adiabatic invariants, this spatial transport is described by a diffusion equation in L-space, whose diffusion coefficient D_{LL} is proportional to the electric field power spectral density at the azimuthal drift frequency ω_D (FALTHAMMAR, 1963). CORNWALL (1968, 1972) evaluated D_{LL} for substorm-like electric fields that rise rapidly and decay exponentially with time constant T

$$D_{LL} = \frac{c^2 \langle E^2 \rangle}{2B_0^2} L^6 \frac{T}{1 + (\omega_D T)^2} \tag{1}$$

where B_0 is the $L=1$ equatorial dipole field strength, and $\langle E^2 \rangle$ is the r.m.s. electric field. $D_{LL} \sim L^6$ for electrons with $\omega_D T \ll 1$ and $D_{LL} \sim L^7$ for (ultrarelativistic) electrons with $\omega_D T \gg 1$.

$\langle E^2 \rangle^{1/2} \simeq 0.1$–$0.3\,\mathrm{mV/m}$ and $T = 1/2$ hour fits the observed radial diffusion of inner zone relativistic electrons (TOMASSIAN *et al.*, 1972). These electric field magnitudes are consistent with observation (MOZER, 1971; CARPENTER *et al.*, 1972) and $T = 1/2$ hour is a natural time scale set by ionospheric line tying for the response of convection to changing solar wind boundary conditions (CORONITI and KENNEL, 1973; SATO and HOLZER, 1973).

Now we turn to the waves responsible for the pitch angle diffusion of energetic electrons. The cold plasma dispersion relation for right hand (RH) circularly polarized whistler waves propagating parallel to a uniform magnetic field is

$$\omega^2/k^2 = C_a^2 \omega/\Omega(1 - \omega/\Omega) \tag{2}$$

where ω is the wave frequency, k the wave number, Ω the electron cyclotron frequency eB/mc and C_a, the Alfvén speed based upon the *electron* mass m:

$$C_a^2 = B^2/4\pi Nm \tag{3}$$

where N is the electron number density.

A resonant particle sees a constant whistler electric field when its parallel velocity Doppler shifts the frequency to its cyclotron frequency. The condition for resonance, $v_{\parallel} = v_R = (\omega - \Omega)/k_{\parallel}$, combined with the dispersion relation, defines the resonant energy E_R

$$E_R = \tfrac{1}{2}mv_R^2 = B^2/8\pi N \frac{\Omega}{\omega}(1 - \omega/\Omega)^2 \tag{4}$$

which is the minimum *total* energy for resonance, since resonant electrons can have arbitrary perpendicular energy.

The parallel whistler mode growth rate γ is (KENNEL and PETSCHEK, 1966)

$$\gamma = \pi\Omega\left(1 - \frac{\omega}{\Omega}\right)^3 \eta(v_R)\left(\bar{A}(v_R) - \frac{1}{\dfrac{\Omega}{\omega} - 1}\right) \tag{5}$$

where $\bar{A}$, the anisotropy,

$$\bar{A}(v_R) = \left.\frac{\displaystyle\int_0^\infty \frac{v_\perp^2 dv_\perp}{v_\parallel}\left(v_\parallel \frac{\partial F_H}{\partial v_\perp} - v_\perp \frac{\partial F_H}{\partial v_\parallel}\right)}{2\displaystyle\int_0^\infty v_\perp dv_\perp F_H}\right|_{v_\parallel = v_R} \tag{6}$$

reduces to $\bar{A} = (T_\perp - T_\parallel)/T_\parallel$, when the hot electron distribution function F_H is an anisotropic bi-Maxwellian. $\bar{A}$ must be positive for instability. $\eta(v_R)$, which measures the number of electrons near cyclotron resonance,

$$\eta(v_R) = 2\pi |v_R| \left.\int_0^\infty v_\perp dv_\perp F_H\right|_{v_\parallel = v_R} \tag{7}$$

should increase with increasing ω/Ω, since E_R decreases. γ consequently increases with increasing ω/Ω, maximizes, and drops to zero at $\omega/\Omega = \bar{A}/(1 + \bar{A})$. Thus, unstable whistlers resonate with electrons whose parallel energy exceeds $B^2/8\pi N \bar{A}(1 + \bar{A})$. These are ordinarily in the tail of the energy distribution, unless the electron β is order unity. Therefore, hot electrons determine the growth rate without affecting the whistler phase and group velocities, which depend upon the cold electron density.

E_R is lowest in the equatorial plane, because it depends strongly on B. Since the observed electron fluxes decrease with increasing energy, the most resonant electrons, and therefore the largest local whistler growth rate, will be at the equator. The equatorial cold electron density therefore determines which electrons resonate with whistlers. Electrons with energies of a few tens of keV resonate with whistlers only just inside the plasmapause and possibly in the diffuse auroral region beyond $L = 6$. Between the plasmapause and $L = 6$, and well within the plasmasphere, $B^2/8\pi N$ is too large for resonance.

If $\gamma l/V_G > 10$, where l is an amplification scale length and V_G is the group velocity, the whistler amplitude will e-fold 10 times as the wave passes through the geomagnetic equator, a rough criterion for the onset of significant quasi-linear diffusion. Because the whistler wave magnetic field exceeds its electric field, resonant electrons diffuse in pitch angle with little change in energy. KENNEL and ENGELMANN (1966) show that quasilinear energy diffusion is much slower than pitch angle diffusion when $\omega/\Omega \ll 1$. Since pitch angle diffusion leads to mirror losses, whistler instability places an upper limit on the energetic electron fluxes that can be stably trapped. KENNEL and PETSCHEK (1966) showed that the observed fluxes of electrons with a few tens of keV energy rarely exceed the limit computed using the above dimensional arguments.

The pitch angle diffusion and loss of the more energetic electrons within the plasmasphere does not conform to the above picture. Except near the

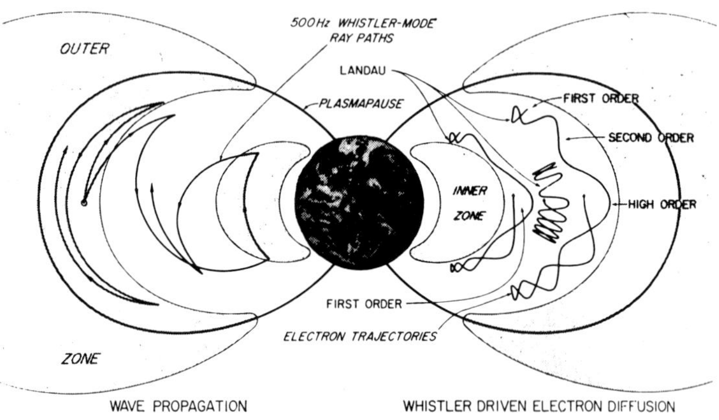

FIG. 5. Plasmaspheric whistler physics. The left hand side illustrates typical ray trajectories for whistler mode waves that are unstable just inside the plasmapause— the boundary between regions of high and low cold electron densities. These cold electrons expand hydrodynamically from the ionosphere until they reach pressure equilibrium inside the plasmasphere. Outside the plasmapause, convection transports ionospheric electrons away before they can reach equilibrium, and the electron density is low. This figure makes plausible the statement that a local unstable region can fill the entire plasmasphere with whistler mode noise (LYONS et al., 1972). The right hand side illustrates how energetic electrons can resonate with plasmaspheric whistler waves. Shown are various trajectories for trapped electrons. All electrons have interactions near their mirror points. Energetic electrons have $n=1$ cyclotron interactions at high latitudes and higher order resonances near the magnetic equator (trajectory farthest to the right). $B^2/8\pi N$ increases with decreasing distance from the earth, and one has a first order ($n=1$) interaction at the equator at best (inmost trajectory). Farther in (inner zone), typical electrons do not resonate at all with the plasmaspheric whistler waves. Here they are scattered into the ionosphere by Coulomb collisions.

plasmapause, $B^2/8\pi N$ and the corresponding resonant energies are so large that the observed electron fluxes are below the stably trapped limit. Only the more detailed analysis outlined in Fig. 5 can explain the losses of plasmaspheric electrons. The left hand panel shows whistler ray paths. A 500 Hz wave, initially amplified in the equatorial plane near the plasmapause, propagates away more or less along field lines and reflects when its frequency equals the local lower hybrid frequency ω_{LHR} (THORNE and KENNEL, 1967). If its propagation vector is canted towards the plasmapause, it will move outward; otherwise it will bounce towards the earth. Since outward moving waves reflect at the plasmapause, all waves eventually move inward and reach a point where their frequency is less than the equatorial lower hybrid frequency. While when $\omega > \omega_{\text{LHR}}$ their group velocities are confined to within $20°$ of the magnetic field direction, when $\omega < \omega_{\text{LHR}}$, they may make any angle to the magnetic field. Similarly, k can make a large angle to B when $\omega < \omega_{\text{LHR}}$. Since whistlers are confined by reflections at the plasmapause and ionosphere, a locally unstable region can fill the entire plasmapause with obliquely propagating whistlers. The whistler mode ELF hiss observed throughout the plasmasphere fits this picture, because it has nearly constant amplitude and frequency. Its few hundred Hz frequency is consistent with unstable amplification by 10–100 keV electrons near the plasmapause.

In contrast to parallel whistlers, oblique whistlers have many wave particle resonances defined by $k_{\parallel}v_{\parallel} = \omega - n\Omega$, where n is any integer. Higher cyclotron resonances, $|n| > 1$, couple to electrons whose parallel energies are order n^2 larger than for $n = 1$ resonance. These interactions scatter electrons in pitch angle. The $n = 0$ Landau resonant electrons have low parallel energies, of order $(\omega/\Omega)B^2/8\pi N$. Landau interactions scatter particles in $v_{\parallel}$ (KENNEL and ENGELMANN, 1966). The right hand side of Fig. 5, which sketches the processes that contribute to the pitch angle diffusion coefficient, shows three typical trapped electron orbits. Because $B^2/8\pi N$ increases away from the equator, high energy electrons have an $n = 1$ resonance at high geomagnetic latitudes, $n = 2$ at intermediate, and high order resonances at the equator. Moreover, the electrons all pass through $n = 0$ Landau resonance near their mirror points, where their pitch angles are near $90°$, so that this interaction amounts to pitch angle scattering as well.

Detailed computations of the pitch angle diffusion coefficient bounce-average the contributions of all wave-particle resonances near and far from the equator. LYONS *et al.* (1971, 1972) modelled the whistler hiss by a Gaussian frequency distribution centered at 600 Hz with a 300 Hz *e*-folding

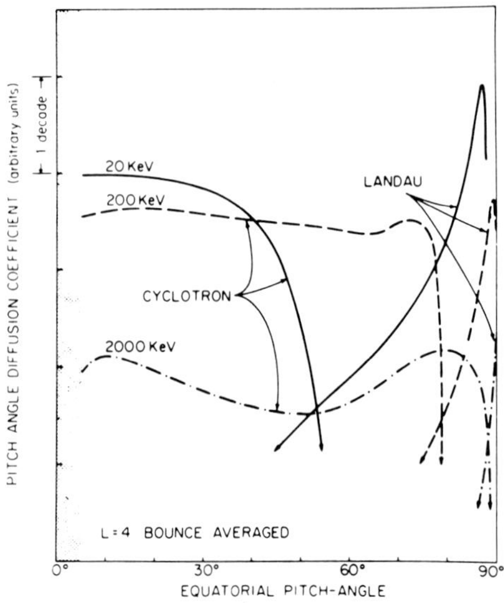

FIG. 6. Bounce-averaged whistler pitch angle diffusion coefficients. These pitch angle diffusion coefficients were computed by LYONS *et al.* (1973), by bounce-averaging over all the pertinent interactions illustrated in the left hand side of Fig. 5. The whistler spectrum was assumed independent of space and time, in rough accord with observation. Landau interactions create the peak near 90° pitch angles, and various cyclotron interactions, the rest. Note the hole in the summed diffusion coefficient between the Landau and cyclotron regions at low energies. The pitch angle distribution should have a strong gradient where the diffusion coefficient minimizes. The hole disappears at high energies.

width and a realatively isotropic distribution of wave energy in K-space. This spectrum was independent of geomagnetic latitude and L-shell. Figure 6 shows their bounce averaged cyclotron and Landau pitch angle diffusion coefficients as a function of equatorial pitch angle at $L=4$ for 20, 200, and 2,000 keV electrons. The overall diffusion coefficient sums these contributions. Note the "hole" in the summed diffusion coefficients between the cyclotron and Landau regions for lower energy electrons.

The pitch angle diffusion coefficients can be incorporated into the quasilinear diffusion equation to solve for the electron pitch angle distribution and the precipitation lifetime. A comparison between theoretical pitch angle distributions and those measured on OGO-5 by H. West is shown in Fig. 7. The low energy flux enhancements observed near 90° result from the slow diffusion predicted in the region between strong cyclotron and

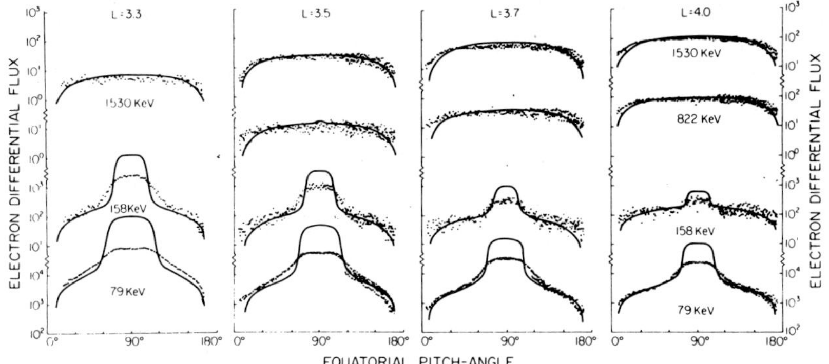

FIG. 7. Theoretical and experimental pitch angle distributions. The pitch angle diffusion equation can be solved for the steady state pitch angle distribution, using the diffusion coefficients shown in Fig. 6. Theory (solid line) is compared with experiment for several electron energies at several L-shells (LYONS et al., 1972). The agreement is excellent. In particular, there is evidence for the strong gradients in the pitch angle distribution corresponding to the "hole" in the diffusion coefficient.

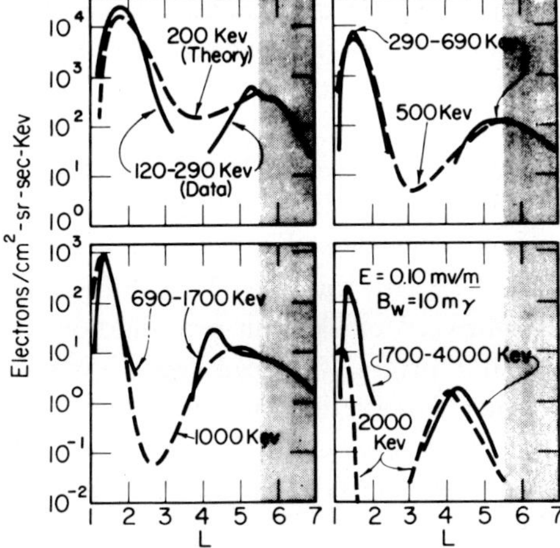

FIG. 8. Equilibrium energetic electron radial profiles within the plasmasphere. Given the precipitation lifetimes due to whistler mode scattering, calculated using the results of Figs. 6 and 7, at large L, and classical Coulomb scattering at small L, the radial diffusion equation may be solved for the radial profile of energetic electron fluxes, given a steady injection source at and beyond the plasmapause (shaded region). Experiment and theory are compared in this figure (LYONS and THORNE, 1973). The excellent agreement rivals any found elsewhere in plasma physics. It is particularly significant that no further physics is needed to explain the sharp flux minima near $L=3$. Erstwhile, it had been possible to wonder whether a different instability occurred there.

Landau interactions. At higher energies, no 90° enhancements are expected and none is observed. The calculated lifetimes are in rough accord with observation.

The precipitation lifetime can be inserted into the radial diffusion equation to solve for the radial profile of plasmaspheric electrons as a function of energy. Theory and observation are compared in Fig. 8 (LYONS and THORNE, 1973). Despite the successive simplifications made in computing the radial profile, the agreement is excellent. In particular, the high energy electron "slot" near $L=3$ emerges naturally from the way whistler and Coulomb scattering combine with radial diffusion. No further plasma physics is needed to understand it.

3.2 The diffuse aurora

3.2.1 Morphology

The human eye does not readily perceive the steady glow of the diffuse aurora. Only sensitive photometric devices and charged particle detectors carried by spacecraft reveal its true extent and importance. The thin auroral arcs that so beguile observers on the ground do not radiate as much total energy as the diffuse aurora, despite their great local intensity. The diffuse aurora is caused primarily by the precipitation of 1–10 keV electrons (LUI and ANGER, 1973; LUI et al., 1973; EATHER, et al., 1976). Zones of soft electron and proton precipitation generally coincide (HULTQUIST, 1975a) and define an auroral oval similar to that deduced from discrete arc observations. While protons typically make less than 30% of the diffuse auroral light (MENDE and EATHER, 1976; HULTQUIST, 1975b), diffuse proton precipitation extends slightly equatorward of electron precipitation in the premidnight sector; only there do proton induced light emissions predominate (FUKUNISHI, 1975). Such observations motivate two theoretical questions. First, where do auroral particles come from and why do they precipitate where they do? Second, what instabilities cause such rapid precipitation of electrons and protons?

3.2.2 Interaction between convection and precipitation

The spatial morphology of diffuse auroral particles suggests they come from the plasma sheet (EATHER and MENDE, 1972a, b; EATHER et al., 1976). In particular, the sharp inner edge of the electron plasma sheet (VASYLIUNAS, 1968) maps magnetically to the equatorward edge of the diffuse aurora. Auroral electrons and protons are usually isotropic in pitch angle when they are observed at high latitudes, and therefore, the fluxes and energy distributions of precipitating electrons are virtually identical to those measured deep in space near the inner edge of the electron plasma sheet (MENG *et al.*, 1979). At present, there is little reason to doubt that the diffuse aurora is due to the direct loss of plasma sheet ions and electrons.

The origin and location of the diffuse aurora was first discussed theoretically by PETSCHEK and KENNEL (1966) and subsequently by KENNEL (1969) and VASYLIUNAS (1969) independently. Their model involves two fundamental statements. First, earthward convection replenishes the electrons and protons lost to the auroral ionosphere. Second, because they are isotropic in pitch angle, their precipitation lifetimes approach the strong diffusion limit T_M. Combining these two statements leads to the following consequences. A flux tube starting in the tail initially loses only a few of its particles as it convects towards the earth, because the minimum lifetimes, at great distances, are much longer than a characteristic flow time. Since the convection is virtually loss free, the electron and proton fluxes increase as the flux tube volume decreases. Because the minimum lifetimes depend on L^4, the flux tube eventually reaches a point where the flow time and electron minimum lifetime are comparable. Electrons are then rapidly precipitated. The difference between electron and proton precipitation fluxes must be compensated by a return flux of ionospheric electrons. Since a hot electron is exchanged for a colder one, the mean energy of the remaining trapped electrons decreases. Thus, this model predicts a sharp decrease of electron temperature at the "inner edge of the electron plasma sheet" whose computed location and scale length agree with observation.

Because protons have a longer minimum lifetime, they should penetrate further than electrons to form their inner edge deeper within the magnetosphere. However, the observed electron and proton precipitation zones more or less coincide. By including in the above model the corotation electric field, coupling via field aligned currents to the conducting ionosphere plasma, and the effects of finite proton pressure, JAGGI and WOLF (1973) and others have shown that protons originally convecting radially inwards can be deflected azimuthally around the earth as soon as they penetrate to L-shells where the high ionospheric conductivity created by electron precipitation has diminished. Thus, on the nightside, one would not expect plasma sheet protons much within the electron inner edge, and the zones of electron and proton precipitation should be similar.

Our first question—about the origin and location of the nightside diffuse aurora—has an answer. Soft auroral electrons and protons are precipitated from flux tubes which convect from the plasma sheet towards the earth's nightside. Current models of magnetospheric convection are much more sophisticated than the above discussion implies. Nonetheless, the simple ideas outlined above do set the stage for our discussion of the plasma physics responsible for soft electron and proton precipitation, our primary interest.

3.2.3   Charge balance and the atmospheric cold electron source

Diffuse auroral precipitation fluxes show no evidence that they have been acted upon by a parallel electric field. Yet, because the diffuse electron fluxes exceed those of the ions, a parallel electric field would develop unless an electron return current flows from the atmosphere to space. This return current can be provided by the ambient cold ($\simeq 1\,\mathrm{eV}$) electrons in the topside ionosphere, or by intermediate energy ($\simeq 100\,\mathrm{eV}$) secondary electrons

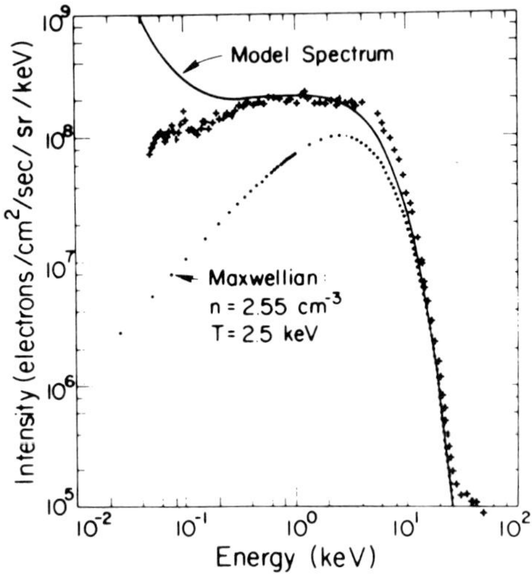

FIG. 9.   Production and injection of secondary electrons by the diffuse aurora. Compared here are observations (DeForest and McIlwain, 1971) of the electron distribution function near the magnetic equator (open dots) and a model spectrum (solid line) computed (Evans and Moore, 1979) according to the following procedure. A Maxwellian spectrum of isotropic plasma sheet electrons (solid dots) was assumed to precipitate upon the atmosphere. These produced secondary electrons whose energy spectrum can be calculated (Evans, 1974) given the primary spectrum. Secondaries replaced primaries on a one-for-one basis. The secondaries are assumed isotropic in pitch angle at the equator in steady state. Knowledge of the cold electron distribution is crucial to understanding the electron cyclotron harmonic instabilities thought responsible for the primary scattering. It has been speculated that these waves can also scatter the cold electrons, perhaps sufficiently to isotropize them as assumed in this model.

created by the interaction of the precipitating electrons with the atmosphere, or both.

Evans and Moore (1979) argue that secondary and backscattered electrons dominantly neutralize the diffuse auroral precipitation. They successfully modelled low altitude observations of diffuse auroral electron distributions using two components: a primary precipitating flux that is isotropic in the downgoing velocity space hemisphere and exhibits the atmospheric loss cone in the upgoing, and completely isotropic distribution of secondaries, *computed using the measured primary flux* by techniques developed by Evans (1974). They also suggested that the upgoing secondary fluxes could be scattered out of the loss cone to populate the magnetosphere with intermediate energy electrons. Figure 9 shows their model of an electron distribution function measured at the geomagnetic equator by DeForest and McIlwain (1971), in which the primary precipitation flux is replaced on a one-for-one basis by secondary electrons calculated self-consistently. The model agreed with observation above 100 eV energy. They conclude that "in terms of electron number densities, those electrons which owe their origin to the earth's atmosphere represent a significant, often dominant portion of the total reservoir of electrons."

In sum, electron precipitation creates its own lower energy electron environment in which the plasma waves responsible for it must grow. The diffuse auroral electron distribution consists of at least two components: a few keV plasma sheet and a few hundred eV secondary component with comparable partial densities. How atmospheric electrons are scattered onto trapped orbits remains an unsolved problem.

3.2.4 *Plasma physical considerations*

The large decrease in electron mean energy observed at the inner edge of the plasma sheet argues that nearly all plasma sheet electrons are eventually precipitated. This suggests that the instability responsible for diffuse auroral electron precipitation occurs near the geomagnetic equator. Waves near the equator can interact with all electrons on the flux tube; conversely, waves at high latitudes cannot empty a flux tube of all its hot electrons. (A similar case can be made for protons, but it is unlikely that all the plasma sheet protons are lost.) Moreover, since electron precipitation is neutralized by ionospheric electrons, interaction between plasma sheet convection and precipitation determines not only the hot plasma, but also the cold electron environment in which the instabilities responsible for the auroral precipitation operate. This environment should permit two instabilities, one for electrons and one for protons, to grow. Finally, since the diffuse aurora is present during quiet as well as disturbed geomagnetic conditions, these instabilities must not require unusual circumstances for their growth. A source of free energy must nearly always be present to drive them.

Can convection generate free energy for plasma instabilities? The possible free energy sources include: electrical currents, fast ion beams, fast electron beams, density and temperature gradients, pitch angle anisotropy and/or a loss cone property. Given the great activity observed on auroral field lines, it is not surprising that all are present at times. Moreover, all are plausible theoretical consequences of the interaction between convection, precipitation, and the ionosphere.

KINDEL and KENNEL (1971) showed that a return flux of cold ionospheric electrons could be unstable to electrostatic ion cyclotron waves at altitudes above 1,000 km. When a net electric current must also flow, the case for instability is even stronger. It is currently thought that ion cyclotron waves may be involved in the formation of regions of strong parallel electric fields, which accelerate the ion and electron beams observed to be associated with field aligned currents on auroral lines of force. These field aligned currents are due in part to the spatially inhomogeneous ionospheric conductivity created by electron precipitation.

Field aligned currents and/or beams probably are not the diffuse auroral free energy source. Field aligned currents are spatially structured, not diffuse. Moreover, our present understanding indicates they will create instabilities far from the geomagnetic equator. Beams associated with strong field aligned currents would also be spatially structured. Thus far, they have been observed mostly near the ionosphere. Finally, beam generated waves should scatter particles out of the beam—from the loss cone onto trapped orbits. Although they may populate auroral flux tubes with accelerated ionospheric plasma, can we also ask them to precipitate plasma back into the ionosphere?

CORONITI and KENNEL (1970) and CHANCE et al. (1973) argued that the gradient in electron temperature that is the inner edge of the electron plasma sheet could be unstable to drift Alfvén waves. However, these long wavelength waves conserve even the proton first adiabatic invariant, while the isotropy of the observed precipitation fluxes indicates that it is violated. Short perpendicular wavelength drift cyclotron and/or hybrid modes are a variant on the basic cyclotron harmonic waves we will discuss shortly; these can violate the first invariant, but require extremely sharp spatial gradients.

This leaves us with the last two, related, free energy sources—thermal anisotropy and a loss cone property. The most familiar example of a loss cone distribution—whence it derives its name—occurs in laboratory mirror machines. There, end losses ensure that the particle distribution function $F(v_\perp, v_\parallel)$ contains virtually no particles with small perpendicular velocities $v_\perp$. In other words, $F(v_\perp = 0) = 0$. At the edge of the loss cone, $\partial F / \partial v_\perp > 0$. Waves with perpendicular phase velocities $\omega / k_\perp$ lying in this positive gradient region can be unstable.

Figure 10 shows the calculated changes in an initially isotropic monoenergetic shell of particles as they convect, conserving their adiabatic invariants, from $L = 10$ to $L = 4$ under the action of a constant electric field (ASHOUR-ABDALLA and COWLEY, 1974). The particle energy increases, the pitch angle distribution becomes anisotropic, with the mean perpendicular energy exceeding the parallel, and a cusp develops near $v_\perp = 0$ where $\partial F / \partial v_\perp$ is positive. Thus, convection can generate both thermal anisotropy and a loss cone property, even in the absence of losses.

Figure 10, while illuminating, oversimplifies the physics. It is important

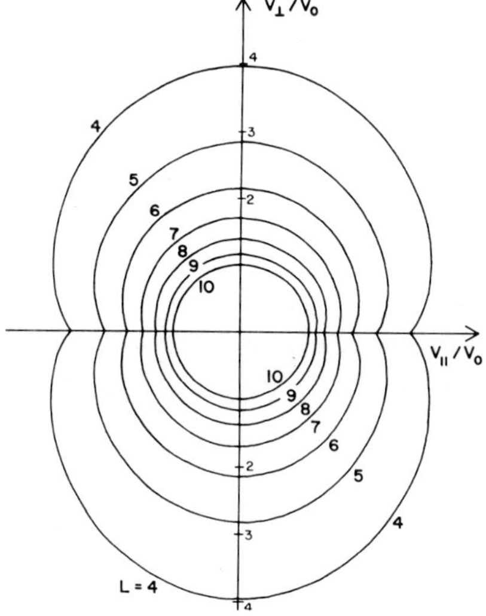

FIG. 10. Production of free energy by steady convection. Shown here is the evolution of an initially monoenergetic isotropic distribution of particles as it convects from $L = 10$ to $L = 4$ in a dipolar magnetic field and steady electric field. The electrons develop a thermal anisotropy, and a cusp at small $v_\perp$ which might be losscone unstable. This makes it plausible that convection itself can develop the free energy for the instabilities responsible for particle precipitation. In the real magnetosphere, the convection is unsteady; the small distribution function structures produced by unsteady convection could also be a free energy source.

to specify the initial energy as well as the initial pitch angle distribution. The reaction of plasma turbulence back upon the distribution function was neglected. For example, the loss cone feature should be weakened by strong pitch angle diffusion. Finally, the true convection electric field is not constant, but varies dramatically with the level of substorm activity, and is highly turbulent (KELLEY and KINTNER, 1978). DEFOREST and MCILWAIN (1971) observed that time variable convection leads to fine structure in the particle velocity distributions. This could also be a free energy source.

The above discussion is summarized in Fig. 11. A series of heuristic arguments has led to thermal anisotropy and a loss cone property as plausible free energy sources for the instabilities causing diffuse auroral precipitation. Both are consequences of magnetospheric convection, both can be present in the equatorial particle velocity distributions of electrons and ions. Now we turn to waves which respond to these free energy sources.

The wave electric field experiment on OGO-5 detected several different types of electrostatic plasma waves with frequencies above the local electron cyclotron frequency. The most common have come to be known as "odd half harmonic" emissions. The facts that, unlike whistlers, odd half harmonic

SOURCES OF FREE ENERGY ON AURORAL FIELD LINES

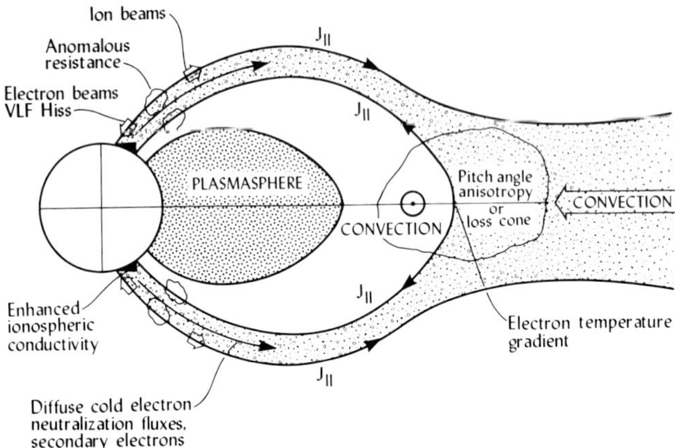

FIG. 11. Production of free energy on auroral field lines. This figure is a schematic cut through the midnight meridian plane whose purpose is to illustrate our contention that the interaction of convection, precipitation, and the ionosphere, creates most of the free energy sources known to plasma physics—field aligned currents, particle beams, spatial gradients, thermal anisotropy, and a loss cone feature. It is generally thought that only the last two are pertinent to the diffuse aurora.

waves resonate with the main part of the electron distribution, and that they were almost always observed near the equator on auroral lines of force with an intensity sufficient for strong diffusion suggested that they cause the diffuse electron aurora (KENNEL *et al.*, 1970). FREDRICKS (1971) showed they could be loss cone unstable.

Theory preceded observation of the waves which may explain diffuse auroral proton precipitation. CORONITI *et al.* (1972) suggested that a quasi-electrostatic mode could be excited at frequencies near but below the ion plasma frequency by an ion loss cone distribution. ASHOUR-ADALLA and THORNE (1977) recalculated its spatial and temporal growth rates, using an ion loss cone distribution modeled upon observations and including cold electrons and ions. They found that the spatial growth rate peaks for low ion harmonics. Just as their results appeared in print, so also did the Hawkeye observations of broadband electrostatic noise (GURNETT and FRANK, 1977). According to these observations, its spectrum peaks at two to three times the ion cyclotron frequency; it is observed on diffuse auroral field lines and its amplitude puts protons on strong diffusion. The fact that Hawkeye's orbit only permitted observations of broadband electrostatic noise at intermediate latitudes ($\lambda=45°$) makes theoretical arguments concerning its free energy source ambiguous. We could extend ionospheric free energy sources (currents, beams) outwards, or extend equatorial sources (loss cone) inwards, to intermediate latitudes. If broadband electrostatic noise causes diffuse electron auroral proton precipitation, it should extend from midlatitudes through the equatorial plane. This remains to be demonstrated.

3.2.5 Summary

We have highlighted a self-consistent conceptual model of the interaction between convecting magnetospheric plasma and the nightside auroral atmosphere (see Fig. 12), which starts with the plasma sheet electron and ion velocity distributions and a given convection electric field and then relates:

1)   Particle precipitation fluxes and charge neutralization return fluxes.

2)   Inhomogeneous ionospheric conductivity profiles.

3)   Interactions of ionospheric conductivity and plasma pressure with the convective flow.

4)   Locating field aligned currents (exclusive of those in smaller scale structures such as arcs).

5)   Implicitly—through field aligned currents—current instabilities, parallel electric fields, and the generation and instability of fast ion and electron streams.

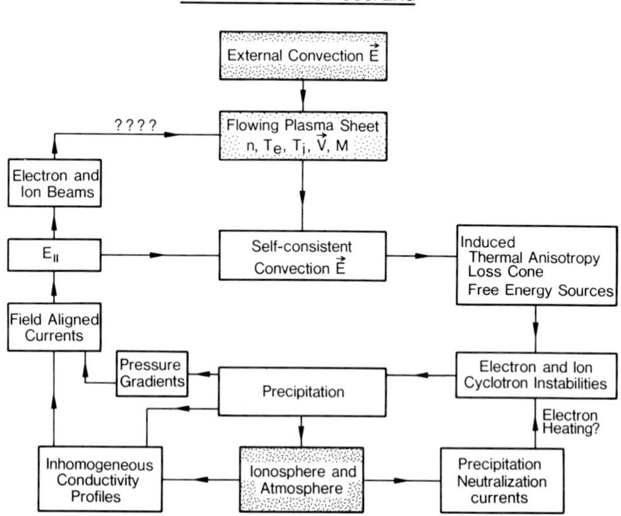

FIG. 12. Flow-precipitation coupling. Illustrated here are some of the physical elements that will eventually be linked together in a complete model of convection and the diffuse aurora. These couplings produce the free energy sources illustrated in Fig. 10. The rest of this paper will concentrate on the three boxes on the far right.

6) Providing equatorial sources of free energy for the instabilities responsible for particle precipitation.

7) Identifying those instabilities which resonate with the main parts of the ion and electron velocity distributions.

8) Estimating self-consistently the hot and cold plasma environment in which these instabilities grow and saturate.

Considerable theoretical and experimental effort has been devoted to every box in Fig. 12 over the past decade, but a complete linkage has yet to be made. Henceforth, we will concentrate upon the right hand side of Fig. 12, where the plasma turbulence fits in, and especially upon electron cyclotron harmonic instabilities.

4. General Properties of Electrostatic Waves in a Magnetized Plasma

4.1 Linear dispersion relation

When the ratio β of plasma to magnetic field energy density is small, there exists a class of plasma waves, whose wavelengths are smaller than the collisionless skin depth, that are longitudinal to a good approximation. Since these waves have little or no magnetic field, the speeds associated with electromagnetic waves, such as the Alfvén speed, or the speed of light, cannot appear in their dispersion relation. Only the thermal velocities of the plasma components remain as characteristic speeds. Therefore, the phase and group speeds of electrostatic waves scale as the thermal speeds. This fact makes electrostatic waves central to the behavior of all low-β plasmas, and understanding them essential for understanding such major issues as energy and particle confinement, since they resonate with the main part of the particle velocity distributions. This statement, no less true for space plasma than for fusion plasmas, is particularly pertinent to the diffuse aurora, in which thermal plasma sheet electrons are precipitated.

The dispersion relation for electrostatic waves in a uniform plasma and magnetic field was first derived by HARRIS (1959).

$$0 = \varepsilon\,(\omega,\,k_\perp,\,k_\parallel) \tag{8a}$$

$$= 1 + \sum_s \frac{\omega_{ps}^2}{k^2} \sum_n \int \frac{G_{ns}(v_\parallel)\mathrm{d}v_\parallel}{\omega - k_\parallel v_\parallel - n\Omega_s} \tag{8b}$$

$$G_{ns}(v_\parallel) = 2\pi \int_0^\infty v_\perp \mathrm{d}v_\perp J_n^2\left(\frac{k_\perp v_\perp}{\Omega}\right)\left(\frac{n\Omega_s}{v_\perp}\frac{\partial f_s}{\partial v_\perp} + k_\parallel \frac{\partial f_s}{\partial v_\parallel}\right) \tag{8c}$$

where $\varepsilon = \varepsilon_R + i\varepsilon_I$ is the plasma dielectric function. The wave frequency $\omega = \omega_R + i\gamma$ is generally complex, and $\gamma > 0$ signifies instability. $k_\perp$ and $k_\parallel$, $v_\perp$ and $v_\parallel$, are the wave vector and particle velocity components perpendicular and parallel to the magnetic field respectively. $\sum_n$ denotes a summation over cyclotron harmonic resonances, and J_n is a Bessel function of integer order n. $\sum_s$ denotes a summation over particle species, and $f_s(v_\perp, v_\parallel)$ is the distribution function of species s. Later, we will use the reduced distribution function $f_s = n_s F_s$, where n_s will denote the number density of species s. The species subscript s can denote different ions, or different components of the same species, such as hot and cold electrons.

Charge neutrality requires that $\sum_s Z_s n_s = 0$, where Z_s is the signed charge of species s in units of the electronic charge e. ω_{ps} and Ω_s denote the plasma and cyclotron frequencies respectively:

$$\omega_{ps} = \left(\frac{4\pi n_s Z_s e^2}{M_s}\right)^{1/2} \tag{8c}$$

$$\Omega_s = \frac{Z_s e B}{M_s c} \tag{8d}$$

where M_s is the mass of the species s, B is the magnetic field strength, and c is the speed of light. CGS units will be used throughout.

4.2 Separable distribution functions

Though neither nature nor man can make purely separable distribution functions, theoreticians find it convenient to use them. When $F_s(v_\perp, v_\parallel) = F_{\perp s}(v_\perp) F_{\parallel s}(v_\parallel)$, the Harris dispersion relation (8) reduces to the illuminating form (GUEST and DORY, 1965)

$$0 = 1 - \sum_s \frac{\omega_{ps}^2}{k^2 A_{\parallel s}^2} \sum_n F_{ns} \tag{9a}$$

$$F_{ns} = C_n(\lambda) Y'(w_n) - \frac{2n\Omega_s}{k_\parallel A_\parallel} \left(\frac{A_\parallel}{A_\perp}\right)^2 D_n(\lambda) Y(W_n) \tag{9b}$$

where

$$C_n = 2\pi \int_0^\infty v_\perp dv_\perp J_n^2\left(\sqrt{2\lambda}\,\frac{v_\perp}{A_\perp}\right) F_\perp(v_\perp) \tag{9c}$$

$$D_n = -\pi A_\perp^2 \int_0^\infty v_\perp dv_\perp J_n^2\left(\sqrt{2\lambda}\,\frac{v_\perp}{A_\perp}\right) \frac{1}{v_\perp} \frac{\partial F_\perp}{\partial v_\perp} \tag{9d}$$

$$\lambda = \frac{k_\perp^2 A_\perp^2}{2\Omega^2} = k_\perp^2 \rho_\perp^2 \tag{9e}$$

$$Y_n(w_n) = \int_{-\infty}^{+\infty} \frac{F_\parallel(u)du}{u - w_n} \tag{9f}$$

$$Y'(w_n) = \int_{-\infty}^{+\infty} \frac{F_\parallel(u)du}{(u - w_n)^2} = \int_{-\infty}^{+\infty} \frac{\partial F_\parallel/\partial u}{u - w_n} \tag{9g}$$

$$w_n = \frac{\omega - n\Omega}{k_\parallel A_\parallel} \equiv \frac{v_n}{A_\parallel} \tag{9h}$$

$A_\perp$ and $A_\parallel$ characterize the widths of the velocity distributions perpendicular and parallel to the magnetic field, and $\rho_\perp$ is the Larmor radius based upon $A_\perp$.

Though plasmas are usually not in thermal equilibrium, theoreticians persist in using Maxwellian distributions and consider it an honorable tradition to model nonthermal distribution functions using sums of

Maxwellians. When $F_\perp = (\sqrt{\pi}\, A_\perp)^{-2} e^{-v_\perp^2 A_\perp^2}$,

$$C_n = D_n \equiv \Gamma_n(\lambda) = e^{-\lambda} I_n(\lambda) \tag{10}$$

where I_n is a modified Bessel function of the first kind. Similarly, when $F_\parallel = (\sqrt{\pi}\, A_\parallel)^{-1} e^{-v_\parallel^2/A_\parallel^2}$, $Y(w_n)$ becomes $Z(w_n)$, the plasma dispersion function (FRIED and CONTE, 1962), which, like the Bessel functions, is not expressible in closed form, but is tabulated.

4.3   Electrostatic waves in a zero temperature plasma

The electrostatic modes in a zero temperature plasma may be found by taking the limits, $A_\parallel \to 0$ in (9f) and (9g), $Y_n \to -1/w_n$, $Y_n' \to 1/w_n^2$, and the limit $A_\perp \to 0$ in (10), whereupon $C_0 = D_0 = 1$, and $C_n \simeq 1/n!(\lambda/2)^n$. Then the dispersion relation (9) becomes

$$\varepsilon = 0 = 1 - \sum_s \left(\frac{\omega_{ps}^2}{\omega^2}\cos^2\theta + \frac{\omega_{ps}^2 \sin^2\theta}{\omega^2 - \Omega_s^2} \right) \tag{11}$$

where $\cos^2\theta = k_\parallel^2/k^2$, $\sin^2\theta = k_\perp^2/k^2$. For one species of electrons, and frequencies sufficiently high that ion motions can be neglected, (11) has the solution

$$2\omega^2 = \omega_{UH}^2 \pm \sqrt{(\omega_{UH}^2)^2 - 4\omega_p^2 \Omega^2 \cos^2\theta} \tag{12}$$

where the subscript denoting electrons has been omitted, and ω_{UH}^2 denotes the upper hybrid frequency

$$\omega_{UH}^2 = \omega_p^2 + \Omega^2 \tag{13}$$

For parallel propagation there is one mode with $\omega = \Omega$ and another with $\omega = \omega_p$. Only the upper hybrid mode propagates perpendicular to the magnetic field.

4.4   Bernstein modes

A pure Maxwellian plasma has no free energy, and no plasma wave in one can grow. In fact, all are damped except those with $k_\parallel = 0$, whose resonant velocities are infinite. The $k_\parallel = 0$, Harris dispersion relation produces the infinite set of undamped modes first studied by BERNSTEIN (1958) for frequencies sufficiently high that ion motions can be neglected:

$$\varepsilon = 0 = 1 - \sum_n \frac{n^2\omega_p^2}{(\omega^2 - n^2\Omega^2)} \frac{2\Gamma_n}{\lambda} \tag{14}$$

where all quantities refer to electrons only. Analytic solutions to (14) for

$\omega_c/\omega_p = 1/5$ and $\omega_{UH} = 5.1\ \Omega$, together with their verification by numerical simulation (KAMIMURA *et al.*, 1978), are shown in Fig. 13.

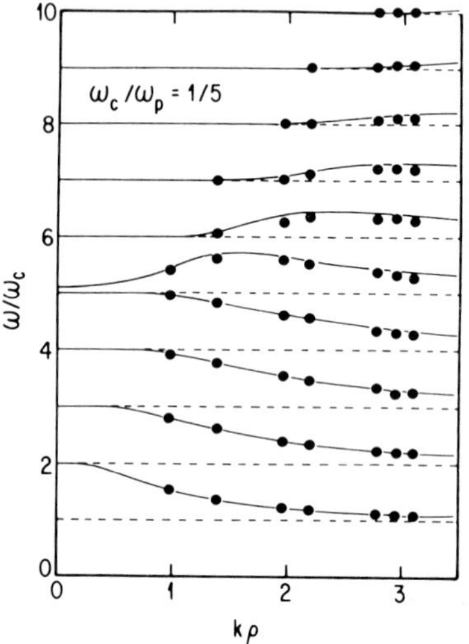

FIG. 13. Analytic calculations of the Bernstein mode dispersion relation and its verification by numerical simulation. The solid lines show the $k_\parallel = 0$ Bernstein mode dispersion relation and the dots, the dispersion characteristics inferred from numerical simulation (KAMIMURA *et al.*, 1978) for a single component electron plasma. ω/ω_c is plotted vs. k_ρ, where ω_c is the electron cyclotron frequency, and ρ is the Larmor radius. The upper hybrid frequency ω_{UH} is 5.09 ω_c. Below the upper hybrid band the modes start at the bottom of their bands, increase to a maximum frequency, and then diminish. In the hybrid band the mode starts at the hybrid frequency, increases to a maximum frequency, and then diminishes. Note that in the absence of cold electrons, the low harmonics have $\partial\omega/\partial k_\perp = 0$ only at $k_\rho \to 0$ and ∞, where finite $k_\parallel$ modes are liable to be stable.

A Bernstein mode propagates in each cyclotron harmonic band, $n\Omega < \omega < (n+1)\Omega$, but how it does so depends upon whether it is above, below, or in the upper hybrid frequency band. The waves in those bands with $\omega < \omega_{UH}$ start at $\omega \simeq (n+1)\Omega$ at long wavelengths (small λ) and descend to $\omega \simeq n\Omega$ when $\lambda \gg n$. Those with $\omega > \omega_{UH}$ start at $\omega \simeq n\Omega$, ascend to a peak near $\lambda \simeq n$, and then return to $\omega \simeq n\Omega$ for $\lambda \gg n$. In these harmonic bands, there is a mode between harmonics with zero group velocity, and therefore a neighboring range of λ for which the group velocity is small. The wave in upper hybrid band is special. It starts at $\omega = \omega_{UH}$, rather than at the next higher cyclotron harmonic, and descends to the next lower cyclotron harmonic frequency as λ increases.

We expect plasma waves with $k_\parallel$ nonzero but smaller than $k_\perp$ to be similar to Bernstein modes. Of course, they would be damped by resonant wave-particle interactions in a Maxwellian plasma. In a non-Maxwellian plasma with free energy for wave growth, Bernstein modes give us a useful picture of the unstable waves' dispersion characteristics, at least insofar as their dependence upon $k_\perp$ is concerned. Bernstein's work first revealed the poverty of the cold plasma approximation. The many collective modes that he discovered are available to a warm magnetized thermal plasma and become channels by which nonthermal plasmas can reduce their free energy with great agility.

4.5 *Bernstein modes in a two-component plasma*

The electrons in the magnetosphere are not usually simple Maxwellians. At the very least, they are a mixture of a hot component that has been heated in space and a cold component that originated in the ionosphere. This fact motivated GAFFEY and LaQUEY (1976) to calculate the Bernstein modes for a two component Maxwellian:

$$f = \frac{n_c}{(\sqrt{\pi}\,a)^3}\,\mathrm{e}^{-V^2/a^2} + \frac{n_H}{(\sqrt{\pi}\,A)^3}\,\mathrm{e}^{-V^2/A^2} \tag{15}$$

where subscripts c and H denote cold and hot, respectively. n_c and n_H are the cold and hot partial electron densities. Lower case a denotes the cold thermal speed, upper case A, the hot.

Several questions immediately come to mind. For example, the upper hybrid frequency divides physical regimes for classical Bernstein modes. Modes with $\omega \gg \omega_{UH}$ have frequencies which typically exceed $n|\Omega_c|$ by a small amount; only those with $\omega \lesssim \omega_{UH}$ can have the frequencies near $\omega \simeq (n+\frac{1}{2})\Omega$ that are most often observed in the magnetosphere. Which is the pertinent hybrid frequency—that based on the total electron density, $n_c + n_H$, or those based upon the partial densities, n_c or n_H?

GAFFEY and LaQUEY (1976) find the expected cold plasma result, a cold upper hybrid mode based upon the total density, when λ_c and λ_H both tend to zero. When $\lambda_c \lesssim 1 \ll \lambda_H$, low harmonic modes behave as though only the cold upper hybrid frequency has an influence, whereas high harmonic modes behave more like classic Bernstein modes. When $n \ll \lambda_c \ll \lambda_H$, the cyclotron harmonic modes act as if only cold electrons existed. An inspection of the relative sizes of the cold and hot Bessel functions makes this behavior clear. Each component makes its largest contribution to the dielectric plasma response for those wavelengths such that the appropriate $\lambda \sim n$.

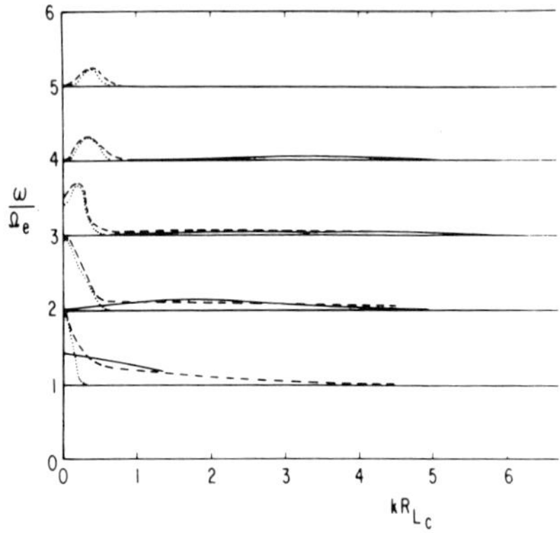

FIG. 14. Bernstein modes in a two-component electron plasma. The frequency normalized to the cyclotron frequency Ω_e is plotted vs. kR_{LC}, where R_{LC} is the cold electron Larmor radius (GAFFEY and LAQUEY, 1976). Plausible magnetospheric plasma parameters ($T_c = 10\,\text{eV}$, $n_c = 0.1\,\text{cm}^{-3}$, $T_H = 1\,\text{keV}$, $n_H = 1\,\text{cm}^{-3}$) were chosen. The cold upper hybrid frequency is 1.4, and the hot 3.3, times the cyclotron frequency. Dotted lines are for hot plasma alone; solid, for cold plasma alone; and dashed, the mixture. Each of the single component calculations resembles that of Fig. 13. In the mixture, cold electrons induce a second peak in the dispersion relation whose kR_{LH} is large. Such second peaks define where $k_\parallel \neq 0$ instabilities have the largest spatial growth rates.

Figure 14 has been adapted from numerical calculations made by GAFFEY and LAQUEY (1976) for $n_c/n_H = 0.1$, $a^2/A^2 = T_c/T_H = 10^{-2}$, and $\omega_{pc}^2 = \Omega_e^2$. For this case, the cold upper hybrid frequency is $\sqrt{2}\,|\Omega_e|$, the hot, $3.3\,|\Omega_e|$, and the total, about $3.5\,|\Omega_e|$. Bernstein modes for hot electrons alone are indicated by dotted lines; for cold electrons alone, solid lines; and for both together, by dashed lines. The behavior described above is immediately apparent.

4.6 Cyclotron harmonic instabilities

Of the two broad classes of electrostatic instabilities—flute modes, which are nonresonant instabilities of the $k_\parallel = 0$ Bernstein modes, and resonant instabilities ($k_\parallel \neq 0$)—we will concentrate on resonant instabilities. We begin to perceive the magnitude of the problem facing us, when, with theoreticians' licence, we assume that the growth rate γ induced by resonant wave-particle interactions is small compared to ω_R, for then Eq. (8) has the approximate solution

$$\varepsilon_R(\omega_R, k_\perp, k_\parallel) \simeq 0 \qquad (16a)$$

$$\gamma = -\varepsilon_I / \partial\varepsilon_R / \partial\omega_R \tag{16b}$$

where $\partial\varepsilon_R/\partial\omega_R$ stands for $\partial\varepsilon_R/\partial\omega$ evaluated at $\omega=\omega_R$. $\gamma>0$ denotes instability. The energy density $w(k_\perp, k_\parallel)$ associated with a given solution of $\varepsilon_R(\omega_R)=0$ is

$$w(k_\perp, k_\parallel) = \frac{\partial\varepsilon_R}{\partial\omega_R} \frac{|E_K|^2}{8\pi} \tag{16c}$$

where E_K is the electric field amplitude. $\partial\varepsilon_R/\partial\omega_R$ measures the energy in oscillating particle motion self-consistently associated with E_K, and $w(k_\perp, k_\parallel)$ may be positive or negative according to the sign of $\partial\varepsilon_R/\partial\omega_R$. The sign of ε_I needed to produce instability changes accordingly.

When $|\gamma/\omega_R| \ll 1$, the resonant denominator in (8a) may be evaluated using the Plemelj relation:

$$\frac{1}{\omega - k_\parallel v_\parallel - n\Omega} = \frac{P}{\omega_R - k_\parallel v_\parallel - n\Omega} - i\pi\delta(\omega_R - k_\parallel v_\parallel - n\Omega) \tag{16d}$$

where P denotes the principal part, δ, the Dirac delta function. The sign of the imaginary term has been chosen to conform to LANDAU's (1946) analysis of resonant instabilities. Using (16d), (16b) reduces to

$$\frac{\partial\varepsilon_R}{\partial\omega_R}\gamma = \pi\sum_s \frac{\omega_{ps}^2}{k^2|k_\parallel|} \sum_n G_{ns}(v_{ns}) \tag{16e}$$

where v_{ns} denotes the nth harmonic parallel resonant velocity

$$v_{ns} = \frac{\omega_R - n\Omega_s}{k_\parallel}. \tag{16f}$$

Thus, not only is there a denumerably infinite set of electron cyclotron harmonic modes, but each mode has a countably infinite set of wave particle resonances associated with it. Then $n=0$ resonance is called the Landau resonance, and all others, cyclotron resonances. $k_\perp=0$ electrostatic waves have only the Landau resonance, but we will not discuss this special case, because of the greater wealth of physical phenomena associated with nonzero $k_\perp$.

For $\partial\varepsilon_R/\partial\omega_R$ positive, a given resonance contributes to instability if $G_{ns}(v_{ns})$ is positive, and overall instability is assured if the partial growth rates from those resonances contributing to growth sum to defeat the damping from the remaining resonances. G_{ns} can only be positive if the species s has a non-Maxwellian distribution with a source of free energy. The kinds of free energy sources explored by plasma research are many and varied. Those who think in terms of separable distributions classify them in terms of parallel and perpendicular free energy sources. Parallel free energy sources include beams (relative parallel drifts) between components of the distribution functions of the same species, currents (relative drifts between electrons and ions), and heat fluxes (zero first parallel moment, but nonzero third moment); perpendicular free energy sources include perpendicular currents, and a loss cone distribution (a peak in $F_\perp$ away from $v_\perp = 0$). Thermal anisotropy is a mixed case, for when $A_\perp > A_\parallel$, one is tempted to say the free energy is in the perpendicular distribution, and *vice versa*. The inclusion of spatial gradients adds density, temperature, and magnetic field gradients to this list.

4.7   *General conditions for instability*

No truly general understanding of what causes a cyclotron instability has emerged. Most research has been individual forays into the Harris dispersion relation with a specific physical situation and distribution function in mind. Many of these were satisfying because they turned up an instability. It is possible, given the many wave mode and particle resonances available, that the situation will remain this diffuse. Nonetheless, several authors (YOUNG *et al.*, 1973; ASHOUR-ABDALLA and COWLEY, 1974) have arrived at rough analytic criteria that permit insight into the causes of cyclotron harmonic instabilities.

Let us first presume that we know a solution to $\varepsilon_R(\omega_R, k_\perp, k_\parallel) = 0$. Inspection of the Bernstein mode dispersion relation, or of Figs. 13 and 14 tells us that we cannot arbitrarily assume that ω_R falls anywhere within a given cyclotron harmonic band, and so we cannot solve for instability without solving $\varepsilon_R = 0$; but given a $\omega_R(k_\perp, k_\parallel)$ we can ask what would make it unstable. We will inspect the stability of separable distributions. For

$$\partial\varepsilon_R/\partial\omega_R > 0 , \qquad -\varepsilon_I = \sum_s (\omega_{ps}/kA_{\parallel s})^2 \sum_n F_{ns}$$

must be positive for instability, and a given resonance contributes to instability if F_{ns} is positive. For instabilities driven by perpendicular free energy, the structure of the parallel velocity distribution is uninteresting, so long as it decreases monotonically for increasing $v_\parallel$ and possesses no free energy itself. We might as well assume it to be Maxwellian.

Take a mode with $m\Omega < \omega_R < (m+1)\Omega$, and write $\omega_R = m\Omega + m\Omega + (\Delta\omega)$; the resonant velocity w_n normalized to the characteristic width $A_\parallel$ of the parallel velocity distribution is

$$w_n = \frac{\Delta\omega + (m-n)\Omega}{k_{\parallel} A_{\parallel}} \tag{17}$$

If $\Delta\omega$ is $0(\Omega)$, the smallest w_n occurs for $n=m$, $m+1$. For example, if

$$\Delta\omega = \frac{1}{2}\Omega, \quad w_n = +\frac{1}{2k_{\parallel}\rho_{\parallel}}, \quad w_{n+1} = -\frac{1}{2k_{\parallel}\rho_{\parallel}},$$

where $\rho_{\parallel} = k_{\parallel} A_{\parallel}/\Omega$ is a fictitious Larmor radius based upon the parallel characteristic velocity. A distribution even in $v_{\parallel}$ has equal numbers of $n=m$ and $m+1$ resonant particles.

It is convenient to recenter the summation over cyclotron harmonics, and rewrite the contribution of the species s to $-\varepsilon_{\mathrm{I}}$ in the form

$$-\varepsilon_{\mathrm{I}s}(m) = \frac{\omega_{ps}^2}{k^2 A_{\parallel s}^2} \lim_{p \to \infty} \sum_{\delta=-p}^{\delta=+p} F_{m+\delta} \tag{18}$$

Increasingly accurate estimates of $\varepsilon_{\mathrm{I}s}(m)$ can be made by increasing p. Our numerical calculations generally take $p=5$. When $k_{\parallel}\rho_{\parallel}$ is small, summing the $\delta=0$ and $\delta=1$ terms yields an approximate necessary condition for instability.

Choosing $F_{\parallel}$ a Maxwellian provides a concrete realization of the above arguments, for then

$$\varepsilon_{\mathrm{I}s} = 2\sqrt{\pi}\left(\frac{\omega_{ps}^2}{k^2 A_{\perp s}^2}\right)\left(\frac{\Omega_s}{k_{\parallel} A_{\parallel s}}\right)\sum_n \left(\frac{\omega}{\Omega_s}\tilde{C}_n + n(D_n - \tilde{C}_n)\right) e^{-w_{ns}^2}\Bigg|_s \tag{19}$$

where $\tilde{C}_{ns} = (A_{\perp s}/A_{\parallel s})^2 C_{ns}$ is always positive. For a mode with $m\Omega < \omega < (m+1)\Omega$, the nth partial contribution to $\varepsilon_{\mathrm{I}s}$ is modulated by

$$e^{-w_{ns}} = \exp\left(\frac{(\Delta\omega/\Omega)+(m-n)}{k_{\parallel}\rho_{\parallel}}\right)^2_s \tag{20}$$

which for $m=n$, peaks near $\Delta\omega=0$ with a halfwidth $k_{\parallel} A_{\parallel s}$. For $n=m+1$, this function peaks about the next higher harmonic frequency with the same half-width. To test for instability we inspect the sign of

$$\varepsilon_{\mathrm{I}s}(m) + \varepsilon_{\mathrm{I}s}(m+1) = 2\sqrt{\pi}\frac{\omega_{ps}^2}{k^2 A_{\perp s}^2}\left(\frac{\Omega_s}{k_{\parallel} A_{\parallel s}}\right)e^{-w_{ms}^2}\Delta_{ms} \tag{21}$$

or equivalently, the sign of Δ_{ms}, where

$$\Delta_{ms} = \tilde{C}_m\left(\frac{\Delta\omega}{\Omega}\right) + mD_n + \left\{\tilde{C}_{m+1}\left(\frac{\Delta\omega}{\Omega}-1\right) + (m+1)D_{m+1}\right\}$$

$$\times \ \exp\left\{\frac{2}{k_{\parallel}^2\rho_{\parallel s}^2}\left(\frac{\Delta\omega}{\Omega}-\frac{1}{2}\right)\right\} \tag{22}$$

When $\Delta\omega/\Omega$ is small, $\Delta_{ms}\simeq mD_m$, so that $D_m<0$ for instability. When $\Delta\omega/\Omega\simeq +1$,

$$\Delta_{ms}\simeq(m+1)D_{m+1}\ \exp\left(\frac{1}{k_{\parallel}^2\rho_{\parallel s}^2}\right) \tag{23}$$

and $D_{m+1}<0$ is needed for instability. In the middle of the harmonic band, $\Delta\omega/\Omega\simeq\frac{1}{2}$,

$$\Delta_{ms}\simeq\tilde{C}-\tilde{C}_{m+1}+mD_m+(m+1)D_{m+1} \tag{24}$$

Once again, negative D_m, D_{m+1} contribute to instability. D_m negative requires that $\partial F_\perp/\partial v_\perp$ be positive somewhere. A similar analysis placing the free energy entirely in the parallel velocity distribution yields the familiar condition that $\partial F_\parallel/\partial v_\parallel$ be positive somewhere.

4.8 Anisotropy instabilities

$F_\perp$ or $F_\parallel$ need not have positive slopes for instability if F is sufficiently anisotropic. One success of the analytic approach in Subsection 4.7 has been to establish when anisotropic bi-Maxwellian plasmas can be unstable to electrostatic waves (HALL and HECKROTTE, 1964; SHIMA and HALL, 1965; YOUNG *et al.*, 1973; ASHOUR-ABDALLA and COWLEY, 1974). In this case, Δ_{ms} is

$$\Delta_{ms}=\Gamma_m\left\{\frac{\Delta\omega}{\Omega_s}\left(\frac{A_{\perp s}}{A_{\parallel s}}\right)^2+m\right\}-\Gamma_{m+1}\left\{\left(\frac{A_{\perp s}^2}{A_{\parallel s}^2}-1\right)-\left(m+\frac{\Delta\omega}{\Omega_s}\right)\right\}$$

$$\times\ \exp\left\{\left(\frac{2}{k_{\parallel}^2\rho_{\parallel s}^2}\frac{\Delta\omega}{\Omega_s}-\frac{1}{2}\right)\right\} \tag{25}$$

Only the second term in Δ_{ms} can be negative, and its magnitude is large only if $\Delta\omega/\Omega_s>\frac{1}{2}$. Thus a necessary condition is

$$\left(\frac{A_{\perp s}^2}{A_{\parallel s}^2}-1\right)>m+\frac{1}{2}, \tag{26}$$

and it is evident that unstable waves have frequencies only *above* odd half harmonics.

Several warnings are now in order. Negative Δ_{ms} does not *guarantee* instability. We must consider the competition with other, possibly damping cyclotron resonances. Moreover, to establish that a mode with the appropriate frequency exists, we must resort to direct calculation. Our arguments above implicitly assumed that $F_{\perp s}$ and $F_{\|s}$ could be characterized by a single speed in each direction, $A_{\perp s}$ or $A_{\|s}$, in effect, that F_s is a one-component distribution function. Since there is a mixture of hot and cold plasmas in the magnetosphere, space physicists cannot avoid such detailed computations.

4.9 Summary

1) A warm magnetized plasma possesses an infinite set of collective modes of oscillation. Each of these cyclotron modes has its own infinite set of wave-particle resonances.

2) By adjusting $k_{\|}$, the wave particle resonance, or the cyclotron harmonic wave mode, plasmas can often find a wave that will resonate with the particles in any given slice in velocity space. Even sources of free energy that are relatively localized in velocity space are therefore potential causes of instability. The very complexity of the linear theory of cyclotron harmonic waves communicates an important physical statement: a collisionless magnetized plasma can find many ways to be unstable and reduce its free energy.

3) The richness of the cyclotron harmonic spectrum suggests that a turbulent plasma can find nonlinear ways to interact with those remaining particles that are inaccessible to linear wave-particle resonance.

4) Because electrostatic waves can resonate with the main part of the velocity distribution, their instabilities can affect the macroscopic properties of the plasma.

5) In many applications, it is important to calculate the spatial growth rate, which requires knowledge of the group velocity. The potential for nonconvective instability is high, because Bernstein modes can have zero perpendicular group velocity. This leaves only the question of whether the parallel and perpendicular group velocities can be small simultaneously.

6) Because the wave propagation characteristics are determined by the main part of the velocity distribution, the structure of the free energy source affects how the waves propagate. Instability and propagation must therefore be calculated together. This contrasts with calculations for electromagnetic waves such as whistlers, where the particles determining propagation and those responsible for instability are different ones.

5. Free Energy Sources for Magnetospheric Electron Cyclotron Harmonic Instabilities

5.1 Introduction

Here we outline historically our developing understanding of magnetospheric cyclotron harmonic instabilities. The community has focussed on two questions. First, what free energy sources drive the observed instabilities? And, second, how do cold electrons affect both instability thresholds and wave propagation? We concentrate here upon the free energy sources, and explore cold electron effects in the next Section. We continue to separate conceptually perpendicular and parallel free energy sources as though nature actually did mimic artifice.

5.2. Perpendicular free energy sources
5.2.1 Generalized loss cone distributions

Distributions for which $\partial f/\partial v_\perp$ is positive somewhere we call generalized loss cone distribution, in contrast with the canonical loss cone distribution, which has zero $\partial f/\partial v_\perp$ at $v_\perp = 0$.

TATARONIS and CRAWFORD (1970) investigated the stability of the $k_\parallel = 0$ flute mode using the simplest loss cone distribution, a ring in velocity space,

$$f_H = \frac{n_H}{2\pi A_\perp} \delta(v_\perp - A_\perp)\delta(v_\parallel) , \qquad (27)$$

in an attempt to understand odd half electron harmonic instabilities in laboratory devices. Shortly after OGO-5 discovered magnetospheric electron cyclotron harmonic waves, FREDRICKS (1971) seized upon the ring distribution, gave it a spread in parallel velocity, and found unstable waves with frequencies between low electron cyclotron harmonics. He thus established a loss cone distribution as the theoreticians' favorite candidate to explain the magnetospheric observations. The unstable waves he found propagated neither perpendicular nor parallel to the magnetic field, thereby ensuring that subsequent instability calculations with more realistic loss cone distributions would have to be numerical.

It has always been clear that the nature does not produce δ-function distributions. The next effort, then, was to see if more gentle loss cones could be unstable. At first, it seemed doomed to failure, for DORY *et al.* (1965), using a canonical loss cone distribution model appropriate to laboratory mirror machines,

$$f_{\text{DGH}} = \left(\frac{n_{\text{H}}}{\sqrt{\pi}\, A_{\|}}\, e^{-v_{\|}^2/A_{\|}^2}\right)\left(\frac{1}{\pi A_{\perp}^2}\frac{1}{j!}\frac{v_{\perp}}{A_{\perp}}\right)^{2j} e^{-v_{\perp}^2/A_{\perp}^2} \equiv n_{\text{H}} F_{\text{DGH}} \tag{28}$$

had found that the low harmonic modes observed in the magnetosphere could not be unstable unless j was unrealistically large ($j \geq 5$). $\varepsilon_{\text{R}}(\omega, \boldsymbol{k}) = 0$ did not produce solutions in those regions of $(\omega, \boldsymbol{k})$ space where ε_{I} had the appropriate (negative) sign for instability. A landmark paper by YOUNG *et al.* (1973), resolved this puzzle and thereby established the basic distribution function model used in most subsequent instability calculations, a mixture of cold electrons and hot electrons with a loss cone free energy source:

$$f = n_{\text{c}} F_{\text{c}} + n_{\text{H}} F_{\text{DGH}} \tag{29a}$$

$$F_{\text{c}} = \left(\frac{1}{\pi a_{\perp}^2}\right)\frac{1}{\sqrt{\pi}\, a_{\|}}\, e^{-(v_{\perp}^2/a_{\perp}^2 + v_{\|}^2/a_{\|}^2)} \tag{29b}$$

where $a_{\perp}$ and $a_{\|}$ are the cold electron perpendicular and parallel thermal speeds respectively.

When F_{c} has no free energy, cold electron cyclotron damping overwhelms warm electron growth, unless $(\omega - n\Omega)/k_{\|}a_{\|}$ is large. When $(\omega - n\Omega)/k_{\|}a_{\|} \gg 1$, the cold electron contribution to the dispersion relation is essentially that for zero temperature. However, even when $n_{\text{c}}/n_{\text{H}}$ is small, cold electrons dominate the real part of the dispersion relation. They reduce the n_{H} threshold for both the flute mode instability and low harmonic finite $k_{\|}$ Harris modes. YOUNG *et al.* (1973) concluded that the weakest DGH distribution is stable to the 3/2 instability for magnetospherically realistic n_{H} unless cold electrons are present. $n_{\text{c}}/n_{\text{H}}$ need not be large, $0(10^{-2})$, to produce first harmonic instability.

After YOUNG *et al.* (1973) explained how low cyclotron harmonics could by unstable at all in the magnetosphere, the first of many parametric surveys of magnetospherically pertinent solutions of the HARRIS (1959) dispersion relation were made by KARPMAN *et al.* (1973, 1975), and Young (1975), using the DGH distribution function. KARPMAN *et al.* found that cold upper hybrid frequency ω_{UHC} determines the unstable frequency range. In particular, first harmonic modes were unstable only below ω_{UHC}. By implication, the cold upper hybrid frequency ought to determine which higher harmonics are unstable, as well.

Up to this point, all authors had computed the temporal growth rate γ. While this enabled them to evaluate instability thresholds, they could not treat amplification to a saturated steady state in the spatially inhomogeneous magnetosphere. The necessary calculations of the *spatial* growth rate K_I, were undertaken by Ashour-Abdalla and Kennel (1976, 1978b). They estimated the magnitude of K_I by

$$K_1 = \frac{\gamma}{|\partial\omega/\partial K|}, \tag{30}$$

a valid approximation if γ is small, which turned out be true *ex post facto* in the cases they investigated. Furthermore, they argued that even if the above approximation fails, it is convenient computationally, and it will certainly indicate the waves that spatially amplify rapidly, which is, after all, the essential physical information needed.

Ashour-Abdalla and Kennel modelled the hot electron loss cone distribution function by a "subtracted Maxwellian," for which C_n, D_n, and Y_n could be evaluated in terms of tabulated functions:

$$f_{\mathrm{H}} = n_{\mathrm{H}} \frac{e^{-v_\|^2/A_\|^2}}{(\pi A_\|^2)^{1/2}} F_{\mathrm{AAK}} \tag{31a}$$

$$F_{\mathrm{AAK}} = \left(\Delta e^{-v_\perp^2/A_\perp^2} + \frac{1-\Delta}{1-\beta} (e^{-v_\perp^2/A_\perp^2} - e^{-v_\perp^2/\beta A_\perp^2}) \right) \frac{1}{\pi A_\perp^2} \tag{31b}$$

where $A_\|^2 = 2T_\|/M$ and $A_\perp^2 = 2T_\perp/M$, $T_\|$ and $T_\perp$ are hot parallel and perpendicular temperatures in energy units, and M is the electron mass. Δ is a loss cone filling parameter; $\Delta = 0$ is an empty loss cone, and $\Delta = 1$ is a stable

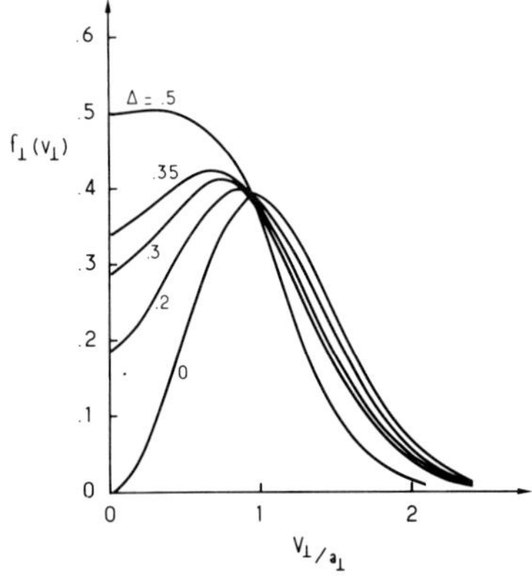

Fig. 15. Subtracted Maxwellian perpendicular distribution. This figure plots the perpendicular distribution f_{AAK} (Eq. (31b)) that we will use in the parametric instability study in Section 6. The distributions corresponding to various values of the loss cone filling parameter Δ are shown.

Maxwellian. F_{AAK} is sketched for various Δ and $\beta=0.9$ in Fig. 15.

The subtracted Maxwellian provides the region of positive slope necessary for loss cone instability. When $\beta\to 1$ and $(1-\beta)v_\perp^2/A_\perp^2\ll 1$, f_{AAK} reduces to the weakest $(j=1)$ DGH distribution function. Since the magnetospheric loss cone is weak, Ashour-Abdalla and Kennel assumed $\beta=0.9$ in all their calculations. F_{AAK} contains two interrelated free energy sources, as may be seen from the hot electron thermal anisotropy

$$\bar{A}_H = \int d^3 v F_{AAK}(v_\perp^2 - v_\parallel)^2 \bigg/ \int d^3 v F_{AAK} v_\parallel^2$$

$$= \Delta\left(\frac{T_\perp}{T_\parallel}-1\right)+(1-\Delta)\left(\frac{T_\perp}{T_\parallel}(1+\beta)-1\right)$$

$$(32)$$

There is no loss cone property if either $\Delta=1$ or $\beta=0$, in which cases $\bar{A}_H$ reduces to the bi-Maxwellian anisotropy. On the other hand, when $T_\perp = T_\parallel$, $\bar{A}_H=(1-\Delta)\beta$.

Using F_{AAK} rather than F_{DGH} proved to modify YOUNG et al.'s (1973) and KARPMAN et al.'s (1973, 1975) conclusions only at the level of numerical detail. However, finite cold electron temperature had a major, unexpected effect on the convective growth rates. A mixture of hot and cold electrons created a region of $k_\perp$ where the perpendicular group velocity was small, and the temporal growth rate, positive. In this region, the parallel group velocity scaled as the cold electron thermal speed $a_\parallel$. If T_c/T_H were small, k_I was large, and if T_c were small enough, nonconvective instability was possible.

ASHOUR-ABDALLA and KENNEL (1976, 1978a) surveyed primarily the first harmonic band. Higher harmonic instabilities were more systematically investigated by HUBBARD and BIRMINGHAM (1978) who used a variant of the AAK distribution function:

$$f_{HB}=n_H F_\parallel F_{AAK} , \qquad (33a)$$

which replaced the parallel Maxwellian by a Lorentzian of order p:

$$F_\parallel = \frac{1}{\pi}\frac{(2p-2)!!}{(2p-3)!!}\frac{(A_\parallel)^{2p-1}}{(v_\parallel^2 + A_\parallel^2)^p} \qquad (33b)$$

$$(0)!! = (-1)!! = 1 \qquad (33c)$$

When p is very large, $f_\parallel$ approaches a δ-function at $v_\parallel = 0$. They chose the Lorentzian because magnetospheric electrons often have a power law high energy tail (ANDERSON and MAEDA, 1977) and because it facilitates computations. In most cases, the Lorentzian and the Maxwellian produced similar results.

HUBBARD and BIRMINGHAM's (1978) convective growth rate calculations generally confirmed those of ASHOUR-ABDALLA and KENNEL (1978) for the first harmonic band. In addition they found the higher harmonic instabilities expected when the cold upper hybrid frequency exceeds twice the cyclotron frequency. Contrary to expectation, they also found instability *above* the cold upper hybrid frequency in this parameter regime. An independent analysis using f_{AAK} showed that *nonconvective* instability was still confined below the cold upper hybrid frequency, although a convective instability was possible above it (ASHOUR-ABDALLA and KENNEL, 1978b; ASHOUR-ABDALLA *et al.*, 1979). Moreover, when n_c/n_H was less than unity, all the harmonic bands at or below the cold upper hybrid frequency had about the same spatial growth rates, and turned convective at roughly the same T_c/T_H. This indicated that several harmonic bands should be observable simultaneously when $n_c/n_H < 1$.

In order to reach large cold upper hybrid frequencies, HUBBARD and BIRMINGHAM (1978), ASHOUR-ABDALLA and KENNEL (1978b), and ASHOUR-ABDALLA *et al.* (1979), all had to increase the ratio n_c/n_H from below unity, a regime pertinent to night side diffuse aurora field lines, to n_c/n_H large, which applies to the dayside of the magnetosphere and plasmapause. They all found that the harmonic band containing the cold upper hybrid frequency was the most unstable when n_c/n_H exceeded unity. This turned out to pertinent to the intense upper hybrid emissions observed by KURTH *et al.* (1979).

5.2.2 Thermal anisotropy

The pure bi-Maxwellian thermal anisotropy ($\beta=0$ or $\Delta=1$, $T_\perp \neq T_\parallel$) produces waves only above half-harmonics, and only when $\bar{A}_H$ exceeds m. On the other hand, the pure loss cone can be unstable below the half-harmonics. Since waves are observed both above and below half-harmonics, the loss cone property is a necessary part of the distribution function model, whereas a bi-Maxwellian anisotropy is not. Furthermore, $\bar{A}_H$ would have to be unreasonably large, for the strong diffusion conditions though to prevail in the magnetosphere, to account for the high-harmonic instabilities that are observed. For this reason, most calculations have taken $T_\perp = T_\parallel$.

5.3. Parallel free energy sources

5.3.1 Anti-loss cone

BERK and GALEEV (1967) found that guiding center drifts preferentially carry $v_{\|} \simeq 0$ particles to the walls of toroidal confinement devices. In contrast with mirrors, this creates an "anti-loss cone" with a deficit of small $v_{\|}$ particles. A similar situation can exist in the magnetosphere. ROEDERER (1968) showed that on the earth's nightside, $v_{\|} \simeq 0$ particles drift to the magnetopause (presumbaly to be lost), whereas $v_{\|} \neq 0$ particles remain on drift orbits closed within the magnetosphere. The resulting anti-loss cone can be larger than the loss cone on auroral lines of force.

KENNEL *et al.* (1970) modelled the hot electron anti-loss cone distribution function as a subtracted Maxwellian

$$f_{\text{KFS}} = n_{\text{H}} \frac{F_{\perp}}{\sqrt{\pi}} \frac{e^{-v_{\|}^2/A_{\|}^2} - e^{-v_{\|}^2/\beta A_{\|}^2}}{A_{\|}(1 - \sqrt{\beta}\,\rho)} \tag{34}$$

where ρ was a depth parameter describing the strength of the anti-loss cone feature, and $F_{\perp}$ was a Maxwellian with a characteristic velocity $A_{\perp}$. They also included cold electrons but only considered parallel propagating waves, which were unstable if $\beta < \rho < \sqrt{\beta}$. They found that cold electrons profoundly affected the range of unstable frequencies, which were near the ion plasma frequency when $n_c = 0$, and near $(n_c/n)\omega_p$ when $n_c \neq 0$, where ω_p is the electron plasma fequency based upon the total electron density n.

NAMBU and WATANABE (1975) replaced the perpendicular Maxwellian in (34) with

$$F_{\text{NW}} = \frac{1}{\pi A_{\perp}^2} \frac{e^{-v_{\perp}^2/A_{\perp}^2} - \bar{\rho}e^{-v_{\perp}^2/\beta A_{\perp}^2}}{(1 - \bar{\rho}\bar{\beta})} \tag{35}$$

Their distribution function, which has been suggested by observation (WEST *et al.*, 1973), is one of the most general thus far investigated. Nambu and Watanabe computed the wave frequencies and growth rates for $\omega_p/\Omega = 2.3$ $A_{\perp}^2/A_{\|}^2 = 10$, $\beta = \bar{\beta} = 0.1$ and $\rho = \bar{\rho} = 1$, and allowed for nonzero $k_{\perp}$. They did not include cold electrons. They found a new instability, with $\omega \simeq 1.2\,\Omega$, associated with the missing anti-loss cone particles, which, however, might be damped by cold electrons. They also studied the electrostatic instability near $\omega = \frac{1}{2}\Omega$ neglected by most other authors except YOUNG *et al.* (1973). Should observations indicate that an anti-loss cone feature is important, a systematic parametric survey of the instabilities it creates would be warranted.

5.3.2 *Electron beams*

A laboratory experiment, may of whose scaled parameters are similar to those in the magnetosphere, showed that a field-aligned energetic electron beam can generate odd half-harmonic emissions (BERNSTEIN *et al.*, 1975). A monoenergetic electron beam, whose energy could be varied between 50 eV and 50 keV, produced odd half-harmonic waves when it was accompanied by countersteaming cold electrons. For this experiment Ω_c/ω_p ranged from 0.2 to 1 and n_c/n_H was $10^{-2}-10^{-3}$, roughly the same values as in the magnetosphere. BERNSTEIN *et al.* (1975) argued that the field-aligned beams of few keV electrons observed by McILWAIN (1975) near the geomagnetic equator at $L=6.6$ could create the odd half-harmonic emissions also observed in this region of space. There has been no theoretical follow-up to this important suggestion.

5.4   Power law energy distributions

BARBOSA and KURTH (1980) extended Hubbard and BIRMINGHAM's (1978) use of a power law distribution from the parallel to the total energy distribution, using a distribution function model previously employed by BARBOSA and CORONITI (1976) in the study of Jovian whistler instabilities:

$$f_{BC} = \frac{1}{2} m^2 E_0^K j_\perp(E_0) \frac{\sin^{2M}\alpha}{E^{K+1}} \tag{36}$$

where E is the electron kinetic energy, E_0 is a threshold energy where the power law distribution first becomes realistic, and K is the spectral index. $j_\perp(E_0)$ is the unidirectional flux of electrons perpendicular to the magnetic field in units of $(\text{cm}^2\text{-s-ster-keV})^{-2}$ and $j_\perp(E)=j_\perp(E_0)(E_0/E)^K$. The omnidirectional integral flux $J(>E)$ is related to $j_\perp(E)$ by

$$J(>E) = \left(2\pi^{3/2}\frac{\Gamma(M+1)}{\Gamma(M+3/2)}\right)(K-1)^{-1}E j_\perp(E) \tag{37}$$

If $K=1$, the term in E_{MAX}/E_0 should be substituted for $(K-1)^{-1}$. At any given pitch angle α, f_{BC} decreases monotonically with increasing energy, while still retaining positive $\partial f/\partial v_\perp$. Unlike f_{DGH} or f_{AAK}, it does not have a nonmonotonic bump in the energy distribution. Relative to a Maxwellian energy distribution, f_{BC} has more high energy (hot) electrons and should produce a broader unstable frequency spectrum.

BARBOSA and KURTH (1980) use f_{BC} to estimate a "limit on stably trapped fluxes" similar to the one derived for whistler instability by KENNEL and PETSCHEK (1966). When n_c/n_H is large and X_{UHC} exceeds $\sqrt{2}$, the growth rate in the lower harmonic bands is roughly

$$\frac{\gamma}{\Omega} \simeq \frac{E j_\perp(E)}{n_c a}, \tag{38}$$

which is proportional to $J(>E)$, and is independent of n_c. A stably trapped limit, found by requiring that a wave e-fold 10 times in propagating a distance such that its normalized frequency changes by 10%, is calculated for the cold upper hybrid mode by BARBOSA and KURTH (1980).

5.5   *Relativistic Bernstein mode instabilities*

The relativistic mass correction enables perpendicularly propagating waves to resonate with particles of finite energy. CURTIS and WU (1979) calculated the growth rates of electrostatic and electromagnetic Bernstein modes supported by a relatively featureless distribution of warm ($\sim$ few keV) electrons. Their source of free energy was a power law energy distribution of energetic electrons ($> 10\,\text{keV}$) with a mild loss cone feature. If their energy distribution has a "shoulder," a region where the energy distribution falls off less rapidly than elsewhere, the gentle loss cone creates odd half harmonic instabilities without cold electrons.

5.6   *Discussion*

The frequencies of cyclotron harmonic instabilities are determined primarily by the Bernstein mode structure of a warm magnetized plasma's dielectric response, and secondarily by the structure of the free energy source. Since different free energy sources can produce odd half-harmonic instabilities, distribution function measurements are needed to guide the further development of theory. In their absence theoreticians have developed the logical, if perhaps irrelevant, consequences of their choices of distribution function model. Nonetheless, their efforts have probably led to valid insights of general applicability, particularly about the effects of cold electrons. Moreover, they have accumulated enough experience with various distribution functions that they are now ready to model the observed distributions.

6.   A Parametric Study of Convective Loss Cone Instabilities

6.1   *Introduction*

A laboratory experimentalist attempts to create the same initial state each time he runs his experiment. If he is studying the nonlinear saturation of an instability, he will inject the same unstable particle distribution into the same background plasma over and over again until he has measured the parameters necessary to understand the behavior he observes. The facts that laboratory experiments are repeatable and may be relatively completely diagnosed motivate laboratory theoreticians to specialize their analyses to a narrow range of parameters. Space physicists, on the other hand, did not create the plasma they work with. They can't even observe all of it. They must cope with observations made at a single point, knowing that the phenomena they detect are affected by things they do not measure that occur elsewhere in the immensity of the magnetosphere. When their spacecraft returns to the same point in space, the magnetosphere will have changed. And the spacecraft encounters along its orbit a variety of plasma environments equivalent to many laboratory experiments. The goal of the theoretician studying plasma waves in space differs therefore from those of his laboratory colleagues. He must arrive at a general understanding of whole classes of related instabilities before he can comprehend space observations. Nowhere is this goal more challenging than with the instabilities contained in the Harris dispersion relation, because of the many ways plasmas can produce waves in essentially the same frequency bands. Systematic instability studies are only the first step in the space theoretician's program. Since waves in space are observed in a saturated state, he must also study how the instabilities saturate in different parameter regions.

The loss cone is the one distribution for which there have been extensive instability studies exploring parameters that are plausibly realistic for the magnetosphere (YOUNG *et al.*, 1973; YOUNG, 1975; ASHOUR-ABDALLA and KENNEL, 1976, 1978a; HUBBARD and BIRMINGHAM, 1978; ASHOUR-ABDALLA *et al.*, 1979a). These have concentrated on two issues. First, can relatively gentle loss cone distributions produce an instability that can persist in strong diffusion conditions? And second, what is the role of cold electrons? The problem of cold electrons underlines the need for a parametric study. We know that a small admixture of cold electrons, $n_c/n_H < 1$, profoundly affects Harris loss cone instabilities. Yet it is just when n_c/n_H is small that the cold electrons are difficult to measure, so that we must investigate the range of situations permitted by our ignorance of their properties. When we can measure the cold electrons, we find that their density varies considerably from the dayside to the nightside of the magnetosphere and from inside to outside the plasmapause. In this case, parametric studies help us understand why the various cyclotron harmonic waves occur where they do. Finally, the distributions we observe in space have already been smoothed by interactions with waves whose instabilities we are studying. We can estimate the effects of wave particle interactions on the growth rates by varying parameters in the general directions expected from quasilinear theory.

In this Section we review only own parametric studies, because we feel that presenting results from different authors that are incommensurate in their choices of parameter and distribution function model needlessly complicates the reader's task. We apologize to the others who have contributed much to the subject.

6.2 *Possibility of nonconvective instability*

We turn to the first of our two questions: how can a gentle loss cone distribution create persistent cyclotron harmonic instabilities? Figure 16 shows a gentle distribution that still leads to strong instability; it is a three dimensional representation of f_{AAK}, Eq. (31), with $A_\perp^2 = A_\parallel^2$, $\beta = 0.9$, $\Delta = 0.2$, and $n_{\mathrm{c}}/n_{\mathrm{H}} = 0.2$. If we choose the ratio X_{UHC} of the cold upper hybrid frequency to the electron cyclotron frequency to be 1.5 (and, with $n_{\mathrm{c}}/n_{\mathrm{H}} = 0.2$, the total X_{UHC} to be 3.8), we may then solve the Harris dispersion relation

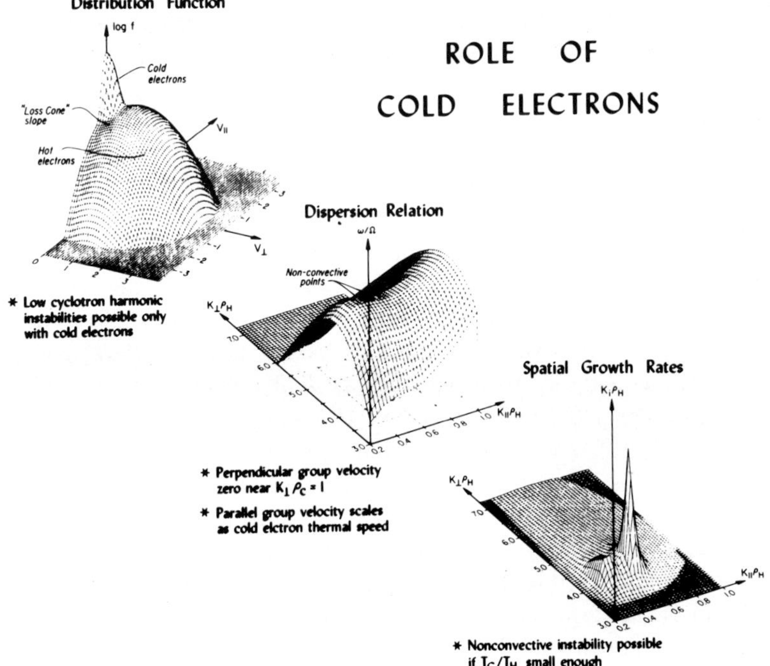

FIG. 16. Role of cold electrons. At the top left is a representation of a gentle loss cone distribution, with $n_{\mathrm{c}}/n_{\mathrm{H}} = 0.2$, $T_{\mathrm{c}}/T_{\mathrm{H}} = 10^{-2}$ and $\Delta = 0.2$, that leads to strong spatial growth. In the center is a three dimensional representation of the first harmonic dispersion relation. Note the two points where the group velocity is zero. At these two points, the spatial growth rate (lower right) maximizes.

numerically and display the results for the first harmonic as in Fig. 16. The real part of the frequency, ω_R, is shown as a function of $k_\perp \rho_H$ and $k_\parallel \rho_H$ at the top and $k_1 \rho_H$, the imaginary part of the spatial growth rate normalized to the hot electron cyclotron radius ρ_H, at the bottom.

The dependence of ω_R upon $k_\perp$ and $k_\parallel$ contains the answer to our question. For a given $k_\parallel \rho_H$, $\omega_R(k_\perp)$ contains the local maximum that one expects from the Bessel function structure of the Harris dispersion relation. It requires no numerical solution to assert that there can be regions where the perpendicular group velocity $\partial \omega / \partial k_\perp$ is zero. However, the numerical solution reveals that the dispersion relation has a saddle shape. In particular, there are two points where $\partial \omega / \partial k_\parallel$ and $\partial \omega / \partial k_\perp$ are zero simultaneously. Here we expect nonconvective instability. Moreover, $\partial \omega / \partial k_\parallel$ is much smaller than A_H elsewhere near the top of the saddle. Therefore, our gentle distribution function produced small temporal growth rates, but large spatial growth rates. This suggests that the even more gentle distributions in the magnetosphere could be strongly unstable. However, we must first ascertain how general the conclusions drawn from this one instability calculation really are.

6.3 *Dependence of nonconvective loss cone instability upon cold electrons*

Given a generalized loss cone free energy source, the properties of the instability depend thereafter on three parameters: n_c/n_H, T_c/T_H, and the magnetic field strength, as expressed for example by X_{UHC}, the ratio of the cold upper hybrid frequency to the cyclotron frequency. We start our explorations in this parameter space by asking how the Bernstein mode structure of a *single* $k_\parallel$ wave changes as T_c/T_H changes. Figure 17 plots $X = \omega_R/\Omega$ vs $k_\perp \rho_H$, for $k_\parallel \rho_H = 0.8$, and $10^{-4} < T_c/T_H < 0.05$. As in Fig. 16, X_{UHC} is again 1.8, and $n_c/n_H = 0.2$, but now $\Delta = 0$. Where the waves are damped, ω_R/Ω is plotted as a solid line; where they are unstable and the parallel group velocity $\partial \omega / \partial k_\parallel$ is positive, as a dotted line; and where they are unstable and $\partial \omega / \partial k_\parallel$ is negative, as a dashed line. $\partial \omega / \partial k_\perp$ is zero at the transition from dashed to dotted lines. If this transition occurs near the local maximum of the dispersion curve, where $\partial \omega / \partial k_\perp$ is zero, the spatial growth rate is large; if $\partial \omega / \partial k_\parallel$ and $\partial \omega / \partial k_\perp$ are zero simultaneously, we have a strong indication of nonconvective instability.

Note first of all that unstable waves have $k_\perp \rho_H > 1.5$ and therefore perpendicular phase velocities $\omega/k_\perp$ smaller than $A_\perp$ that lie in the region of positive slope of the $\Delta = 0$ distribution in Fig. 15, a rough necessary condition for loss cone instability (ROSENBLUTH and POST, 1965). It is

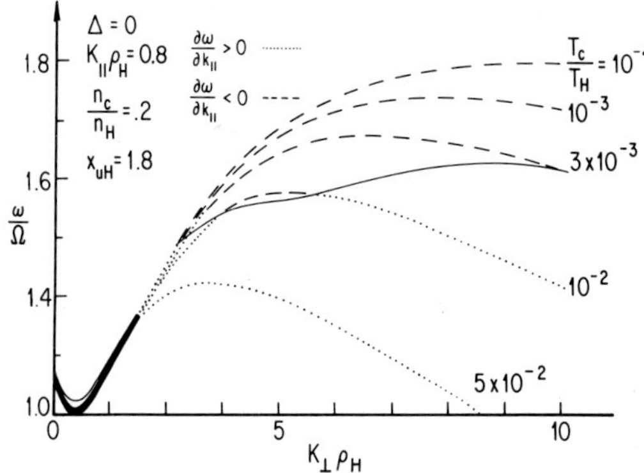

FIG. 17. Effect of cold electron temperature on the Harris mode dispersion relation. ω/Ω is plotted vs. $k_\perp\rho_H$ for $n_c/n_H=0.2$, $X_{UHC}=1.8$, $\Delta=0$, a single $k_{\parallel}\rho_H=0.8$, and various T_c/T_H. Where $\partial\omega/\partial k_{\parallel}>0$ and the mode is unstable, the lines are dotted; where $\partial\omega/\partial k_{\parallel}<0$, dashed; where the mode is damped, the lines are solid. When T_c/T_H is small, the frequency approaches the cold upper hybrid frequency for large $k_\perp\rho_H$. In general, the dispersion relation peaks and $\partial\omega/\partial k_\perp=0$ near $k_\perp\rho_c\simeq1$. For T_c/T_H small enough, the parallel group velocity is small near these peaks, and the spatial growth rate is large. For $T_c/T_H>10^{-2}$, in this case, the parallel group velocity has no zeroes.

precisely when $k_\perp\rho_H$ exceeds unity that a small density of cold electrons can affect the dispersion relation, since the Bessel functions of the dominant harmonic terms contributing to the hot electron dielectric response are already diminishing. On the other hand, the contributions of the $n\neq0$ cold electron terms increase until the perpendicular wavelength decreases to the cold electron cyclotron radius, or $k_\perp\rho_c\simeq1$. When $T_c=0$, this never happens and the dispersion relation approaches that for a cold plasma electrostatic wave propagating perpendicular to the magnetic field as the coupling to hot electrons diminishes with increasing $k_\perp\rho_H$. In other words $X\rightarrow X_{UHC}$. $T_c/T_H=10^{-4}$ corresponds to this limit. On the other hand, our knowledge of Bernstein modes tells us that the frequency of the mode in the cold upper hybrid band diminishes with increasing $k_\perp\rho_c$. This effect becomes increasingly important as T_c/T_H increases. For example, when $T_c/T_H=.05$ in Fig. 17, the wave frequency reaches a local maximum where $k_\perp\rho_H\simeq5$ and $k_\perp\rho_c\simeq1$. Thus, by the time that hot electron effects have diminished enough for the dispersion relation to feel the cold electrons, the cold electrons produce a mode whose frequency peaks below X_{UHC}.

The spatial growth rate can be large near the local maximum where $\partial\omega/k_\perp=0$ if $\partial\omega/\partial k_{\parallel}$ is small as well. When $T_c/T_H<10^{-2}$ in Fig. 17, $\partial\omega/\partial k_{\parallel}$

changes sign near where $\partial\omega/\partial k_\perp=0$, suggesting that a neighboring mode with a slightly different $k_\parallel$ would be nonconvective. It turns out that $\partial\omega/\partial k_\parallel$ scales as the cold electron thermal speed, and is much smaller than A_H. Taken together, the above arguments indicate that a nonconvective instability is possible below the cold upper hybrid frequency when T_c/T_H is sufficiently small. Increasing T_c/T_H eventually removes the nonconvective property, as in the case $T_c/T_H=0.5$ of Fig. 17. When $T_c/T_H=1$, there is no difference between cold and hot electrons (except for the hot loss cone free energy) and the first harmonic mode is stable (YOUNG *et al.*, 1973).

What about higher harmonic modes? Since the total upper hybrid frequency is 3.8, and the hot upper hybrid frequency 3.5, times the electron cyclotron frequency, we might also expect instabilities in the next two harmonic bands. We have calculated, for the same parameters as in Fig. 17, the second harmonic dispersion relation on an expanded scale. When $T_c/T_H<.05$, there can be temporal instability, but since the parallel group velocity does not change sign, there can be no nonconvective instability. The frequencies do not rise much above 2Ω when $k_\perp\rho_H>1$, because the cold electron Bernstein modes above the cold upper hybrid frequency do not do so. From this argument, we tentatively conclude that nonconvective instability occurs only at or below the cold upper hybrid frequency.

To determine the generality of the conclusions we drew above from the solutions for $k_\parallel\rho_H=0.8$, we solve the Harris dispersion relation for all unstable $k_\parallel$ and $k_\perp$ to find the maximum spatial growth rates, or point(s) of nonconvective instability, for a given n_c/n_H and T_c/T_H. We then repeat this procedure for many n_c/n_H and T_c/T_H. Figure 18 presents a complete parameter search of the first and second harmonic bands for $X_{UHC}=3$ and a partially filled loss cone with $\Delta=0.2$. We have shaded the regions of nonconvective instability in a $(n_c/n_H, T_c/T_H)$ parameter space; outside them, we have contoured the spatial growth rates (solid lines) and frequencies (dotted lines) of the maximally growing modes. Nonconvective instability occurs for a finite range of n_c/n_H, between 0.1 and 2 in this case. Moreover, both harmonic bands turn convective at the same $T_c/T_H=.05$, when $n_c/n_H<1$. The third harmonic band was convective throughout.

When n_c/n_H is sufficiently large, we expect nonconvective instability only in the harmonic band containing the cold upper hybrid frequency. In this limit, the cold electrons dominate the real part of the dispersion relation everywhere, and the hot electrons only affect the growth rate. The cold electron Bernstein modes in the bands below the upper hybrid frequency start at the cyclotron harmonics at small $k_\perp\rho_c$. Here loss cone instability is

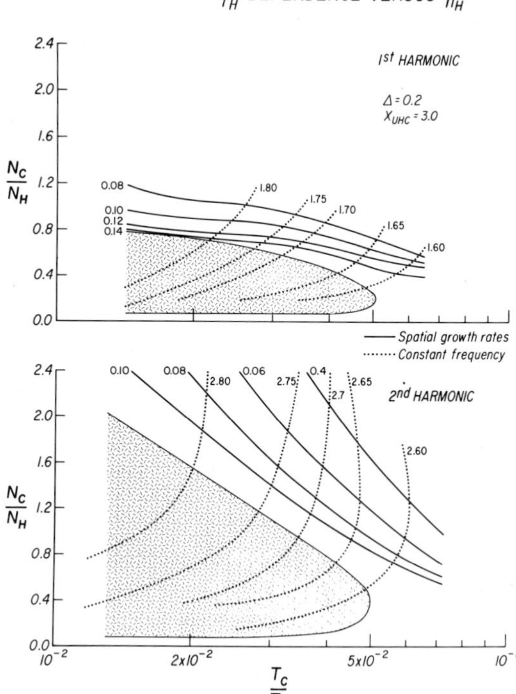

Fig. 18. Survey of T_c/T_H dependence versus n_c/n_H. Shown here are the results of a complete parameter search. Keeping $X_{\mathrm{UHC}}=3$ and $\Delta=0.2$, the entire unstable region was calculated for each n_c/n_H and T_c/T_H, and the nonconvective or fastest spatially growing modes were identified. The nonconvective zones are shaded. Outside, the solid lines are contours of constant maximum $k_\perp\rho_H$. The dotted lines show the frequencies of the fastest growing or nonconvective modes. The first harmonic is above, and the second is below. The third harmonic always remained convective. Both the first and second harmonics turned convective at about the same T_c/T_H. At the convective transition, the frequencies were slightly above the odd half harmonics.

swamped by cold electron cyclotron damping. In the hybrid band the cold electron mode can be between cyclotron harmonics, so that cold cyclotron damping can be avoided. Moreover, $\partial\omega/\partial k_\perp$ is small unitl $k_\perp\rho_c\simeq 1$, and so large spatial growth rates are possible. Figure 18 did not test this possibility since there the cold upper hybrid frequency was precisely three times the cyclotron frequency. In Fig. 19, we keep T_c/T_H fixed at .05, vary n_c/n_H and

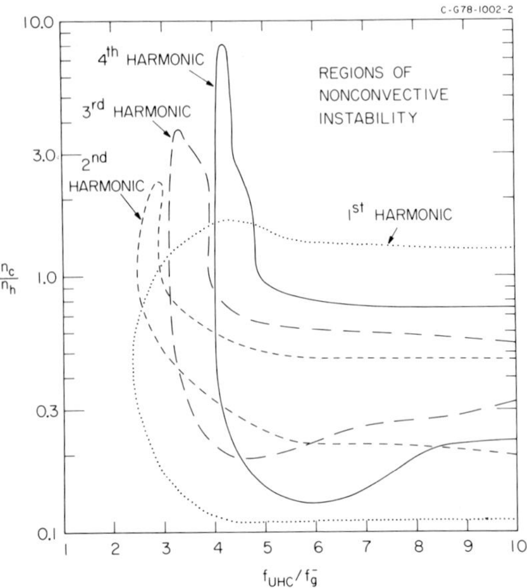

FIG. 19. Regions of nonconvective instability. Here we plot the regions of nonconvective instability in n_c/n_H, $X_{UHC} = f_{UHC}/f_g$—parameter space, for the first four harmonic bands. When $n_c/n_H < 1$, each of the harmonics can be nonconvectively unstable so long as its frequency is below the cold upper hybrid frequency. When $n_c/n_H > 1$, a given harmonic can be nonconvective only if it contains the cold upper hybrid frequency. Generally speaking, we expect $n_c/n_H < 1$ in the diffuse aurora, and $n_c/n_H > 1$ near the plasmapause. This figure has been used to interpret observations such as in Fig. 26, where upper hybrid emissions near the plasma pause intensified at odd half harmonic frequencies.

X_{UHC}, and plot the boundaries of the nonconvective regions of the first four harmonic bands. We choose $\Delta = 0$. This figure reveals an important difference between $n_c/n_H < 1$ and $n_c/n_H > 1$ regimes. When $n_c/n_H < 1$, a multiharmonic nonconvective instability, involving all the bands at or below the cold upper hybrid frequency, is possible. When n_c/n_H exceeds 1.6, only that harmonic band containing the cold upper hybrid frequency is nonconvective. When n_c/n_H is sufficiently large, nonconvective instability disappears altogether.

6.4 Location of growing modes in K-space

We have chosen the extreme case displayed in Figs. 20a, b to make our point. There we contour the spatial growth rates in $(k_\perp, k_\parallel)$ space, for $T_c/T_H = .079$, $n_c/n_H = 0.3$, $\Delta = 0$, and the large ratio $X_{UHC} = 5$. We shade

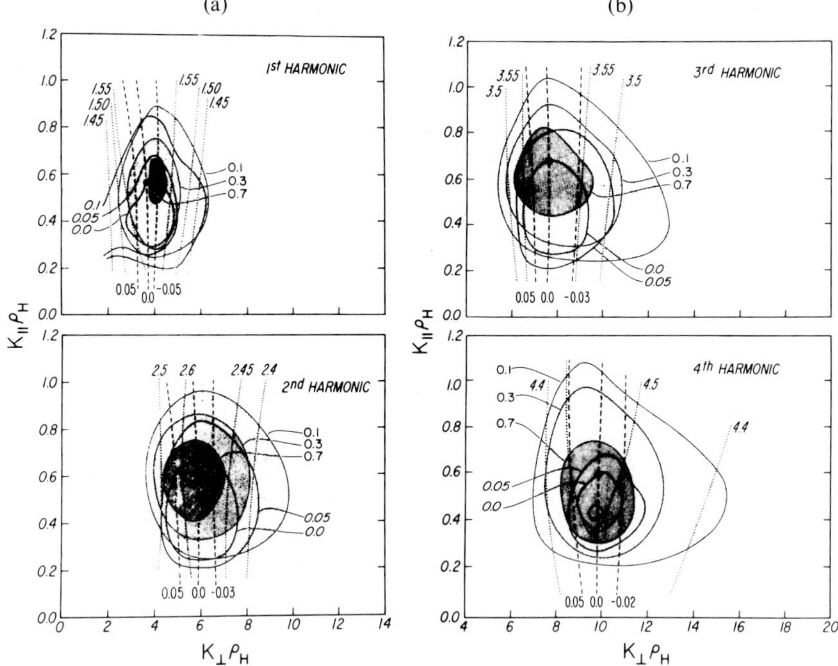

FIG. 20. Cyclotron harmonic instabilities in K space. We have shaded regions of $k_I\rho_H = 0.1$, 0.3, and 0.7 in $(k_\perp, k_\parallel)$ space for $T_c/T_H = .078$, $n_c/n_H = 0.3$, $\Delta = 0$, and $X_{UHC} = 5$. The dotted lines denote constant frequency. The parallel (solid lines) and perpendicular (dashed lines) group velocities normalized to the hot electron thermal speed are also contoured. Where the group velocity components are simultaneously zero, the nonconvective point is marked by a large dot. The first two harmonics are shown in (a) and the third and fourth, in (b). The fifth harmonic remained convective. The most rapidly growing modes all have $k_\parallel\rho_H \simeq 0.6$, and so resonate with roughly the same electrons. However, the spatial growth rates peak near $k_\perp\rho_c \simeq \sqrt{m}$, where m is the harmonic number.

regions of spatial growth rate $k_I\rho_H = 0.1$, 0.3, 0.7. Dotted lines denote contours of constant frequency; solid lines, parallel group velocity; and dashed lines, perpendicular group velocity. Where there is a simultaneous zero of $\partial\omega/\partial k_\perp$ and $\partial\omega/\partial k_\parallel$, the corresponding nonconvective point is marked by a black dot. The fifth harmonic band, while unstable, remained convective, and we chose not to present it here. The peak growth rates occur roughly at $k_\perp\rho_H = (2mT_H/T_c)^{1/2}$, near the maxima of the cold electron Bessel functions. Because each harmonic band is most unstable for roughly the same $k_\parallel\rho_H$, the velocities of the resonant particles that make the dominant contributions to each harmonic's growth rate are roughly equal. Each harmonic feeds off the same region of velocity space.

6.5 Discussion

Here we summarize the qualitative implications of our parameter searches. First, nonconvective instability is possible when T_c/T_H is sufficiently small. This suggests that distribution functions with smaller free energy sources than those used in our numerical calculations can still be strongly unstable. Since we used essentially the same free energy source in all our calculations, where the spatial growth rates peak depends primarily upon the properties of the cold electrons.

Concerning the effects of the cold upper hybrid frequency on multiharmonic instabilities, we found:

1) The cold upper hybrid frequency determines the harmonic bands that can be nonconvectively unstable.

2) When $n_c/n_H \lesssim 1$, nonconvective instability occurs in all the bands at and below the cold upper hybrid frequency.

3) When $n_c/n_H \gtrsim 1$, nonconvective instability occurs only in that band containing the cold upper hybrid frequency.

Concerning the effects of T_c/T_H on multiharmonic instabilities, we found that:

1) Nonconvective instability is possible when T_c/T_H is sufficiently small.

2) When $n_c/n_H \lesssim 1$, all harmonic bands turn convective at about the same T_c/T_H.

3) When $n_c/n_H > 1$, the different bands turn convective at different T_c/T_H.

Concerning the wavenumbers and frequencies of individual instabilities, we found that:

1) The fastest spatially growing, or nonconvective modes, have increasing $k_\perp \rho_H$ with increasing harmonic number.

2) The fastest growing, or nonconvective modes have about the same $k_\parallel \rho_H$, implying that waves in each harmonic band resonate with roughly the same electrons.

3) The most unstable waves have frequencies nearest half-harmonic frequencies near the convective-nonconvective transition at large T_c/T_H.

ASHOUR-ABDALLA *et al.* (1979) showed that filling in the loss cone by increasing Δ leaves the instabilities nonconvective, though it reduces the size of the nonconvective region in parameter space. We emphasize that we do not believe that the electrostatic instabilities whose consequences we observe in space will *necessarily* be nonconvective. Since the spatial growth rates remain appreciable near but outside the regions of nonconvective instability in parameter space, the nonconvective regions indicate where instability is important.

Parameter searches reveal physical trends that can guide the qualitative interpretation of magnetospheric observations. However, specific numerical predictions, such as the T_c/T_H for convective-nonconvective transition, may not be applicable to a given observation. Only use of measured hot and cold electron distributions tests the theory reliably. Nonetheless, these trends help to sort out the phenomenology of the waves observed in the magnetosphere. For example, cyclotron harmonic waves should not occur inside the plasmasphere because n_c/n_H is much too large; but just beyond the plasmapause, we expect upper hybrid emissions because n_c/n_H is moderately large. We expect first harmonic emissions on nightside diffuse auroral lines of force because n_c/n_H should be small and X_{UHC} not too large. Moreover, those observed at high latitudes where X_{UHC} can be near unity should be nearer the cyclotron frequency than those at the geomagnetic equator where X_{UHC} probably is largest on given flux tube.

7. Observations of Magnetospheric VLF Electric Fields

7.1 *Introduction*
Early research established the phenomenology of VLF electric field emissions in earth's magnetosphere. More recent parameteric studies of electron cyclotron harmonic instabilities provided the background necessary to classify the observed wave spectra. Examples of each type of terrestrial emission have recently been found in Jupiter's magnetosphere. We will discuss these developments in order.

7.2 *OGO-5 observations*
The VLF electric field instrument onboard OGO-5, the first to operate successfully beyond the plasmapause, detected several classes of narrow band, apparently electrostatic waves whose frequencies exceeded the local electron cyclotron frequency (KENNEL *et al.*, 1970). The most common were odd-half harmonic emissions, with frequencies between cyclotron harmonics. OGO-5 detected multiple half harmonic emissions more rarely than the basic 3/2 emission. At the top of Fig. 21 is a frequency-time spectrogram from June 16, 1968 when OGO-5 was near the local dawn

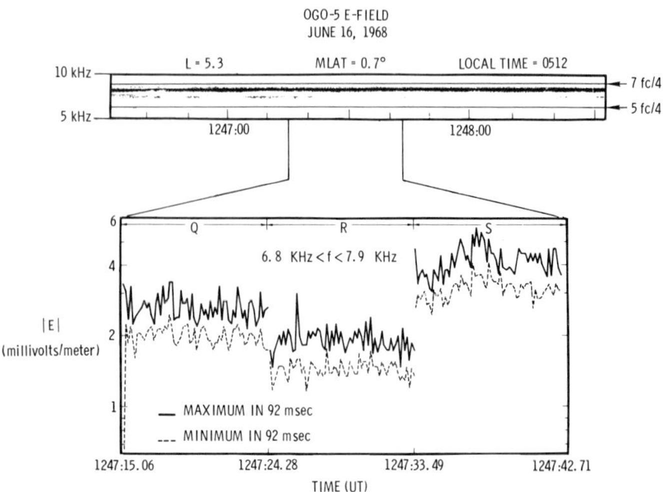

FIG. 21.   Class Ib (3/2) emissions observed on OGO-5. Here we show simultaneous broad-band analog data (top inset, in a frequency-time display) and narrow band digital data for an emission near $3f_c/2$. The interference lines from the Rb-magnetometer on board OGO-5, at odd quarter harmonics of f_c, enabled a quick visual determination of the local cyclotron frequency. The 15% narrow band channel centered at 7.35 kHz registered a quasi-steady electric field amplitude of about 2.5 mV/m. The peak and minimum amplitudes measured in 92 ms differed by about 50%. Thus this particular event was not impulsive. Q, R, and S indicate the periods during which individual directional elements of the orthogonal triad of E-field antennas were sampled (KENNEL *et al.*, 1970).

geomagnetic equator at $L=5$. Interference lines at odd quarter cyclotron harmonics from the JPL/UCLA magnetometer define the local magnetic field strength. The 8 kHz emission is between 3 fce/2 and 7 fce/4 (fce $=\Omega/2\pi$). The bottom inset shows that the calibrated electric field amplitude measured in a narrow band channel was 2–4 mV/m, depending upon the vector component (Q, R, S) of the electric field sampled. Not all electric field emissions were as unstructured as this one. Figure 22 shows "3/2" emissions whose intensity was intermittent, or whose frequency alternated between two values, or whose frequency drifted with time. The local magnetic field remained constant in all three cases. OYA (1972) attempted an explanation of such complex time dependences in terms of nonlinear wave-wave couplings.

The intense emissions observed on OGO-5 were probably locally generated. The emission in Fig. 23 kept its frequancy near 3/2 fc even when the local magnetic field strength changed by ten percent in ten seconds. Since

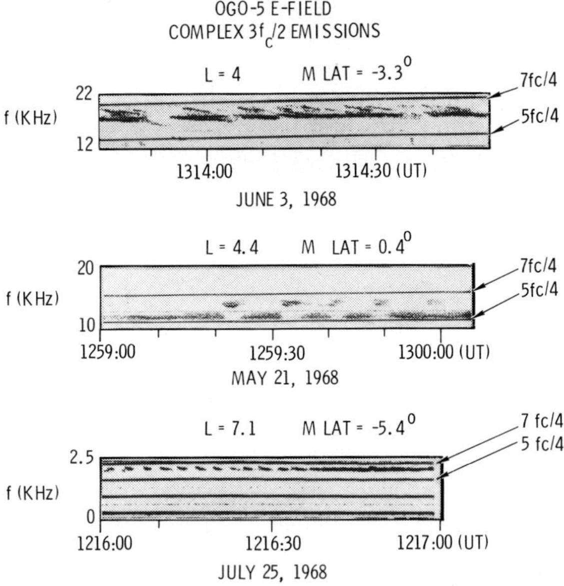

FIG. 22.   Complex class Ib (3/2) emissions observed on OGO-5. The top inset shows an event with considerable structure above $3f_c/2$, but little below; in the middle, the frequency alternates from below to above $3f_c/2$; at the bottom there is a periodically modulated 3/2 emission that was associated with a periodic oscillation in the magnetic field. We present this picture to remove the impression that magnetospheric cyclotron harmonic waves always resemble smooth steady hiss bands.

its bandwidth was about 10% of f_c, the waves cannot have propagated more than the distance in which the magnetic field changes by ten per cent, or about 0.2 R_E.

The first OGO-5 study found that intense electrostatic waves were localized within a few degrees of the geomagnetic equator. However, this result was biassed, since only the data between $-10° < \lambda_m < 10°$ was sampled completely, and the magnetometer interference set a variable intensity threshold of about 1 mV/m. FREDRICKS and SCARF (1973), using later OGO-5 measurements which were more sensitive because the magnetometer interference was absent, found electric field emissions at latitudes well above 10°. Recent studies on ISEE and GEOS, whose intensity thresholds were about 1 μV/m, continue this trend. Diffuse low intensity bands of electrostatic waves can be found nearly everywhere beyond the plasmapause. On the other hand, it now appears that some classes of intense electrostatic

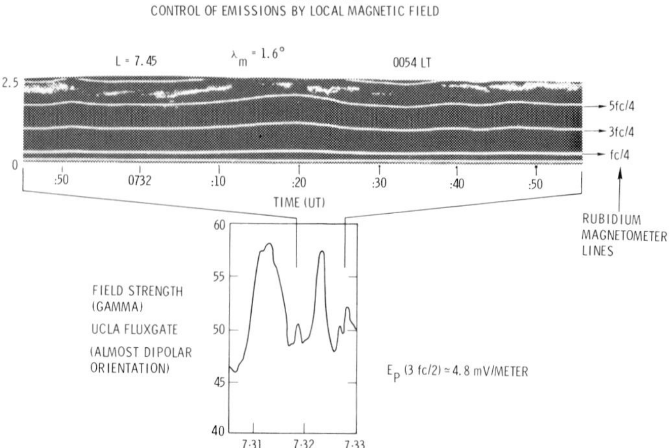

Fig. 23. Control of 3/2 emissions by local magnetic field. This figure was one of the first to suggest that 3/2 emissions are locally generated near the spacecraft because their center frequency tended to follow the local variations in the geomagnetic field.

emissions are equatorially localized. Despite the fact that OGO-5 was in a position to observe intense electrostatic waves relatively infrequently, a majority of equator crossing beyond the plasmapause had detectable activity between 0000–1200 LT. Thus, KENNEL *et al.* (1970a) argued that intense electrostatic waves were present near the equator at least half the time on auroral lines of force. FREDRICKS and SCARF (1973) extended this conclusion to 1800 LT for the higher sensitivity data.

7.3   *Types of magnetospheric electron cyclotron harmonic emission*
7.3.1   *Introduction*

In retrospect OGO-5 identified nearly all the magnetospheric wave emissions with amplitudes exceeding about 0.1 mV/m that were found with subsequent, more sensitive experiments on spacecraft in different orbits. However, at that time, there did not exist a sufficient understanding of linear instability theory to know how to distinguish between theoretically meaningful classes of waves. For example, KENNEL *et al.* (1970a) grouped together all the emissions with $f \gg f_c$, whereas we now know that there are three different types of waves in this frequency range. The observational and theoretical classification of magnetospheric cyclotron harmonic emissions went hand in hand over the next decade.

7.3.2   *Emissions with $f \simeq f_c$*

Figure 24 shows an emission with frequency near the cyclotron frequency. These were observed much more rarely on OGO-5 than 3/2 emissions, because of the equatorial bias of the OGO-5 data sample. However, they were the emissions most frequently observed when IMP-7 was more than 20° from the geomagnetic equator (HUBBARD *et al.*, 1979).

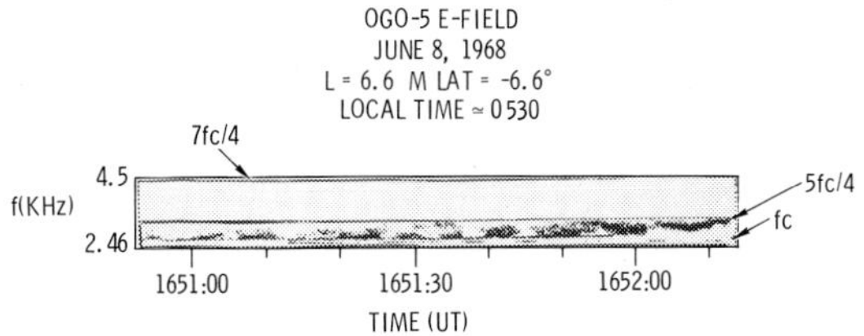

FIG. 24.   A class Ia emission observed by OGO-5. Shown here is a structured class Ia emission with $f \simeq 1.25 f_c$ observed by OGO-5 slightly off the geomagnetic equator. OGO-5 also detected multiharmonic emissions at freqeuncies well above the local cyclotron frequency, but we do not show these results.

7.3.3   *Diffuse bands*

High sensitivity electric field experiments on IMP-6, Hawkeye, ISEE 1 and 2, and GEOS frequently detect low intensity electrostatic waves beyond the plasmapause in several cyclotron harmonic bands at and below the local cold upper hybrid frequency. They differ from strong multi-harmonic emissions (Section 7.3.5) in two significant ways. Their integrated amplitude is nearer $1 \, \mu V/m$ than the 0.1–$10 \, mV/m$ of multiharmonic emissions, and their bandwidth is broader than the 10% characteristic of intense emission—thus the term diffuse band. Figure 34 shows an example of diffuse bands measured on ISEE-1.

CHRISTIANSEN *et al.* (1978a, b) suggested that diffuse bands were enhanced thermal fluctuations (see also ASHOUR-ABDALLA *et al.*, 1978b). The incoherent Cerenkov and cyclotron harmonic noise radiated by individual electrons can rise several orders of magnitude above the thermal level if the electrons have some residual free energy, even if it is insufficient for instability. Christiansen *et al.* argue that a stronger, unstable free energy source would soon destroy itself quasilinearly and could not last the long periods of time that diffuse bands are observable.

The thermal fluctuation amplitude E_{TH} may be crudely estimated by

$$\frac{E_{TH}^2}{8\pi} \simeq \frac{nT_e}{n\rho^3} \tag{39}$$

where $\eta\rho^3$ is the number of electrons in a Larmor cube. With $n \sim 1\,\mathrm{cm}^3$, $T_e \simeq 1\,\mathrm{keV}$, $B \simeq 100\,\gamma$, (39) yields $E_{TH} \approx 0.2\,\mu\mathrm{V/m}$. Therefore, waves with amplitudes of a few $\mu\mathrm{V/m}$ could be enhanced thermal fluctuations. This also indicates that present electric field detectors have a sensitivity adequate to measure all the cyclotron harmonic waves of interest in the outer magnetosphere.

7.3.4 Upper hybrid emissions

GURNETT and SHAW (1973), MOSIER et al. (1973), and SHAW and GURNETT (1975) first identified electrostatic noise near the upper hybrid frequency at or slightly beyond the plasmapause. The emission frequency was observed to track the upper hybrid frequency from just inside the plasmapause, across the sharp density decrease at the plasmapause, and into the outer magnetosphere in about two thirds of all the IMP-6 plasmapause crossings (KURTH, 1979). Ordinarily these upper hybrid emissions are part of the diffuse band system; they have relatively low amplitudes, about a few $\mu\mathrm{V/m}$. Occasionally, they rise to amplitudes of $1\text{--}20\,\mathrm{mV/m}$, making them among the most intense electrostatic waves observed in the magnetosphere (KURTH, 1979; KURTH et al., 1979a). KURTH (1979) found 145 examples of upper hybrid emissions above a $1\,\mathrm{mV/m}$ threshold in 1500 IMP-7 and Hawkeye plasmapause crossing observed over five and one-half years, and GURNETT (1975), CHRISTIANSEN et al. (1978a), HUBBARD and BIRMINGHAM (1978), and HUBBARD et al. (1979) have discussed isolated examples. They occur at all local times and in the latitude range $-50° < \lambda_m < 50°$ (KURTH, 1979). They do not correlate strongly with magnetic storms or auroral substorms.

The great intensity of upper hybrid emissions makes it possible to study their polarization (KURTH et al., 1979a). Figure 25 shows the electric field amplitude measured in logarithmically spaced frequency channels during a Hawkeye plasmapause and magnetic equator crossing. Analysis of the spin modulation of the amplitude with respect to the local magnetic field direction enables identification of regions where the wave electric field is parallel (shaded) and perpendicular (unshaded) to the magnetic field. The emission at $31.1\,\mathrm{kHz}$ had an electric field vector that was inclined more than $80°$ to the magnetic field, consistent with the expected polarization of upper hybrid waves. The waves with $E\|B$ are most likely low-level continuum radiation in the $[L, 0]$ mode, which we will discuss shortly. The intense upper hybrid waves have a measured magnetic amplitude of $1/30\text{--}1/40$ the electric field amplitude, so that they are nearly longitudinal.

Upper hybrid waves often intensify when the upper hybrid frequency

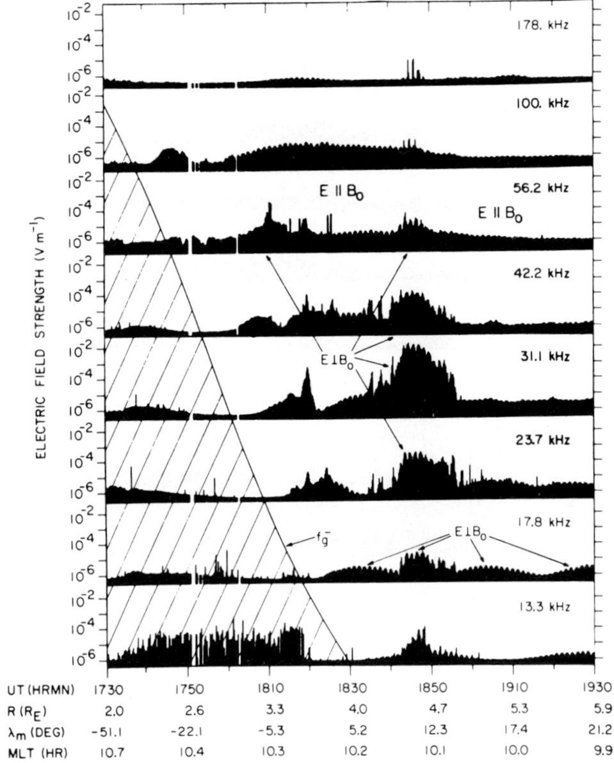

UT (HRMN)	1730	1750	1810	1830	1850	1910	1930
R (R_E)	2.0	2.6	3.3	4.0	4.7	5.3	5.9
λ_m (DEG)	−51.1	−22.1	−5.3	5.2	12.3	17.4	21.2
MLT (HR)	10.7	10.4	10.3	10.2	10.1	10.0	9.9

FIG. 25.   Intense upper hybrid (class 4) emissions observed by Hawkeye. A display of plasma wave measurements (KURTH, 1979) of an intense upper hybrid event with a summary of polarization measurements superimposed over the data. The spin modulation apparent in many of the features has been analyzed to determine the orientation of the wave electric field with respect to the geomagnetic field. Shaded are regions of frequency and time where all waves are polarized such that $E\|B_0$, consistent with the polarization of nonthermal continuum radiation. The unshaded regions denote frequencies and times, including the intense electrostatic event at 31.1 kHz, where waves with measurable polarization have $E\perp B_0$, consistent with the expected polarization of upper hybrid waves. The hatched region contains whistler mode turbulence.

equals an odd half harmonic of the electron cyclotron frequency. Lines of constant $(n+\frac{1}{2})f_c$ are superimposed on the amplitude data in Fig. 26. The emission at 31.1 kHz intensifies when $7/2\,f_c \simeq f_{\mathrm{UHR}}$. Successive intensifications each time $(n+\frac{1}{2})f_c \simeq f_{\mathrm{UHR}}$ are commonly observed as the spacecraft moves across the plasmapause (W. S. Kurth, 1979, private communication).

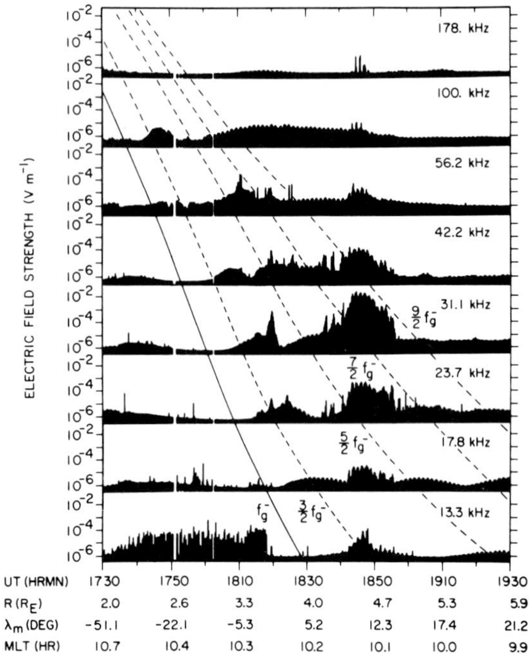

Fig. 26. Intensification of upper hybrid emissions at odd half harmonics. A display of plasma wave measurements (Kurth, 1979) of the same event shown in Fig. 25, here overlaid with lines denoting the cyclotron frequency, f_g^-, and lower order $(n+\frac{1}{2})f_g^-$ bands using the ordinate as a frequency scale. A large number of the features lie at or near the $(n+\frac{1}{2})f_g^-$ harmonics. In particular, the intense event of 31.1 kHz lies directly at the $7f_g^-/2$ harmonic. This event may be interpreted using Fig. 20 as follows. As Hawkeye moved outward beyond the plasmapause, the local cold upper hybrid frequency decreased. The most intense emission occurred when $f_{\mathrm{UHC}} \simeq 7/2\, f_g^-$. Lower harmonic emissions were also observed. A similar event was observed by ISEE-1 on Nov. 5, 1977 (see Fig. 34, 35, and 37).

The earth's magnetosphere contains a low level of nonthermal electromagnetic radiation, whose frequency exceeds the local plasma (or upper hybrid) frequency, that is trapped between high plasma density reflection points at the plasmapause and magnetopause (Gurnett and Shaw, 1973; Gurnett, 1975). Kurth (1979) suggests that the continuum radiation may originate, by mode-conversion or mode coupling, from the intense upper hybrid electrostatic emissions. In some cases, the intensity of the continuum radiation appears to fall off as the inverse square of the distance from a plasmapause crossing at which upper hybrid noise is observed. In other cases, the relation between the two is less clear.

7.3.5 Multiharmonic emissions

Figure 27 shows three hours of data taken when ISEE-1 was at magnetic latitudes of 17–22°, well beyond the plasmapause, and moving outbound around magnetic local dawn on November 22, 1977 (D. Gurnett and W. Kurth, 1980, private communication). Several odd half harmonic emissions are observed simultaneously. We note that all the odd half harmonics are present up to an upper cutoff, and they have a tendency to intensify simultaneously, viz., near 2000 UT. This upper cutoff is often the lower cutoff of the continuum radiation, and is close to the upper hybrid frequency. Fredricks (private communication) has remarked that the first harmonic is nearly always present when one or more higher harmonics are observed.

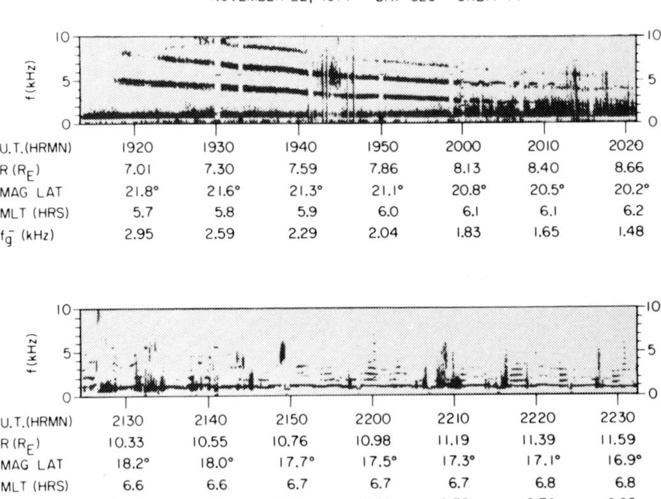

U.T.(HRMN)	1920	1930	1940	1950	2000	2010	2020
R (R_E)	7.01	7.30	7.59	7.86	8.13	8.40	8.66
MAG LAT	21.8°	21.6°	21.3°	21.1°	20.8°	20.5°	20.2°
MLT (HRS)	5.7	5.8	5.9	6.0	6.1	6.1	6.2
f_g^- (kHz)	2.95	2.59	2.29	2.04	1.83	1.65	1.48

U.T.(HRMN)	2130	2140	2150	2200	2210	2220	2230
R (R_E)	10.33	10.55	10.76	10.98	11.19	11.39	11.59
MAG LAT	18.2°	18.0°	17.7°	17.5°	17.3°	17.1°	16.9°
MLT (HRS)	6.6	6.6	6.7	6.7	6.7	6.8	6.8
f_g^- (kHz)	0.86	0.80	0.78	0.72	0.75	0.72	0.68

FIG. 27.   A multiharmonic (class 2) emission observed on ISEE-1. This figure shows a long lasting multiharmonic event observed on an outbound pass of ISEE-1 near local dawn, well beyond the plasmapause (W. Kurth and D. Gurnett, 1979, private communication). Filling of the dayside magnetosphere by the ionosphere is expected to provide a cold electron density such that f_{UHC} exceeds $(n+\tfrac{1}{2})f_g^-$. Note that the lower harmonics, and particularly 3/2, are nearly always present, whereas the highest harmonics come and go, perhaps reflecting density structures in the outer magnetosphere.

TABLE 1. Electrostatic emissions in the outer magnetosphere.

Class	Names (b)	Frequency range	Intensity	Observational comments	Theoretical comments
1a	Low 3/2 or cyclotron waves	$1 < f/f_c < 1.2$	$> 1\,\mathrm{mV/m}$	Most common class for $20 < \lambda < 55°$	Requires $f_c \lesssim f_{UHC} \ll 2 f_c$
1b	3/2	$1.2 < f/f_c < 2$ $\Delta f/f_c \lesssim 0.2$	$> 1\,\mathrm{mV/m}$	Possibly most common near equator	Requires $f_{UHC} < 2 f_c$
2	$n + \frac{1}{2}$ or mutiharmonic emissions	$\Delta f/f_c \approx 0.1$ in each band $f_{max} \lesssim f_{UHR}$	$> 1\,\mathrm{mV/m}$ in each band	Less common than 3/2 emission; nearly always include a 3/2 component	Requires $f_{UHC} > 2 f_c$
3	Diffuse bands	$f_{max}/f_c \lesssim 4$ $\Delta f/f_c \simeq 0.5$ in lowest band	Few $\mu\mathrm{V/m}$	Common on dayside at all latitudes	May not be unstable
4	$f \sim f_{UHR}$ intense upper hybrid emission	$f \ll f_{max} \sim f_{UHR}$ $\Delta f/f_c \approx 0.1$	$1\text{–}20\,\mathrm{mV/m}$	Observed near but beyond plasmapause intensify when $(n+\frac{1}{2})f_c \simeq f_{UHR}$	Requires $n_c \gg n_H$

7.4   Classification and spatial distribution of cyclotron harmonic emissions

Table 1 has been adapted from HUBBARD and BIRMINGHAM (1978) who used this format to summarize the available observational and theoretical knowledge about cyclotron harmonic emissions.

Figures 28, 29, and 30 present statistical surveys of one year of IMP-6 cyclotron harmonic wave observations (HUBBARD et al., 1979). Figure 28 shows that class 1 emissions were almost always present on nightside diffuse auroral lines of force, and were often found on the dayside as well. According to Fig. 29, class 2 emissions were more likely to be on the dayside. The occasions when continuum radiation was also present are indicated by heavy lines. Class 3 (light shading) and class 4 (dark shading) were largely dayside phenomena. Figure 30 displays occurrence probability distributions as a function of geomagnetic latitude for classes 3 and 4 (top panel) and 1 and 2 (bottom panel). IMP-6 spent very little time at low latitudes and so probably did not test as well as OGO-5 did the equatorial occurrence of class 1b (3/2) emissions; HUBBARD and BIRMINGHAM (1978) found that the class 1a emissions, observed infrequently in the OGO-5 equatorially biassed sample, dominated the IMP-7 observations above $\lambda_m \gtrsim 20°$.

The electrostatic wave experiment team on GEOS-2 (GOUGH et al., 1979) has arrived at a classification scheme similar to that of HUBBARD and

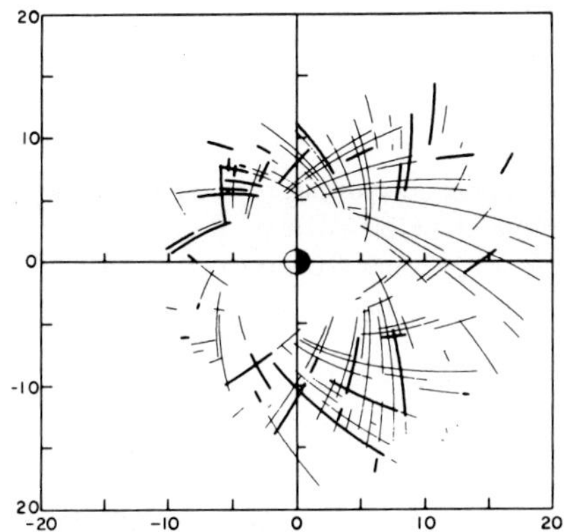

FIG. 28.   Spatial distribution of class 1 events from IMP-6. Plotted as a function of local time (measured counterclockwise from noon at the left) is the radial position of IMP-6 when class I emissions were observed (HUBBARD et al., 1979). The period studied was Feb. 5, 1972 to Feb. 4, 1973. The heavy trace indicates the simultaneous observation of continuum radiation. Note that class I emissions were present over large portions of the IMP-6 orbits in the night side diffuse auroral region. A similar but less complete spatial distribution was obtained by the early OGO-5 experiment.

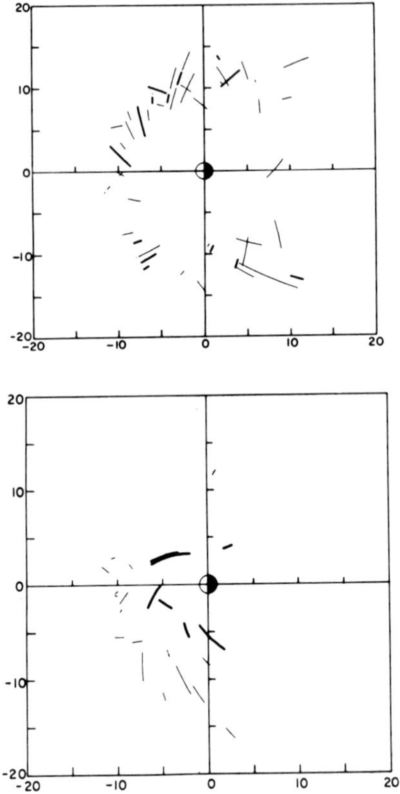

FIG. 29.   Spatial distribution of classses 2, 3 and 4 events from IMP-6. The top inset shows, for the same time period covered in Fig. 28, the spatial distribution for multiharmonic (class 2) events (HUBBARD *et al.*, 1979). Again, simultaneous observation of continuum radiation is denoted by a heavy line. Multiharmonic emissions are primarily a dayside phenomenon. The bottom trace shows the spatial distributions of class 3 diffuse bands (dark trace) and class 4 intense upper hybrid emissions (light trace).

BIRMINGHAM (1978), using wave measurements and active and passive measurements of the cold plasma density:

Class I: A single frequency band near but above the electron cyclotron frequency found near local midnight. Here $f_{pc} < f_c$.

Class II: Multiharmonic emissions, with the most intense component in the first harmonic band, observed near local dawn. Here $f_{pc} \simeq f_c$.

Class III: Multiharmonic emissions, with the most intense component

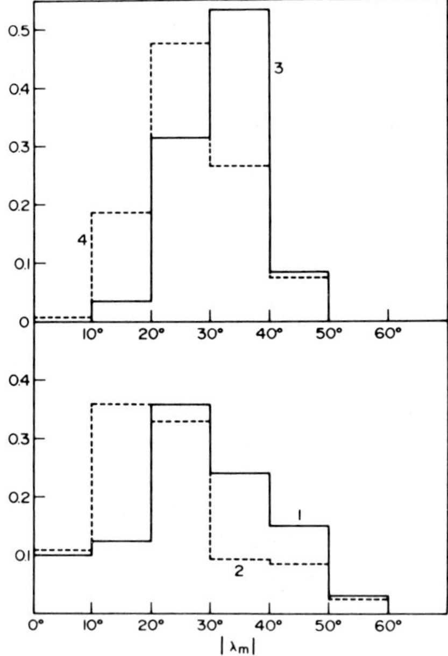

FIG. 30. IMP-6 distribution of electrostatic emissions with geomagnetic latitude λ_m. The occurrence probability of class 1 and 2 (bottom inset) and classes 3 and 4 (top inset) waves (HUBBARD *et al.*, 1979). Each histogram is normalized to unity; note there were only 7 events in class 3. The data are biassed by the small amount of time IMP-6 spent at small $|\lambda_m|$. These data supplement the original OGO-5 latitudinal distributions, which were biased towards small $|\lambda_m|$. More experiment and theory are clearly needed to sort out the latitude distribution.

in the cold upper hybrid band. Here $f_{pc} > 2f_c$.

Class IV: A weak cold upper hybrid emission, unaccompanied by detectable emissions in the lower harmonic bands.

Although the experimental community should settle this issue itself, it appears to us that class I of GOUGH *et al.* (1979) may correspond to classes 1a and/or 1b of HUBBARD and BIRMINGHAM (1978); classes II and III may be parts of class 2, and class IV may be related to diffuse bands (class 3). GEOS-2 may not have encountered the plasmapause sufficiently often to detect many intense upper hybrid emissions.

Class III events were recorded during the vast majority of GEOS-2 equator crossings between $L = 4$ and 7, when the wave analyzers were in the appropriate detection mode. They were found at nearly all local times (0000–0600 MLT, 0900–1800 MLT) where the GEOS-2 orbit crossed the magnetic equator. Approximately 75% of the emissions were less than 4° in latitudinal extent, and 75% were centered between $-2° < \lambda_m < +3°$. Wrenn *et al.* (1979) and Gough *et al.* (1979) have found a remarkable correlation between equatorially confined electrostatic wave emissions and highly anisotropic fluxes of 20–500 eV electrons. According to Wrenn *et al.* (1979) a pancake distribution $f(\alpha) \sim \sin^p \alpha$, $p > 3$, is generally observed by GEOS-2 on the morning side. This contrasts with local midnight, the diffuse auroral region, where the electron distribution is more nearly isotropic and peaks at an energy above their 500 eV detector limit. During the morning hours, a high anisotropy index ($p > 3$), combined with low density ($< 10/\text{cm}^3$) cold plasma, is associated with a wave intensity above $3 \mu\text{V}/\text{m}$ measured in the first harmonic band. Gough *et al.* (1979) showed that it is the intensity of class III emissions that is related to the anisotropy index. When the integrated amplitude in all the bands above the cyclotron frequency rises above $300 \mu\text{V}/\text{m}$, p increases to 5–7. Whether the waves cause the anisotropic electrons, or *vice versa*, is unclear. In any case, both theory and observation now strongly indicate the importance of equatorial studies of electrostatic waves and the electrons they interact with.

7.5　Discussion

Modern plasma wave instruments can detect all electron cyclotron harmonic waves at or below the upper hybrid frequency whose amplitudes exceed the thermal fluctuation level. They are found throughout the magnetosphere beyond the plasmapause. They can be divided into four distinct classes by frequency and amplitude: 3/2, multiharmonic, and intense upper hybrid emissions, and diffuse bands. The diffuse bands' low amplitudes distinguish them from the other three; these may be enhanced thermal fluctuations whereas the others probably are saturated instabilities. Single band 3/2 emissions tend to occur on the nightside, multiharmonic emissions primarily on the dayside. Diffuse bands and low 3/2 (or cyclotron) waves occur more at high latitudes; intense 3/2 or multiharmonic emissions are confined to within a few degrees of the geomagnetic equator. Intense upper hybrid emissions are associated with the plasmapause and may generate electromagnetic continuum radiation.

Since intense electrostatic waves were often present near the geomagnetic equator on auroral field lines, and their 1–10 mV/m amplitudes could put the main part of the 1–10 keV plasma sheet electron distribution in strong diffusion, Kennel *et al.* (1970a) suggested that electrostatic cyclotron harmonic, and particularly 3/2, waves were responsible for the diffuse aurora. More recent studies support this conclusion; for example, HUBBARD *et al.* (1979) found that 3/2 waves occur almost always near local midnight, so that they could account for the ever-present diffuse aurora.

7.6 Electrostatic waves in the Jovian magnetosphere

The plasma wave and radio astronomy instruments onboard Voyager 1 and 2 have discovered cousins of terrestrial electrostatic wave emissions in Jupiter's magnetosphere (SCARF *et al.*, 1979; WARWICK *et al.*, 1979; GURNETT *et al.*, 1979; KURTH *et al.*, 1980). KURTH *et al.* (1980) present Jovian examples of each of the four basic classes of emissions in Table 1. The Jovian emissions are therefore completely analogous to those in the earth's magnetosphere, though their spatial morphology naturally differs.

Figure 31 presents Voyager 2 observations in Jupiter's outer magnetosphere that reveal a relationship between intense upper hybrid emissions and electromagnetic continuum radiation identical to that found at earth. The lower frequency cutoff of the continuum radiation is near the upper hybrid frequency. The narrow band upper hybrid emission (class 4) have an amplitude that exceeds $60\,\mu\text{V/m}$. The upper hybrid emissions in Fig. 31 change frequency discontinuously in steps of about the electron cyclotron frequency, $180\,\text{Hz}$, a behavior similar to that observed at earth.

There is little or no evidence for class 2 multiharmonic emissions in Jupiter's outer magnetosphere. However, the Voyager plasma wave receivers detected odd half harmonic emissions on virtually every crossing of the magnetic equator within 23 R_J. Figure 32 illustrates a multiharmonic event

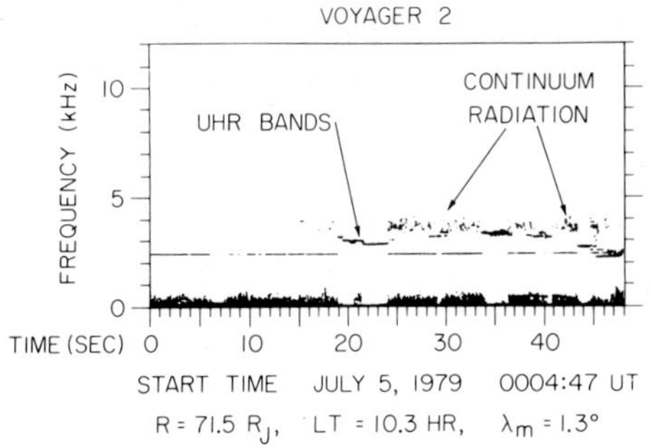

FIG. 31.   Upper hybrid (class 4) emissions detected at Jupiter. Shown is an example of upper hybrid emissions observed in Jupiter's outer magnetosphere. Upper hybrid emissions and continuum radiation bear the same relationship to one another at Jupiter that they do at earth. We thank D. A. Gurnett, W. S. Kurth, and F. L. Scarf for providing us with the Voyagers 1 and 2 figures shown in this article.

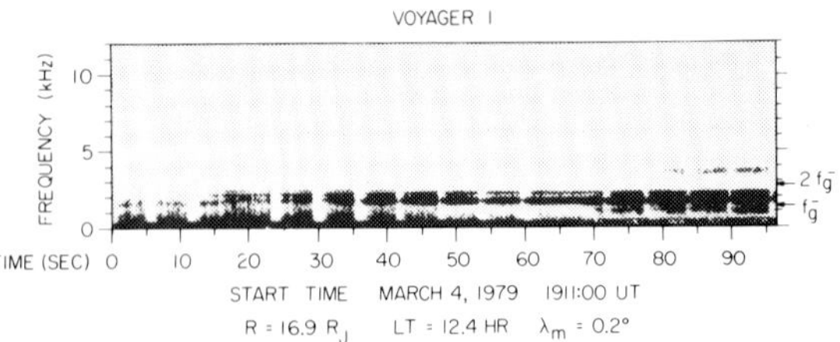

Fig. 32. A class I and class II event observed in Jupiter's magnetosphere. Shown here is an intense 3/2 event very near Jupiter's magnetic equator (0.2°) that develops a 5/2 component towards the end. Care must be taken in relating these events to their analogs at earth, because at Jupiter n_c/n_H exceeded unity, whereas it is thought to be small at earth during such events. Also the Jovian events are much more strongly localized to the equator than the terrestrial ones.

detected by the Voyager 1 plasma wave instrument near the local noon magnetic equator at a distance of 16.9 R_J. A 3/2 emission is present for the entire 96 seconds of data shown, and a 5/2 emission appears at the end of the interval. These emissions are somewhat more complex than those found at earth; the first harmonic band has an intense amplitude peak near $3/2 f_c$ with a more diffuse emission covering much of the rest of the band. (The gap near 2.4 kHz is due to a notch filter that suppresses interference from the spacecraft power supply. The semi-regular interruption of the signal is an AGC effect due to periodic interference from a low energy charged particle instrument's stepper motor.) KURTH et al. (1980a) report that the mean magnetic latitude at which eight such events were detected was 1.1°; the mean width of the emission region was 0.28 R_J. The peak amplitudes were typically in the 3/2 and upper hybrid bands, and increased with decreasing radial distance to a maximum of a few mV/m.

Figure 33 shows a class Ia emission observed at $\lambda_m = -1.3°$ near Voyager 2 periapsis. Note the intense short duration bursts, a feature not emphasized by observers of terrestrial electrostatic waves. Figure 5 of KURTH et al. (1980a), which we do not present here, shows an example of class 3 diffuse bands.

The Planetary Radio Astronomy experiments on Voyager 1 and 2 (BIRMINGHAM et al., 1980) have detected intense upper hybrid emissions and odd half harmonic emissions both below and above f_{UHR} in the Io plasma torus. The upper hybrid emissions were nearly always present, but the half harmonic emissions varied with position in a manner consistent with an increase of a factor 4–5 in T_c/T_H from the inner ($L \simeq 5\ R_J$) to outer ($L \simeq 9\ R_J$) portions of the torus. Since n_c/n_H was probably large throughout the torus,

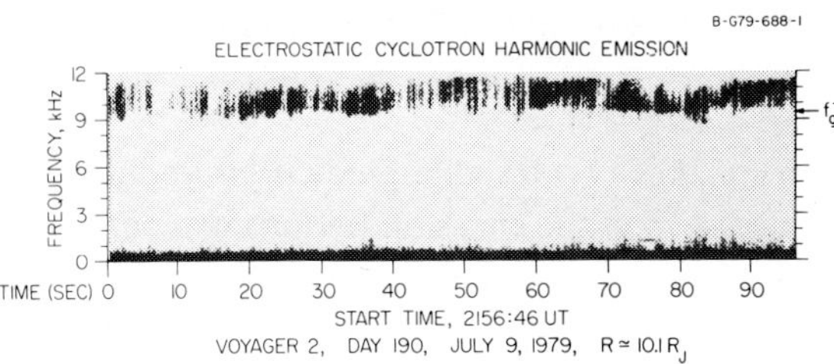

Fig. 33.   Fine structure of a 3/2 event in Jupiter's magnetosphere. 90 seconds of a Voyager 2 electric field measurement near the outer boundary of the Io torus,reveals a highly impulsive emisssion just above the local electron cyclotron frequency. F. Scarf (private communication, 1980) has suggested that Doppler shifting from co-rotation could play a role in the interpretation of such observations.

the instabilities were probably predominantly convective. Although the analogy should not be carried too far, the Io plasma torus and the earth's plasmapause are similar insofar as their electrostatic waves are concerned.

The Voyager 2 plasma wave instrument detected whistler mode emissions near $f_c/2$ at a magnetic equator crossing near the outer edge of the Io plasma torus (CORONITI et al., 1980). High resolution spectral information together with whistler instability theory permitted an estimate of the density $(n_H \sim 2.5\,\mathrm{cm}^{-3})$, energy (few keV) and omnidirectional energy flux $(10^2$ ergs/cm^2-sec) associated with the hot electrons. Presumably, these hot electrons also interact with the electrostatic waves, although BIRMINGHAM et al.'s (1980) theoretical analysis was insensitive to, and could not determine, n_c/n_H.

CORONITI et al. (1980) argue that pitch angle scattering from a combination of whistler and electrostatic waves produces a strong diffusion precipitation energy flux within a factor two of the 50 ergs/cm^2-sec required for the aurora associated with the Io torus (BROADFOOT et al., 1979). Thus, a model of the Jovian aurora similar to that of the terrestrial diffuse aurora may be feasible.

An experimentalist might yield to a moment's regret that the Voyagers did not discover a strikingly new class of electrostatic emissions; nonetheless, the similarity between the terrestrial and Jovian observations is an important assurance that, as plasma and magnetospheric physicists, we are on the right track.

8. Comparisons between Linear Theory and Observation

8.1 *Introduction*

Linear theory makes two testable predictions about electron cyclotron harmonic instabilities. First the emission frequency is controlled by the ratio of hot to cold electron densities and by the cold upper hybrid frequency, more or less independently of the free energy source residing in the hot electrons. Secondly, plausible free energy sources are a generalized loss cone feature and thermal anisotropy. We discuss tests of each prediction in turn.

8.2 *Dependence upon cold electrons*

The spatial morphology of terrestrial electrostatic emissions can be interpreted in terms of what is generally known about the distribution of hot and cold electron densities in the magnetosphere. For example, both theory and observation agree that electrostatic cyclotron harmonic waves should not survive well within the plasmasphere. Moreover, intense upper hybrid emissions occur near but beyond the plasmapause because they require $n_c > n_H$. Since the density of cold electrons of ionospheric origin is larger on the dayside than on the nightside, multiharmonic emissions, which require the cold upper hybrid frequency to exceed twice the cyclotron frequency, ought to be more frequent on the dayside. Finally, if cold electrons stream outward from the ionosphere with little pitch angle scattering, their density would be proportional to the magnetic field strength; then the ratio of the cold upper hybrid frequency to the local electron cyclotron frequency would decrease away from the geomagnetic equator along a field line, potentially explaining the predominance of $f \simeq f_c$ over 3/2 emissions at geomagnetic latitudes above $20°$.

A prerequisite to a more detailed test of theory is an experimental definition of hot and cold electrons. The theoretical meaning of hot and cold is clear: hot electrons have free energy, cold electrons fix the unstable frequency range. We have argued that magnetospheric convection both heats and produces free energy in the hot electron distribution. Cold electrons can have at least three sources. Electrons in the topside ionosphere, with about 1 eV energy, flow into space in response to pressure and charge imblances. A flux of $10^8/cm^2$-sec of 10 eV photoelectrons streams outward on field lines whose feet are in sunlight. Few hundred eV secondary electrons can be backscattered into space from regions of intense precipitation of more energetic electrons. The existing instability studies do not consider multicomponent cold electron distributions, nor do the measurements clearly distinguish between components. Theoretically, we expect the pertinent cold upper hybrid frequency to be based upon the sum of the various cold densities, whereas the group velocities should depend upon the structure of the cold electron distribution function.

In the absence of measurements of perpendicular wavelength, linear theory is blind to the values of T_c and T_H and is sensitive only to their ratio. Thus, what is meant by T_c and T_H may also depend upon the location of the observing spacecraft. For example, beyond $L=6$ on the earth's nightside, we expect hot electrons to have plasmasheet energies of about 1 keV. As far as theory is concerned, 100 eV electrons could be "cold." On the other hand, in the GEOS orbit ($L<7$), which is predominantly within the plasmasheet inner boundary, the "hot" electrons could have hundred eV energies, and the cold electrons, ionospheric energies.

Experimentally, it easy to infer the *total* electron density from the observed plasma frequency; one also easily measures the electron distribution of energies above about 10 eV photoelectron energies, but it is unclear whether all of these electrons should be counted in the "hot" distribution. As a first approximation, HUBBARD *et al.* (1979) considered all the electrons measured above the 13.3 eV threshold of the IMP-6 plasma analyzer to be hot and subtracted this density from the total density to estimate n_c. They then classified the observed electron cyclotron harmonic waves according to the inferred n_c/n_H. 3/2 emissions, multiharmonic emissions, diffuse bands, and intense upper hybrid emissions corresponded to $n_c/n_H \simeq 0.2$, 0.55, 6 and 1.5, respectively. A much larger fraction of diffuse bands and upper hybrid emissions had $n_c/n_H \geq 1$ than did 3/2 or multiharmonic emissions. HUBBARD *et al.* (1979) also present evidence that the effective T_c/T_H for 3/2 and multiharmonic emissions is about 10^{-1}, whereas it is $0(10^{-2})$ for diffuse bands and upper hybrid emissions. The specific values of n_c/n_H deduced are model dependent, because they could depend on T_c/T_H. It may prove difficult to calculate exact values of n_c/n_H or T_c/T_H from wave observations, therefore. Nonetheless, the general theoretical expectations about the dependence of cyclotron harmonic emissions upon the cold electron density have been confirmed.

8.3   Dependence upon hot electron free energy source

The most primitive requirement from linear theory, that cyclotron harmonic waves intensify when the fluxes of hot electrons increase, has been known to be satisfied for some time. SCARF *et al.* (1973) showed that the intensity of 3/2 emissions increases greatly during substorms. ANDERSON and

MAEDA (1977) found that 3/2 emissions observed on Explorer 45 intensify (disappear) when the fluxes of 1–10 keV electrons increase above (diminish below) 10^6/cm^2-sec-sr-keV. However, the angular and energy distributions of the hot electrons measured on OGO-5 and Explorer 45 did not have sufficient resolution to identify the free energy source.

HUBBARD *et al.* (1979) present some evidence that the electrons above 13.3 eV have a second peak in their energy distribution when intense electrostatic waves are observed. Two-dimensional electron velocity distribution measurements onboard Hawkeye-1 (KURTH *et al.*, 1979a) provided the first detection of a loss cone free energy source made simultaneously with cyclotron harmonic waves. However, there were only a few occasions when Hawkeye's spin vector was aligned appropriately to the magnetic field to permit an unambiguous identification of the free energy source.

Three-dimensional ISEE 1 and 2 and GEOS electron velocity distribution measurements identify the free energy source sufficiently clearly to permit comparison with linear theory, although their time resolutions still do not permit detection of short-lived strongly unstable features in velocity space. RONNMARK *et al.* (1978) report a GEOS study of an intense upper hybrid emission accompanied by 3/2, 5/2, 7/2 and 11/2 bands. They measured the local magnetic field directly, and the total electron density both actively and passively. They fit the observed 300 eV–20 keV electron velocity distribution, averaged over three minutes, with an analytical model which had a loss cone feature. In the absence of more detailed observations, they modelled the cold electrons by a 5 eV isotropic Maxwellian. Their calculations revealed convective instability in the four odd half harmonic bands, whose amplitudes were about $30\,\mu$V/m, and a nonconvective instability in the upper hybrid band, whose amplitude was about $3\,$mV/m. Although there is no generally accepted nonlinear theory of wave saturation, absolute instabilities should saturate at larger amplitudes than convective ones. Thus, RONNMARK *et al.* (1978) concluded that the measured amplitudes qualitatively agreed with linear theory. The predicted and observed frequencies were in reasonable agreement.

Figure 34 shows an ISEE-1 Swept-frequency-receiver (SFR) color spectrogram of plasma wave data recorded on the Nov. 5, 1977 inbound pass (GURNETT *et al.*, 1978). On this day, ISEE-1 penetrated the magnetopause at 1720 UT near local noon at about 20°N geomagnetic latitude. Sporadic bursts of 3/2 and multiharmonic emissions were observed to occur until 1900 UT, at which time the intensity of electrons with energies exceeding

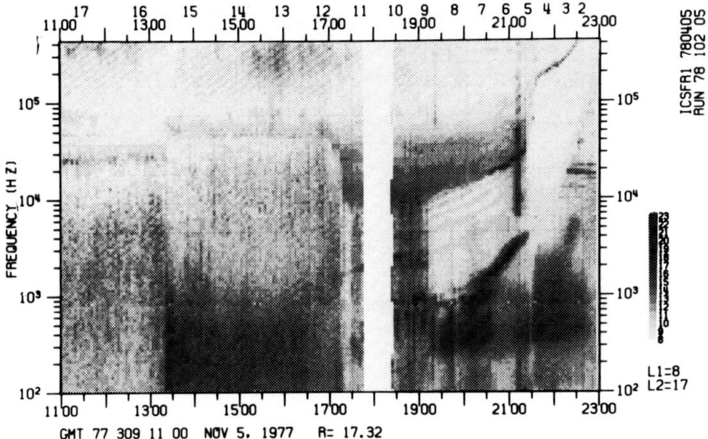

FIG. 34.   A survey of the plasma waves observed by ISEE-1 on Nov. 5, 1977. ISEE-1 encountered the bowshock at about 1320 UT, and the magnetopause near 1726 UT on an inbound pass on Nov. 5, 1977. These features may be discerned by sharp changes in the frequency-time display produced by the sweptfreqeuncy receiver. The local electron plasma frequency is discernible as a line at the lower frequency boundary of the continuum radiation. The local plasma frequency increases sharply during the plasmapause entry between 2115–2130 UT. Intense upper hybrid emissions occurred just beyond the plasma pause. Weak diffuse bands were observed between about 1915 UT and 2115 UT, and a 3/2 emission between the magnetopause crossing and 1915 UT. The low frequency waves are whistler chorus between the magnetopause and plasmapause, and ELF whistler hiss inside the plasmasphere.

230 eV measured by the LEPEDEA instrument decreased abruptly, and the emissions ceased. A period of sustained 3/2 activity, with amplitudes of 0.1–1 mV/m occurred from 1825–1837 UT. Low intensity diffuse bands were observed between 1900 and 2030 UT. An intense upper hybrid emission dominated the wave activity from 2030 UT until ISEE-1 crossed the plasmapause at 2125 UT. A burst with peak amplitude of 7 mV/m occurred between 2108 and 2118 UT (GURNETT *et al.*, 1979b).

Figure 35 shows three dimensional LEPEDEA measurements (SENTMAN *et al.*, 1979) of the electron distribution function above 230 eV taken between 1829–1837 UT (3/2 emissions), 2011–2019 UT (diffuse bands) and 2111–2119 UT (upper hybrid emission). These distributions are surprisingly featureless. A weak feature near the geometric loss cone, indicated by shading, does appear in the "3/2 distribution" (top inset). The thermal anisotropy is about 0.4, too small to account for 3/2 waves. The "diffuse band distribution" (middle) consists of two distinct populations,

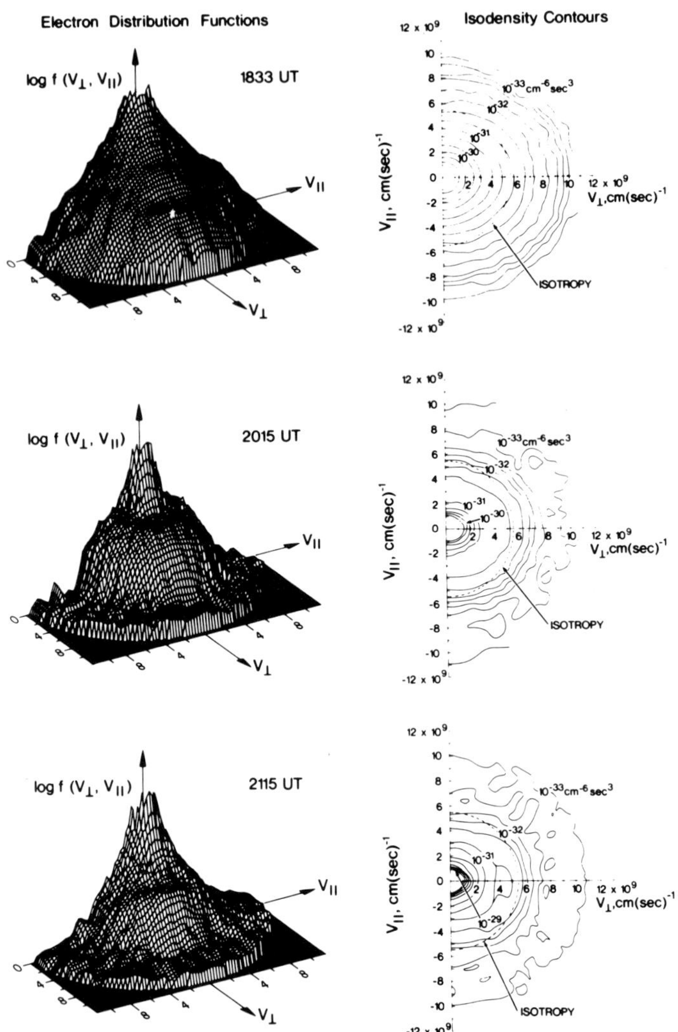

Fig. 35a.   Electron distribution functions measured during ISEE-1 wave events on Nov. 5, 1977. Two respresentations of the distribution function for a class 1 (3/2) event are shown at the top, a diffuse band (class 3) event at the middle, and a class 4 (upper hybrid event) at the bottom (SENTMAN *et al.*, 1979). These distributions are relatively featureless, at least averaged over the 8 minutes during which they were compiled.

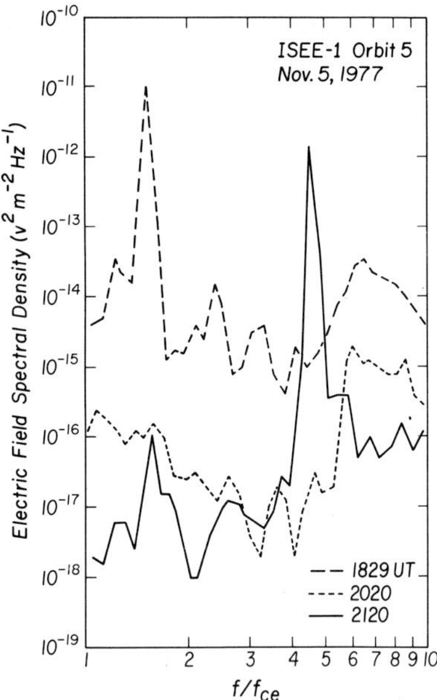

FIG. 35b.   Electric field spectral distributions measured during ISEE-1 wave events on Nov. 5, 1977. Shown here are the spectral distributions for the class 1 (1829 UT), class 3 (2020 UT) and the class 4 (2121 UT) wave events corresponding to the electron distributions in Fig. 35a (SENTMAN *et al.*, 1979). The emissions above $f/f_{ce} = 5$ are due to electromagnetic continuum radiation.

each of which has an anisotropy of about 0.8, with a relatively broad plateau around 4×10^9 cm/sec. The "upper hybrid distribution" is shown in the bottom inset. The relatively broad plateau in the "diffuse band distribution" has been reduced in size to a small region near 90° pitch angle, and the temperature anisotropy diminished to about 0.3.

For the events above, the total electron density was inferred from the lower cutoff of the continuum radiation, and the densities corresponding to the electrons measured above 230 eV were calculated using a two-component bi-Maxwellian fit, which proved necessary for the "diffuse band" distribution. The hot density was defined to be the sum of these two densities. n_c/n_H was 4, 15, and 30 for the 3/2 emission, diffuse bands, and upper hybrid emission respectively, a trend in general agreement with HUBBARD *et al.* (1979).

It is disconcerting that there was no persuasive evidence for a loss cone feature in the "upper hybrid distribution" at 2115 UT. The free energy

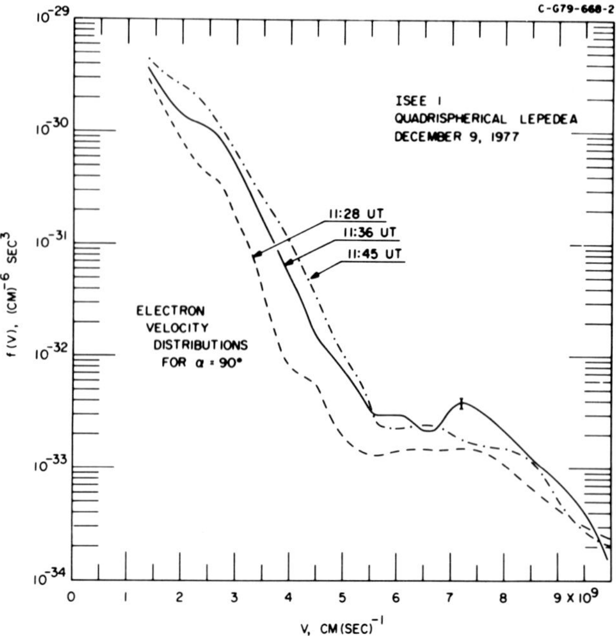

F_IG. 36. Projected perpendicular velocity distributions during an ISEE-1 upper hybrid event on Dec. 9, 1977. In this case, the free energy source was probably detected. The solid line shows the perpendicular velocity distribution compiled during the upper hybrid event, while dashed lines show the distributions when no waves were present (K_URTH *et al.*, 1979c).

source for these emissions may have been smeared by a combination of quasilinear diffusion and the finite time resolution (8 min) of the measurement, or may have been at energies below 230 eV. A similar upper hybrid event, recorded by ISEE-1 on Dec. 9, 1977, has been studied by K_URTH *et al.* (1979c). Figure 36 shows three cuts of the three dimensional velocity distribution for 90° pitch angle measured in 8 minute intervals before and after the upper hybrid emission (dotted), and during (solid). A small loss cone feature appears only when the hybrid emission was present; in addition the detector happened to be sampling 90° pitch angles when the waves intensified.

K_URTH *et al.* (1979b) compared the Nov. 5, 1977 observations of the intense upper hybrid emissions with linear theory. Figure 20 predicts an intensification of the upper hybrid emission each time $(n + \frac{1}{2})f_c \simeq f_{\mathrm{UHR}}$, an effect observed during this event. Figure 37 shows the maximum spatial

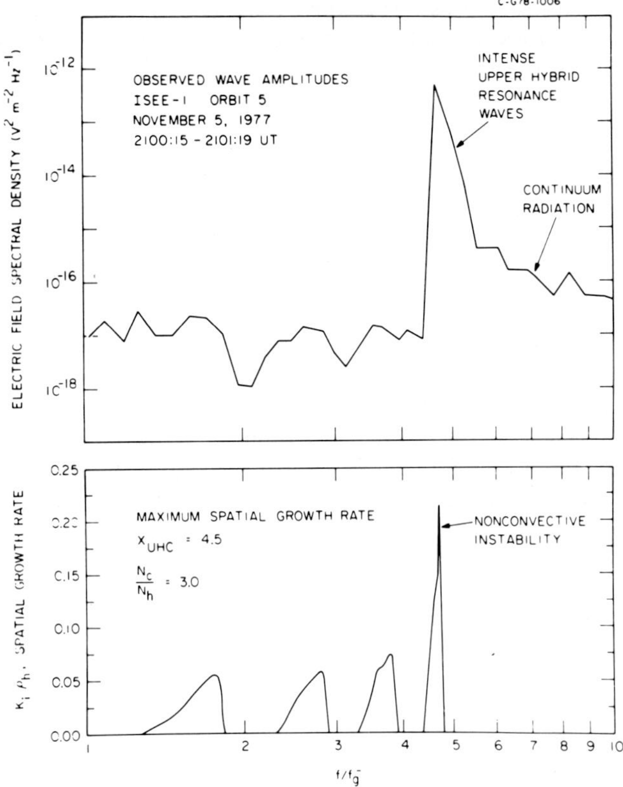

Fig. 37. Wave amplitudes and spatial growth rates for the ISEE-1 upper hybrid event of Nov. 5, 1977. The top inset shows the electric field spectral density as a function of frequency normalized to electron cyclotron frequency f_c. The intense upper hybrid peak is apparent, and there is evidence for weaker multiharmonic emissions below the hybrid frequency. The bottom inset shows the spatial growth rate normalized to the hot electron Larmor radius, $K_i\rho_h$, calculated using f_{AKK} for the parameters indicated. The peaks in spatial growth rate and spectral density do roughly correspond (Kurth et al., 1979b). A similar event observed by GEOS yielded a similar conclusion (Ronnmark et al., 1979).

growth rate as a function of frequency calculated for the empty loss cone distribution (f_{AAK} with $\Delta = 0$), $X_{UHC} \simeq 4.5$ and $n_c/n_H \simeq 3$. The cyclotron harmonic bands are convectively unstable, while the upper hybrid band is nonconvective. Thus, Ronnmark et al. (1978) and Kurth et al. (1979b) arrive at similar conclusions from two independent studies of intense upper hybrid emission accompanied by diffuse odd half harmonic bands.

The apparent success with the Nov. 5, 1977 upper hybrid event illustrates how elusive comparison between linear theory and observations of Harris modes can be. The value of n_c/n_H used in the computation was a factor ten smaller than that inferred from the continuum cutoff. Moreover, the empty theoretical loss cone was a much stronger free energy source than the one found on Dec. 9, 1977 or the one that probably failed to survive averaging on Nov. 5, 1977. The saving feature may be the prevalence of strong spatial growth. For example, underestimating n_c/n_H overestimated the number of resonant electrons. We subsequently recalculated the spatial growth rates for the Nov. 5, 1977 event, using a three component electron distribution consisting of a $T=0$ cold electrons whose n_c/n_H was inferred from the continuum cutoff, warm electrons with $T_W/T_H=.05$ and $n_W/n_H \simeq 3$ (in rough accord with LEPEDEA measurements) and hot electrons with $\Delta=0$. There was little qualitative change in the frequencies of peak spatial growth. Furthermore, parametric studies generally indicate that spatial instability persists as the loss cone is filled in, though the growth rates again decrease in magnitude. Thus the computations may have found the correct frequency range—not surprising given the structure built into the Harris dispersion relation—and overestimated the growth rates. This emphasizes the need to compare observations with nonlinear theory, which can predict the magnitude of the growth rate that accounts for the observed amplitudes.

8.4   Summary

Detailed comparisons of measured wave and electron distributions with theory are in their infancy. Electron measurements are hampered by a time and velocity space resolution that is often barely adequate to detect the weak free energy sources thought responsible for cyclotron harmonic waves, as well as inadequate definition of cold electron properties. Nonetheless, a promising beginning has been made for intense upper hybrid emissions. The next task is to extend these efforts to other emission classes, and particularly to nightside 3/2 diffuse auroral emissions. Moreover, the types of free energy sources that destabilize magnetospheric cyclotron harmonic waves need to be classified observationally. On the theoretical side, parametric studies of nonlinear saturation that are similar in scope to those for linear theory are needed.

9. Some Nonlinear Considerations

9.1 *Introduction*

In this chapter we approach the frontier of our subject. The corresponding complexities of the linear theory and of the observational phenomenology have prevented us from achieving a general insight into the nonlinear behavior of magnetospheric cyclotron harmonic waves. We have therefore chosen to review velocity space diffusion, where there are results of general validity, more extensively than saturation, where present efforts are truly preliminary.

9.2 *General properties of quasilinear diffusion*

The quasilinear diffusion driven by electrostatic waves in a magnetized plasma (BERNSTEIN and ENGELMANN, 1966; KENNEL and ENGELMANN, 1966; YOUNG, 1975) is described by

$$\frac{\partial f}{\partial t} = \sum_n \int dk_\parallel \hat{G}_n \{D_n \hat{G}_n\} f \tag{40a}$$

where

$$\hat{G}_n = \frac{n\Omega}{v_\perp} \frac{\partial}{\partial v_\perp} + k_\parallel \frac{\partial}{\partial v_\parallel} \tag{40b}$$

$$D_n = 2\pi \int_0^\infty k_\perp dk_\perp \frac{e^2 E^3}{M^2 K^2} J_n^2\left(\frac{k_\perp v_\perp}{\Omega}\right) R_n \tag{40c}$$

$$R_n = \frac{\Gamma_K}{(\omega - k_\parallel v_\parallel - n\Omega)^2 + \Gamma_K^2} \tag{40d}$$

and Γ_K is the sum of the linear growth rate γ_K and the resonance broadening frequency shift η_K due to a finite amplitude wave spectrum. In the resonant limit, $\Gamma_K \to 0^+$,

$$R_n \to \pi\delta(\omega - k_\parallel v_\parallel - n\Omega) = \frac{\pi}{\left|\dfrac{\partial\omega}{\partial k_\parallel} - v_\parallel\right|} \delta\left(k_\parallel - \frac{\omega - n\Omega}{v_\parallel}\right) \tag{41}$$

where δ is the Dirac delta function. Equation (40) also takes the form

$$\frac{\partial f}{\partial t} = \frac{1}{v_\perp} \frac{\partial}{\partial v_\perp} v_\perp \left(D_{\perp,\perp} \frac{\partial f}{\partial v_\perp} + D_{\perp\parallel} \frac{\partial f}{\partial v_\parallel}\right) + \frac{\partial}{\partial v_\parallel}\left(D_{\parallel\perp} \frac{\partial f}{\partial v_\perp} + D_{\parallel,\parallel} \frac{\partial f}{\partial v_\parallel}\right) \tag{42a}$$

where

$$\hat{D} = \begin{pmatrix} D_{\perp\perp} & D_{\perp\|} \\ D_{\|\perp} & D_{\|\|} \end{pmatrix} = \sum_n \int dk_\| \frac{\Omega^2}{v_\perp^2} D_n \hat{\Lambda}_n \tag{42b}$$

and

$$\hat{\Lambda}_n = \begin{pmatrix} n^2 & \dfrac{nk_\| v_\perp}{\Omega} \\ \dfrac{nk_\| v_\perp}{\Omega} & \left(\dfrac{k_\| v_\perp}{\Omega}\right)^2 \end{pmatrix} \tag{42c}$$

Equations (40) and (42) indicate that in the resonant limit particles diffuse as if their diffusion tensor were a sum of individual tensors, one for each $k_\|$ and n. If one $k_\|$ dominated the spectrum, and the particles of interest had only an nth order resonance with that wave, they would diffuse along a surface defined by the characteristics of the operator $\hat{G}_n$:

$$\frac{k_\| v_\perp^2}{2n\Omega} + v_\| = \text{constant} \qquad (n \neq 0) \tag{43a}$$

$$v_\perp = \text{constant} \qquad (n = 0) \tag{43b}$$

These characteristic lines in $(v_\perp^2, v_\|)$ space become surfaces because azimuthal symmetry in Larmor phase about the magnetic field is maintained if the diffusion time scale exceeds the Larmor period (KENNEL and ENGELMANN, 1966). In general, the wave distribution has a spread in $k_\|$, several harmonic modes can be excited, and the same particle can have different order resonances with different harmonics. Nonetheless, the characteristic surfaces provide intuitive information about how diffusion proceeds. Landau particles, with $n = 0$, diffuse only in $v_\|$. When $|k_\| v_\perp/n\Omega|$ is small, cyclotron resonant particles diffuse primarily in $v_\perp$; when $|k_\| v_\perp/n\Omega|$ is large, they diffuse primarily in $v_\|$ like Landau particles.

We elucidate the nature of quasilinear diffusion in the magnetosphere with an *ad hoc* model of the wave spectrum based upon magnetospheric observations and linear theory. The observed spectra generally consist of components in each harmonic band from the first to the cold upper hybrid band. The unstable waves in each band generally span the same range of $k_\|$, but a different range of $k_\perp$ that increases with increasing harmonic number m. We therefore model the wave distribution by a function, separable in $k_\|$ and $k_\perp$, that distinguishes between harmonic bands:

$$E^2(k_\perp, k_\parallel) = \sum_{m=-1}^{m=+Q} E_m^2(k_\parallel) \frac{k^2 e^{-k_\perp^2/k_{\perp m}^2}}{\pi k_{\perp m}^4} \tag{44a}$$

where Q denotes the number of excited harmonics, and $k_{\perp m}$ increases with increasing $|m|$. The partial diffusion tensor $\hat{D}_{nm}$, for the nth order resonance with the mth harmonic, is then

$$\hat{D}_{nm} = \int dk_\parallel d_{nm} \hat{\Lambda}_n \tag{44b}$$

$$d_{nm} = \frac{e^2}{M^2} E_m^2 R_n \frac{\Gamma_n(\lambda_m)}{\lambda_m} \tag{44c}$$

where Γ_n is defined in Subsection 4.2, and $\lambda_m = k_{\perp m}^2 v_\perp^2 / 2\Omega^2$.

9.2.1 Diffusion inside the loss cone

d_{nm} peaks, for a given $v_\parallel$, at the maximum of Γ_n/λ_m and is therefore largest for non-zero $v_\perp$. On the other hand, whether a given interaction (n, m) can maintain strong diffusion depends upon the magnitude of the diffusion coefficient inside the loss cone where $v_\perp$ is sufficiently small that $\lambda_m \ll 1$. Since $k_\parallel v_\parallel / \Omega$ then is also small, the largest component of the diffusion tensor is $D_{\perp\perp}^{nm}$, which in the resonant limit is approximately

$$D_{\perp\perp}^{mn} \simeq \frac{\pi n}{2(n-1)!} \frac{e^2 E_m^2(K_R)}{M^2 |v_\parallel|} \left(\frac{k_{\perp m}^2 v_\perp^2}{2\Omega^2}\right)^{n-1} \tag{45}$$

where we have assumed that $\partial\omega/\partial k_\parallel \ll v_\parallel$ and defined the resonant parallel wave number K_R

$$K_R \simeq \frac{\omega - n\Omega}{v_\parallel}. \tag{46}$$

Since $n \neq 0$ interactions create little diffusion in $v_\parallel$ when $|k_\parallel v_\perp / n\Omega|$ is small, it is convenient to transform to approximate pitch angle coordinates, $v_\perp \simeq v_\parallel \alpha$, whereupon an approximate diffusion equation, including a loss term is, assuming that the (m, n) interaction dominates at a given $v_\parallel$,

$$\frac{1}{\alpha} \frac{\partial}{\partial\alpha} \left(\alpha D_{\alpha\alpha}^{mn} \alpha^{2(n-1)} \frac{\partial f}{\partial\alpha}\right) - \frac{f}{T_B} = 0 \tag{47}$$

T_B is the quarter bounce time, and

$$D_{\alpha\alpha}^{mn} = \frac{\pi n}{2(n-1)!} \frac{e^2 E_m^2(K_R)}{M^2 |v_\parallel^3|} \left(\frac{k_{\perp m}^2 v_\parallel^2}{2\Omega^2}\right)^{n-1}. \tag{48}$$

For shorthand we will call $D_{\alpha\alpha}^{mn}$, D_0, with the dependence upon the indices m and n understood.

The pitch angle distribution, normalized to unit diffusion flux across the edge of the loss cone at $\alpha = \alpha_0$, has been found by KENNEL (1969), in terms of modified Bessel functions of the first kind I_p:

$$h(\alpha) = \left\{ \frac{1}{D_0} \left(\frac{D_0 T_{\rm B}}{\alpha_0^2} \right)^{1/2} \alpha^{-(n-1)} \frac{I_{(n-1)/(2-n)}\left(\frac{1}{2-n} \left(\frac{\alpha^{2(2-n)}}{D_0 T_{\rm B}} \right)^{1/2} \right)}{I_{1/(2-n)}\left(\frac{1}{2-n} \left(\frac{\alpha_0^{2(2-n)}}{D_0 T_{\rm B}} \right)^{1/2} \right)} \right\} \tag{49}$$

except for $n = 2$, which has a power law solution

$$h(\alpha) = \frac{1}{D_0} \left\{ \frac{1}{\alpha_0^2} \frac{\left(\frac{\alpha}{\alpha_0} \right)^{\beta-1}}{(\beta-1)} \right\} ; \qquad \beta = \left(1 + \frac{1}{D_0 T_{\rm B}} \right)^{1/2} . \tag{50}$$

$n = 2$ separates two distinct types of solutions. In the weak diffusion limit, $D_0 T_{\rm B} \ll 1$, the pitch angle distribution grows exponentially from small α to the edge of the loss cone for $n \neq 2$, and as a high power of α for $n = 2$. In the strong diffusion limit, $D_0 T_{\rm B} \gg 1$, the small argument limit of the Bessel functions lead to:

$$h(\alpha) \simeq \frac{1}{D_0} \left(\frac{2 T_{\rm B} D_0}{\alpha_0^2} \right) = T_{\rm M}, \qquad \text{for all} \quad n \tag{51}$$

where $T_{\rm M} = 2 T_{\rm B}/\alpha_0^2$ is the minimum lifetime. Thus wherever the small argument limit is valid, the pitch angle distribution is isotropic. For $n \leq 2$, the pitch angle distribution is isotropic throughout the loss cone. When $n > 2$ and diffusion at $\alpha = \alpha_0$ is strong, the distribution will be isotropic part way into the loss cone and then plunge exponentially to zero at $\alpha = 0$. This may be understood physically as follows. The time t^* it takes a particle to diffuse from some small pitch angle α^* to $\alpha = 0$ is roughly $t^* \simeq \alpha^{*2(2-n)}/D_0$. For $n > 2$, $t^* \to \infty$ as α^* decreases, and the pitch angle distribution has a hole at $\alpha = 0$. For $n \leq 2$, isotropy can be maintained *throughout* the loss cone.

The wave amplitude for strong diffusion at the edge of the loss cone is given by

$$D_0(\alpha_0)\alpha_0^{2(n-1)} T_m \geq 1 . \tag{52}$$

We relate $E_m^2(K_{\rm R})$ to the total integrated wave amplitude ε_m^2 using the approximation $E_m^2(K_{\rm R}) \simeq \varepsilon_m^2/|K_{\rm R}|$; furthermore, linear theory indicates that

$K_R A/\Omega \simeq \frac{1}{2}$ for all harmonics, where A is the hot electron thermal speed. For the earth's dipole field we may use

$$B \simeq \frac{1}{3L^3} \quad \text{Gauss} \tag{53a}$$

$$\alpha_0^2 = \frac{2}{L^3} \tag{53b}$$

$$T_M = \frac{L^4 R_E}{v_\|} \tag{53c}$$

to estimate ε_{mn}^*, the critical electric field amplitude for the nth order resonance with the mth harmonic to create strong diffusion at the edge of the loss cone:

$$\varepsilon_{mn}^* \simeq 0.1 \ mv/M \left(\frac{(n-1)!}{n!}\left(\frac{6}{L}\right)^7\left(\frac{A}{10^9}\right)^3\right)^{1/2}\left(\frac{v_\|}{A}\right)^2\left(\frac{L^3}{k_{\perp m}^2 v_\|^2/\Omega^2}\right)^{(n-1)/2} \tag{54}$$

For the (m, n) interaction, we estimate that

$$\frac{v_\|}{A} \simeq \left| m - n + \frac{1}{2} \right| \tag{55}$$

and since the $k_\perp$ where the spatial growth rate peaks is controlled in part by cold electrons, that

$$\frac{k_\perp^2 a^2}{2\Omega^2} \simeq m \tag{56}$$

where a is the cold electron thermal speed. These approximations enable us to compute the ratios, displayed in Table 2, of the critical field for the (m, n) interaction to that for $(1, 1)$ at $L=6$. Estimates of $v_\|/A$ are shown in parentheses. (Note that for $T_c/T_H = a^2/A^2$ sufficiently small, λ_m is no longer small at the edge of the loss cone.) Putting together the above arguments and estimates we conclude

   1)   Only $|n|=1$, 2 interactions can create the isotropy *throughout* the loss cone observed in diffuse auroral precipitation. This argument holds even if λ_m is not small at the edge of the loss cone, since the pitch angle diffusion coefficient depends on a high power of α at its center if $n>2$.

   2) $|n-m| \simeq 0$, 1 interactions involve resonant electrons in the main part of the hot electron distribution. Together with point 1) above, this suggests that only 3/2 and 5/2 waves can account for diffuse auroral strong diffusion.

TABLE 2. $\varepsilon_{mn}^{*}/\varepsilon_{11}^{*}$ at $L=6$.

$m=$	1	2	3	4
$n<1$	1 (1)	9 (3)	25 (5)	49 (7)
2	$3.6\left(\dfrac{T_c}{T_H}\right)^{1/2}$ (1)	$2.6\left(\dfrac{T_c}{T_H}\right)^{1/2}$ (1)	$6.4\left(\dfrac{T_c}{T_H}\right)^{1/2}$ (3)	$9.2\left(\dfrac{T_c}{T_H}\right)^{1/2}$ (5)
3	$36\dfrac{T_c}{T_H}$ (3)	$18\dfrac{T_c}{T_H}$ (1)	$12\dfrac{T_c}{T_H}$ (1)	$9\dfrac{T_c}{T_H}$ (3)
4	$119\left(\dfrac{T_c}{T_H}\right)^{3/2}$ (5)	$70\left(\dfrac{T_c}{T_H}\right)^{3/2}$ (3)	$114\left(\dfrac{T_c}{T_H}\right)^{3/2}$ (1)	$76\left(\dfrac{T_c}{T_H}\right)^{3/2}$ (1)

3) The critical amplitude for 3/2 strong diffusion is about 0.1 mV/m at $L=6$, if the hot electron thermal speed is 10^9 cm/sec. This critical amplitude scales as the hot electron temperature to the 3/4 power and as $(6/L)^{7/2}$.

4) The above calculation of ε_{mn}^{*} implicitly assumed that waves existed along an appreciable fraction of the field line. A more accurate calculation would bounce average the diffusion equation. If the waves are localized near the geomagnetic equator, the effective diffusion coefficients are reduced by τ/T_{B}, where τ is the time a particle spends in the wave region. ε_{11}^{*} should therefore be increased by $(T_{\mathrm{B}}/\tau)^{1/2}$ and is probably several tenths of a mV/m.

5) When the wave amplitudes in any of the harmonic bands exceed the 3/2 critical value, strong diffusion can spread to electrons with energies larger than thermal.

6) By fixing the $k_{\perp}$ of maximum spatial growth, T_c/T_H plays a significant role in determining the strength of pitch angle diffusion in the loss cone due to higher harmonic excitations.

9.2.2 *Energization of hot electrons*

The energy and pitch angle diffusion driven by cyclotron harmonic waves proceed at comparable rates. Also, even if only one harmonic band is excited, a particle interacts with it via several successive resonances in diffusing from large to small pitch angles. And the particle may encounter several different regions of localized wave turbulence as it bounces in the geomagnetic field. For these reasons, there is no simple analytic description of magnetospheric quasilinear diffusion outside the loss cone. To elucidate its nature, LYONS (1974) numerically calculated bounce averaged diffusion coefficients using a model wave distribution.

Quasilinear diffusion in spherical velocity space coordinates involves three diffusion coefficients that are related to one another by

$$D_{\alpha\alpha} = \sum_n D_{\alpha\alpha}^n \; ; \qquad D_{\alpha v} = \sum_n D_{\alpha\alpha}^n \Gamma_n \; ; \qquad D_{vv} = \sum_n D_{\alpha\alpha}^n \Gamma_n^2 \qquad (57a)$$

$$\Gamma_n = -\left(\frac{\sin\alpha \, \cos\alpha}{\dfrac{2n}{3} + \sin^2\alpha} \right) \qquad (57b)$$

if only 3/2 waves are excited. LYONS (1974) modelled the electric field spectral density by

$$E^2(k, \omega) = E_0^2 \frac{\delta(k_\perp - k_{\perp 0})}{k_{\perp 0}} \exp \left. -\left\{ \left(\frac{|k_\parallel| - k_{\parallel 0}}{\delta k_\parallel} \right)^2 \right\} \delta\left(\omega - \frac{3}{2}\Omega \right) \right. \qquad (58)$$

and chose $k_{\parallel 0}\rho = 0.5$, $k_{\perp 0}\rho = 2$, and $\delta k_\parallel \rho = 0.25$. He assumed that the wave energy density scales inversely as the magnetic field strength along a field line, so that E_0^2 is proportional to B_{eq}/B. Figure 38 shows Lyons' bounce-averaged $D_{\alpha\alpha}$ and D_{vv} as a function of equatorial pitch angle α_0 and energy K normalized to the thermal energy K_{th}. These diffusion coefficients scale

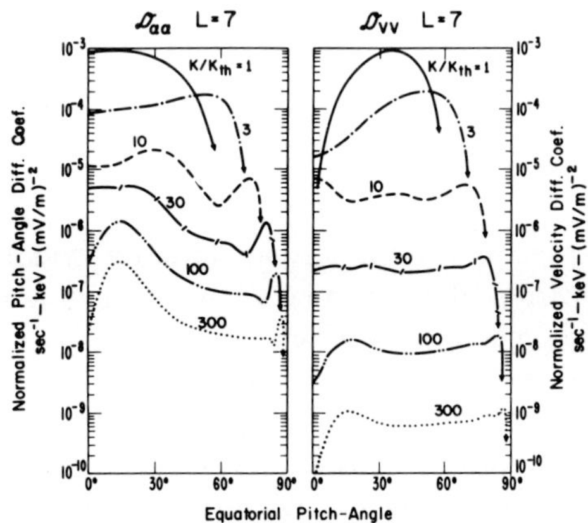

FIG. 38.   Pitch angle and speed diffusion coefficients for 3/2 waves. Normalized coefficients for diffusion in pitch angle ($D_{\alpha\alpha}$) and speed (D_{vv}) are plotted as a function of equatorial pitch angle at $L=7$ for various values of the electron energy K normalized to the thermal energy K_{th} (LYONS, 1974). Note that the two diffusion coefficients are comparable for $K/K_{th} < 10$. Above pitch angle diffusion dominates. For high energy electrons, the competition with whistler mode diffusion must be taken into account, since whistlers are often present simultaneously. The actual diffusion coefficients may be obtained by multiplying $D_{\alpha\alpha}$ and D_{vv} by E_0^2/K_{th}, with E_0 in mV/m and K_{th} in keV.

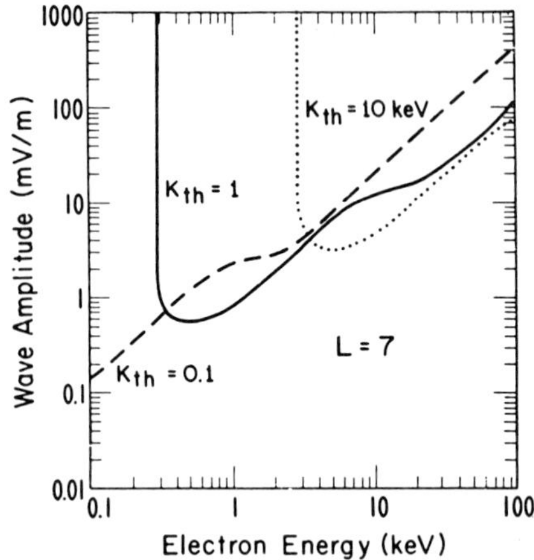

FIG. 39. Energy range for which electrons are put into strong diffusion by 3/2 waves. The minimum equatorial wave amplitude required for strong pitch angle diffusion at $L=7$ is shown as a function of electron energy for thermal energies $K_{th}=0.1$, 1, and 10 keV (LYONS, 1974). For amplitudes of 1–10 mV/m, electrons between 1 and 10 keV, the diffuse auroral energy range, are in strong diffusion for $1 < K_{th} < 10$ keV.

according to the equatorial wave energy density. $D_{\alpha\alpha}$ and D_{vv} are comparable for $K/K_{th} < 10$, whereas at large energies D_{vv} scales as $(K_{th}/K)^{5/2}$ and is smaller than $D_{\alpha\alpha}$, which scales as $(K_{th}/K)^{3/2}$. The diffusion rates maximize at $K \simeq K_{th}$.

LYONS (1974) estimated the strong diffusion critical amplitude for $K = K_{th} = 1$ keV electrons at $L=7$ to be a few tenths of a mV/m, consistent with Table 2. A few mV/m puts 10 keV electrons on strong diffusion (Fig. 39). By calculating Lyons' diffusion coefficients, SENTMAN *et al.* (1979) showed that the upper hybrid event observed by ISEE-1 on Nov. 5, 1977 was intense enough to smooth features extending over tens of degrees in pitch angle in a few minutes, possibly accounting for the featureless electron distribution that was observed, upon averaging for several minutes.

Magnetospheric electron cyclotron harmonic waves are often narrow band in frequency. This suggests, but does not prove, that perhaps only a few wave modes are excited at any given point. Should their coherence time be long, the random phase approximation underlying the quasilinear theory is invalid. Using techniques developed by AAMODT and BODNER (1969) and AAMODT (1970a, b), LYSAK *et al.* (1980) calculated the resonant ion heating rate when a few ion cyclotron waves dominate the spectrum. Since their arguments are kinematic, they apply equally well to electron cyclotron harmonic waves.

A single electron interacting with one constant amplitude Bernstein mode will oscillate back and forth in one of the potential wells created by the Bessel function structure of the basic mode. When several waves are present, the particle can increase its perpendicular energy $W_\perp$ by jumping from well to well. Transposing Lysak et al.'s Eq. (26) to electrons, we find

$$\frac{d}{dt}\left(\frac{4}{5}\left(\frac{W_t}{T_c}\right)^{5/4}\right) = \frac{e\phi_0}{T_c}(\pi^2 k_\perp^2 \rho_c^2)^{-1/4}\Omega \simeq \frac{e\phi_0\Omega}{T_c}(2m\pi^2)^{-1/4} \tag{59}$$

where ϕ_0 is the potential amplitude associated with the largest amplitude wave. According to Lysak et al., Eq. (59) is valid if $k_\parallel\rho_H < 1$, a condition satisfied for the most rapidly growing electron cyclotron harmonic waves. The heating rate is much faster than quasilinear, because it depends on ϕ_0 rather than ϕ_0^2.

When the wave electric field exceeds the critical value E_0,

$$E_0 \simeq \frac{1}{4}\left(\frac{\omega}{\Omega}\right)^{2/3}\frac{aB}{c}(k_\perp\rho_c)^{-1} \simeq \frac{1}{4(2m)^{1/2}}\left(\frac{\omega}{\Omega}\right)^{2/3}\frac{aB}{c} \tag{60}$$

electron heating becomes stochastic, because the particle interacts with many harmonics simultaneously (SMITH and KAUFMAN, 1975; KARNEY, 1978). The above E_0 is roughly the amplitude at which resonance broadening in a multimode wave spectrum destroys cyclotron harmonic structure (YOUNG, 1975). Wave growth would certainly saturate at this point. For $T_c = 100\,\text{eV}$, $m = 1$, $\omega/\Omega = 3/2$, $E_0 \simeq 200\,\text{mV/m}$ at $L = 6$.

In sum, electron cyclotron harmonic waves can heat resonant electrons at a rate comparable to which they diffuse them in pitch angle up to energies of at least ten times thermal. Little is known about the energy distribution resulting from such heating. Some general arguments for quasilinear heating have been presented by KENNEL (1969). When the wave amplitude is below strong diffusion, the electron energy distribution should arrive at a characteristic form, because the energization and loss times scale together with the electric field amplitude. Above the strong diffusion threshold, heating can be very rapid because the loss time is fixed. Here particles are heated to energies where they are no longer in strong diffusion. The heating rate is bounded by the rate of production of free energy for wave growth, whose ultimate source for diffuse auroral precipitation is convection.

9.2.3 Cold electron heating

Cold electrons are by definition nonresonant. When the wave

distribution changes with time, the quasilinear diffusion equation describes an adiabatic adjustment of the cold electrons to the changing wave amplitude. The effective nonresonant diffusion coefficient D_{NR} is of order γ/Ω times the resonant diffusion coefficient (ASHOUR-ABDALLA and KENNEL, 1978a). Nonresonant diffusion is completely reversible, and the rearrangement of the cold electron distribution ceases when the wave spectrum arrives at a steady state. However, high order nonlinear processes can couple the waves to the cold electrons. Suppose such processes induce an irreversible cold diffusion coefficient D_{c}. The time τ_{c} to reconstruct the cold electron distribution function would then be roughly $\tau_{\mathrm{c}} \simeq a^2/D_{\mathrm{c}}$, whereas the hot diffusion time τ_{H} is A^2/D_{H} so that if $D_{\mathrm{c}}/D_{\mathrm{H}}$ exceeds $T_{\mathrm{c}}/T_{\mathrm{H}}$, cold electrons diffuse more rapidly relative to their thermal width than hot electrons. This suggests that cold electrons could be heated by cyclotron harmonic waves. The nonlinear heating of nonresonant particles is a problem unusual to laboratory plasma physics, but pertinent to magnetospheric electrostatic waves, for an increase in $T_{\mathrm{c}}/T_{\mathrm{H}}$ can change an instability from nonconvective to convective, whereupon it can saturate by propagation. Of course, there must be sufficient free energy in the hot electrons to heat the cold electrons. If ΔE_{H} is the free energy per hot electron in the "initial" distribution, the final maximum cold electron temperature $T_{\mathrm{c}}^{\mathrm{max}}$ is bounded by $(n_{\mathrm{H}}/n_{\mathrm{c}})\Delta E_{\mathrm{H}}$. Thus, when $n_{\mathrm{c}}/n_{\mathrm{H}}$ is small, cold electrons could be strongly heated. This is precisely the parameter regime to which the question whether the instability is convective or nonconvective is most pertinent. Finally, even when cold electrons are not strongly heated, they can still be scattered out of the loss cone.

ΔE_{H} is maintained by the balance of the production of free energy by external agents such as convection and its destruction by velocity space diffusion. The cold electron temperature is regulated by the balance of nonlinear heating and the exchange of heat with the ionosphere. Because of the delicacy of these balances, measurements, rather than calculations, are the most direct way to determine $T_{\mathrm{c}}/T_{\mathrm{H}}$. The above theoretical arguments do suggest that the cold electron distribution function in space should not necessarily resemble that in the ionosphere.

9.3 Saturation

We have no general understanding of how electron cyclotron instabilities saturate in the magnetosphere. Perhaps this is because each of the at least four emission classes may saturate at different levels in different ways. But we are uncertain even about how to approach the question of saturation, because we do not know whether to treat the linear instability as convective or nonconvective. It is not simply our ignorance of T_c/T_H that creates uncertainty, for even if we know the waves are officially convective, their group velocities can still be small. Should we expect smooth quasisteady hiss bands characteristic of convective saturation over appreciable distances, or do most measurements average over the impulsive bursts often found when nonconvective instabilities grow to saturation?

Our present knowledge of linear theory permits a few general remarks about the convective-nonconvective question. Since the frequencies of magnetospheric cyclotron waves always scale with the local cyclotron frequency, the waves do not propagate large distances to the observing spacecraft. Because a wave cannot propagate through cyclotron harmonics without being heavily damped, it must come from that volume of space defined by where its frequency equals the next higher and next lower cyclotron harmonic frequency. However, all observed emissions but diffuse bands usually fill only about 10% of a harmonic frequency band. Are they therefore nonconvective? Not necessarily. For example, for the T_c/T_H for which odd half harmonic waves grow in a 10% frequency bandwidth, the instability is convective. Some electrostatic emissions are confined to the magnetic equator of earth, and nearly all at Jupiter. Are they made nonconvectively unstable by highly anisotropic electrons that are also equatorially localized? Again, not necessarily. According to BARBOSA and KURTH (1980), for the n_c/n_H large limit pertinent to Voyager observations, the odd half harmonic waves are convectively unstable, but the preferentially amplified waves bounce back and forth across the equator in a small range of latitudes. In fact, equatorial localization can be an argument that the waves propagate. How else, if not by propagation, can they be sensitive to the spatial symmetry of the equator?

In 9.2.4 we proposed that nonresonant heating of cold electrons could increase T_c/T_H and thereby change the instability from nonconvective to convective. We would then be much more likely to observe the instabilities in the steady state they approach after they turn convective. This is unlikely to happen for intense upper hybrid emissions, because they require n_c/n_H large, and it is energetically difficult to reconstruct the cold electron distribution. On the other hand, odd half harmonic waves can be unstable for n_c/n_H small, and their 10% observed bandwidths are consistent with the range of unstable frequencies for T_c/T_H at the nonconvective-convective transition.

Can cold electrons be nonresonantly heated? ASHOUR-ABDALLA *et al.* (1980a) attempted to answer this question by numerical simulation.

Location of Particles in
Perpendicular Velocity Space
Time = 0.0
50 ω_{pe}^{-1}
100 ω_{pe}^{-1}
4.0
0.0
-4.0
V_z
V_y
-4.0

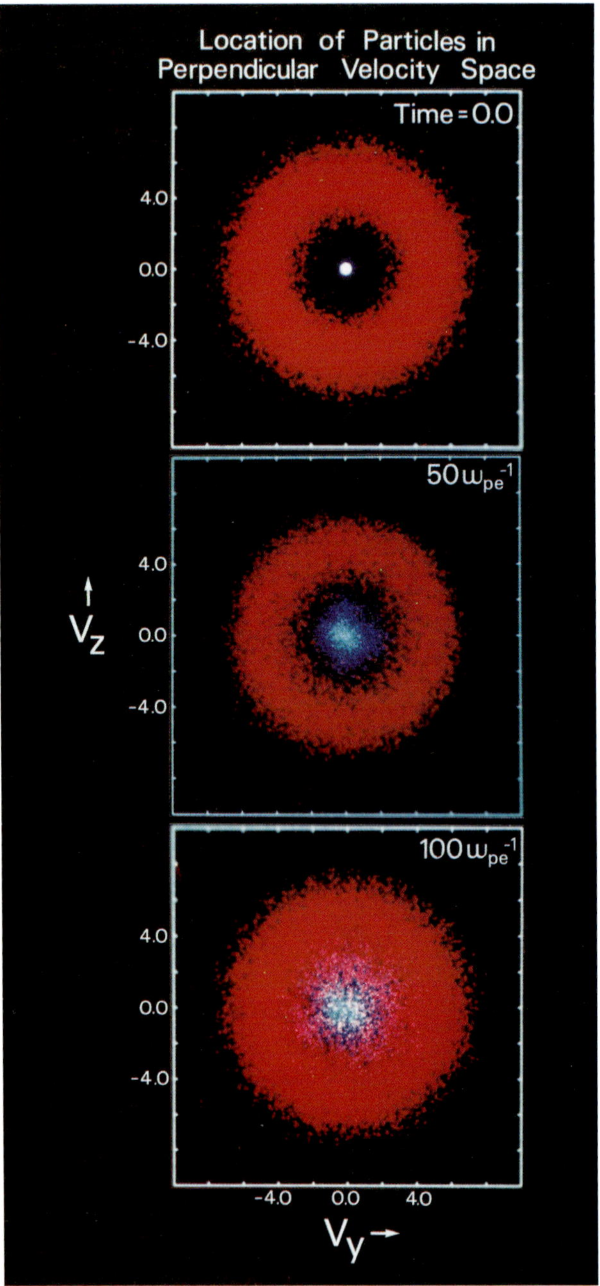

Fig. 40

Figure 40 displays the electron distribution in $v_\parallel (V_z)$ and $v_\perp (V_y)$ at three different times in a simulation with $\Omega = 0.3\ \omega_p$, $X_{\mathrm{UHC}} = 1.95$, $n_c/n_H = \frac{1}{4}$ and $T_c/T_H = 10^{-2}$ initially. For reasons of economy, the hot electron free energy source—a ring distribution—was chosen unrealistically strong, so that the waves would grow from a thermal fluctuation background in a reasonably short computer run. Two modes, with $\omega \sim 3/2\Omega$ grew until $t = 50\ \omega_p^{-1}$, saturated, and decayed back to thermal levels. At saturation, the hot electrons still had free energy, and a calculation using the smoothed hot electron distribution at saturation, and the initial cold electron distribution still showed linear instability. Saturation was therefore not due to quasi-linear scattering of hot electrons. It occurred because the cold electrons acquired a quiver velocity large enough to bring them into cyclotron resonance, at which point they began to heat rapidly, and the waves to saturate. The cold electron perpendicular temperature increased more than the parallel, so this kind of saturation could explain in principle the observations of highly anisotropic electrons associated with equatorial multiharmonic emissions by WRENN et al. (1979). Saturation occurred at an amplitude of several hundred mV/m, scaled to the magnetosphere.

The saturation mechanisms in the literature that are based upon the destruction of the original particle orbits that support the instability—nonlinear cold electron heating (ASHOUR-ABDALLA et al., 1980a), onset of stochastic heating (Eq. (60)), or turbulent resonance broadening (YOUNG, 1975)—have magnetospheric saturation amplitudes in the 100 mV/m range. They are related mechanisms calculated in different ways. Yet such large amplitudes have been reported for only one event at the onset of a large substorm (SCARF et al., 1973), and typical amplitudes are 1 mV/m or less. Therefore: either we have not yet found the right saturation mechanisms, or the measurements average over many impulsive events which grow to the 100 mV/m range.

Let us consider the above two options further. It is unlikely that saturation based upon orbit destruction will occur at an amplitude of 1% of

---

FIG. 40. Cold electron heating observed in a numerical simulation of 3/2 instabilities. The top inset shows the initial ring distribution of hot electrons, colored red, whose free energy caused a strong 3/2 instability. The cold electrons are blue. By $t = 50\ \omega_p^{-1}$, the instabilities saturated; at this point, the cold electrons were heating rapidly. As the wave amplitudes died out, cold electrons continued to heat and gradually lost their identity. The saturation amplitude was several hundred mV/m scaled to the magnetosphere (ASHOUR-ABDALLA et al., 1980).

that calculated using the present approaches. Thus, if the observed waves are truly steady, we must turn to other nonlinear mechanisms or to convective saturation. A completely different nonlinear mechanism is wave-wave coupling. Although they could redistribute wave energy within a harmonic band (OYA, 1972) or between harmonic bands, or couple electron and ion modes together (RONNMARK, 1975), there is no persuasive evidence that wave-wave couplings saturate the observed emissions. This leaves linear convective saturation. Let us now suppose the emissions are unsteady. Bursts would grow on a time scale of $10^{-2} f_c$ (a few milliseconds for the terrestrial magnetosphere and less for Jupiter's). Since orbit destruction destroys the wave's coherence, they would cease growing on the wave coherence time, which scales inversely as the frequency, after threshold is attained. The average amplitude would then depend upon the burst repetition rate, which in turn depends upon the rate of resupply of free energy to the small region of space in which the burst grew. We would therefore expect the observed range of average saturation amplitudes to be broader than the range of peak amplitudes. With convective saturation, the average and peak amplitudes should be more nearly equal; the range of amplitudes observed could still be broad, because it depends upon the linear growth rate. Finally, we note that the high time resolution Voyager 2 measurements in Fig. 33 revealed a highly impulsive class Ia emission in Jupiter's magnetosphere.

Let us now propose a research strategy that may help resolve our present uncertainties. On the experimental side, high time resolution wave measurements and better measurements of cold electrons are needed. On the theoretical side, further studies of nonlinear saturation and propagation that explore the parameters encountered in the earth's and Jupiter's magnetospheres are called for. While numerical simulations have proven to provide much insight into nonlinear plasma physics, they must be carefully tailored to suit magnetospheric problems. They should model not only the temporal growth to saturation of a strong instability, but also attempt to simulate the magnetosphere's delicate web of quasi-steady processes—the balance of particle injection and loss that leads to smooth distributions and relatively low average amplitudes. Because of the magnetosphere's great size, propagation studies have proven invaluable in interpreting space measurements. Moreover, theoreticians need to know whether a few or many modes are excited. Because wavelength measurements are difficult, and because linear theory tells us that many modes with different wavelengths are crowded into the same frequency range, this question may not be answerable by observation. Propagation studies can tell us how rich the wave-number spectrum of amplified waves can be.

10. Concluding Remarks

We began Section 2 by asserting that "Our understanding of radial diffusion and weak pitch angle scattering . . . of energetic electrons within the plasmasphere sets a standard for our future comprehension of the strong pitch angle diffusion of . . . diffuse auroral electrons. . . ." Much of this article has been a comment on why our understanding of strong diffusion is not as yet quantitative. In virtually every important plasma physical aspect, the diffuse aurora and the radiation belts are at opposite extremes. ELF whistlers propagate throughout the plasmasphere, whereas electrostatic electron waves propagate short distances at best. The smooth ELF spectrum satisfies the constraints of quasilinear theory, and it is easy to construct bounce averaged diffusion coefficients. Diffusion is largely in pitch angle and it is weak. Cold electrons play a passive role. None of these simplifications pertain to magnetospheric cyclotron waves. Cold electrons are crucial; we do not know whether the wave amplitudes are smooth or bursty and whether quasi-linear diffusion strictly applies. If it does, the pitch angle diffusion is strong, and there is significant electron heating. We do not know the spatial distribution of the waves very well, and so there is no simple way to construct bounce-averaged diffusion coefficients. Yet these complications make strong diffusion much more interesting. We will have to plumb the depths of modern plasma turbulence theory to understand it. In the meantime, we can be satisfied that our experimental and theoretical work on magnetospheric stong diffusion is comparable in scope and depth to that on laboratory mirror trapping, and that both are paradigm problems that lead the development of contemporary plasma physics.

We express our intellectual debt to F. V. Coroniti, with whom we have discussed all these problems over the years. We are also deeply indebted to D. A. Gurnett, W. S. Kurth, and F. L. Scarf, with whom we have collaborated, and who provided many of the figures in this paper. We regret that we cannot acknowledge all of our debts. This work was supported by NASA NGL-05-007-190 and NSF-19558.

REFERENCES

AAMODT, R. E., Non-linear evolution of multiple finite-amplitude resonant loss-cone modes, *Phys. Fluids*, **13**, 2147, 1970a.

AAMODT, R. E., Particle motion in the presence of three-dimensional finite amplitude cyclotron harmonic waves, *Phys. Fluids*, **13**, 2341, 1970b.

AAMODT, R. E. and S. E. BODNER, Non-linear dynamics of a single harmonic loss-cone flute mode, *Phys. Fluids*, **12**, 1471, 1969.

ANDERSON, R. R. and K. MAEDA, VLF emissions associated with enhanced magnetospheric electrons, *J. Geophys. Res.*, **82**, 135, 1977.

ASHOUR-ABDALLA, M. and S. W. H. COWLEY, Wave-particle interactions near the geostationary orbit, in *Magnetospheric Physics*, edited by B. M. McCormac, p. 241, D. Reidel, Dordrecht, Holland, 1974.

ASHOUR-ABDALLA, M. and C. F. KENNEL, Convective cold upper hybrid instabilities, in *Magnetospheric Particles and Fields*, edited by B. M. McCormac, p. 181, D. Reidel, Dordrecht, Holland, 1976.

ASHOUR-ABDALLA, M. and C. F. KENNEL, Nonconvective and convective electron cyclotron harmonic instabilities, *J. Geophys. Res.*, **83**, 1531, 1978a.

ASHOUR-ABDALLA, M. and C. F. KENNEL, Multiharmonic electron cyclotron instabilities, *Geophys. Res. Lett.*, **5**, 711, 1978b.

ASHOUR-ABDALLA, M. and C. F. KENNEL, Diffuse auroral precipitation, *J. Geomag. Geoelectr.*, **30**, 239, 1978c.

ASHOUR-ABDALLA, M. and R. M. THORNE, The importance of electrostatic ioncyclotron instability for quiet-time proton auroral precipitation, *Geophys. Res. Lett.*, **4**, 45, 1977.

ASHOUR-ABDALLA, M. and R. M. THORNE, Towards a unified view of diffuse auroral precipitation, *J. Geophys. Res.*, **83**, 4775, 1978.

ASHOUR-ABDALLA, M., C. F. KENNEL, and W. LIVESEY, A parametric study of electron multiharmonic instabilities in the magnetosphere, *J. Geophys. Res.*, **84**, 6540, 1979a.

ASHOUR-ABDALLA, M., C. F. KENNEL, and D. D. SENTMAN, Magnetospheric multiharmonic instabilities, in *Astrophysics and Space Sciences Book Series*, edited by P. J. Palmadesso, D. Reidel, Dordrecht, Holland, 1979b.

ASHOUR-ABDALLA, M., J. N. LEBOEUF, J. M. DAWSON, C. F. KENNEL, and R. W. HUFF, Simulation of electrostatic cyclotron harmonic instabilities, in *Proceedings of International Conference on Plasma Physics*, Vol. 1, p. 133, 1980a.

ASHOUR-ABDALLA, M., J. N. LEBOEUF, J. M. DAWSON, and C. F. KENNEL, A simulation study of cold electron heating by loss-cone instabilities, *Geophys. Res. Lett.*, **7**, 889, 1980b.

BARBOSA, D. D.and F. V. CORONITI, Relativistic electrons and whistlers in Jupiter's magnetosphere, *J. Geophys. Res.*, **81**, 4531, 1976.

BARBOSA, D. D. and W. S. KURTH, Superthermal electrons and Bernstein waves in Jupiter's inner magnetosphere, *J. Geophys. Res.*, **85**, 1980.

BERK, H. L. and A. A. GALEEV, Velocity space instabilities in a Toroidal Geometry, *Phys. Fluids*, **10**, 441, 1967.

BERNSTEIN, I. B., Waves in a plasma in a magnetic field, *Phys. Rev.*, **109**, 10, 1958.

BERNSTEIN, I. B. and F. ENGELMANN, Quasi-linear theory of plasma waves, *Phys. Fluids*, **9**, 937, 1966.

BERNSTEIN, W., H. LEINBACH, H. COHEN, D. S. WILSON, T. N. DAVIS, T. HALLINAN, B. BOKER, J. MARTZ, R. ZEIMKE, and W. HUBER, Laboratory observations of RF emissions at ω_{pe} and $(n+\frac{1}{2})\omega_{ce}$ in electron beam-plasma and beam-beam interactions, *J. Geophys. Res.*, **80**, 4375, 1975.

BIRMINGHAM, T. J., J. K. ALEXANDER, M. D. DESCH, R. F. HUBBARD, and B. M. PEDERSEN, Observations of electron gyroharmonic waves and the structure of the Io torus, NASA Goddard SFC Tech. Memo. 80696, to be published, *J. Geophys. Res.*, 1980.

BROADFOOT, A. L., M. J. S. BELTON, P. Z. TAKACS, B. R. SANDEL, D. E. SHEMANSKY, J. B. HOLBERG, J. M. AJELLO, S. K. ATREYA, T. M. DONAHUE, H. W. MOOS, J. L. BERAUX, J. E. BLAMONT, D. F. STROBEL, J. C. McCONNEL, A. DALGARNO, R. GOODY, and M. B. McELROY, Extreme ultraviolet observations from Voyager 1 encounter with Jupiter, *Science*, **204**, 929, 1979.

CARPENTER, D. L., K. STONE, J. C. SIREN, and T. CRYSTAL, Magnetospheric electric fields deduced from differing whistler paths, *J. Geophys. Res.*, **77**, 2819, 1972.

CHANCE, M. S., F. V. CORONITI, and C. F. KENNEL, Finite β drift Alfvén instability, *J. Geophys. Res.*, **78**, 7521, 1973.

CHRISTIANSEN, P., M. P. GOUGH, G. MARTELLI, J.-J. BLOCK, N. CORNILLEAU, J. ETCHETO, R. GENDRIN, D. JONES, C. BÉGHIN, and P. DECREAU, GEOS 1, identification of natural magnetospheric emissions, *Nature*, **272**, 682, 1978a.

CHRISTIANSEN, P. J., M. P. GOUGH, G. MARTELLI, J. J. BLOCH, N. CORNILLEAU, J. ETCHETO, R. GENDRIN, C. BÉGHIN, P. DECREAU, and D. JONES, GEOS-1 observations of electrostatic waves and their relationship with plasma parameters, *Space Sci. Rev.*, **22**, 838, 1978b.

CORNWALL, J. M., Diffusion processes influenced by conjugate-point wave phenomena, *Radio Sci.*, **3**, 740, 1968.

CORNWALL, J. M., Radial diffusion of ionized helium and protons: a probe for magnetospheric dynamics, *J. Geophys. Res.*, **77**, 1756, 1972.

CORONITI, F. V. and C. F. KENNEL, Auroral micropulsation instability, *J. Geophys. Res.*, **75**, 1863, 1970.

CORONITI, F. V. and C. F. KENNEL, Can the ionosphere regulate magnetospheric convection?, *J. Geophys. Res.*, **78**, 2837, 1973.

CORONITI, F. V. and R. M. THORNE, Magnetospheric electrons, *Ann. Rev. Earth Planet. Sci.*, **1**, 107, 1973.

CORONITI, F. V., R. W. FREDRICKS, and R. W. WHITE, Instability of ring current protons beyond the plasmapause during injection events, *J. Geophys. Res.*, **77**, 6243, 1972.

CORONITI, F. V., F. L. SCARF, C. F. KENNEL, W. S. KURTH, and D. A. GURNETT, Detection of Jovian whistler mode chorus: implications for the Io torus aurora, *Geophys.Res. Lett.*, **7**, 45, 1980.

CURTIS, S. A. and C. S. WU, Electrostatic and electromagnetic gyroharmonic emissions due to energetic electrons in magnetospheric plasmas, *J. Geophys. Res.*, **84**, 2057, 1979.

DeForest, S. E. and C. E. McIlwain, Plasma clouds in the magnetosphere, *J. Geophys. Res.*, **76**, 3587, 1971.

Dory, R. A., G. E. Guest, and E. G. Harris, Unstable electrostatic plasma waves propagating perpendicular to a magnetic field, *Phys. Rev. Lett.*, **14**, 131, 1965.

Eather, R. H. and S. B. Mende, Systematic auroral energy spectra, *J. Geophys. Res.*, **77**, 660, 1972a.

Eather, R. H. and S. B. Mende, High latitude particle precipitation and source regions in the magnetosphere, in *Magnetosphere-Ionosphere Interactions*, edited by K. V. Foldestad, p. 139, Universities Forlaget, Oslo, 1972b.

Eather, R. H., S. B. Mende, and R. J. R. Judge, Plasma injection at synchronous orbit and spatial and temporal auroral morphology, *J. Geophys. Res.*, **81**, 2805, 1976.

Evans, D. S., Precipitating electron fluxes formed by a magnetic field-aligned potential difference, *J. Geophys. Res.*, **79**, 2853, 1974.

Evans, D. S. and T. E. Moore, Precipitating electrons associated with the diffuse aurora: evidence for electrons of atmospheric origin in the plasma sheet, *J. Geophys. Res.*, **84**, 6451, 1979.

Falthammar, C.-G., Effects of time-dependent electric fields on geomagnetically trapped radiation, *J. Geophys. Res.*, **70**, 2503, 1963.

Fredricks, R. W., Plasma instability at $(n+\frac{1}{2})f_c$ and its relationship to some satellite observations, *J. Geophys.*, **76**, 5344, 1971.

Fredricks, R. W. and F. L. Scarf, Recent studies of magnetospheric electric field emissions above the electron gyrofrequency, *J. Geophys. Res.*, **78**, 310, 1973.

Fried, B. D. and S. Conte, *The Plasma Dispersion Function*, Academic Press, New York, 1961.

Fukunishi, H., Dynamic relation between proton and electron auroral substorms, *J. Geophys. Res.*, **80**, 553, 1975.

Gaffey, J. D. and R. LaQuey, Upper hybrid resonance in the magnetosphere, *J. Geophys. Res.*, **81**, 595, 1976.

Gough, M. P., P. J. Christiansen, G. Martelli, and E. J. Gershuny, Interaction of electrostatic waves with warm electrons at the geomagnetic equator, *Nature*, **279**, 515, 1979.

Guest, G. E. and R. A. Dory, Microinstability of a mirror confined plasma, *Phys. Fluids*, **11**, 1775, 1963.

Gurnett, D. A., The earth as a radio source: the nonthermal continuum, *J. Geophys. Res.*, **80**, 2751, 1975.

Gurnett, D. A. and L. A. Frank, A region of intense plasma wave turbulence on auroral field lines, *J. Geophys. Res.*, **82**, 1031, 1977.

Gurnett, D. A. and R. R. Shaw, Electromagnetic radiation trapped in the magnetosphere above the plasma frequency, *J. Geophys. Res.*, **78**, 8136, 1973.

Gurnett, D. A., F. L. Scarf, R. W. Fredricks, and E. J. Smith, The ISEE-1 and ISEE-2 plasma wave investigation, *IEEE Trans. Geosci. Electr.*, **GE-16**, 225, 1978.

GURNETT, D. A., W. S. KURTH, and F. L. SCARF, Plasma wave observations near Jupiter: initial results from Voyager 2, *Science*, **206**, 987, 1979a.

GURNETT, D. A., R. R. ANDERSON, F. L. SCARF, R. W. FREDRICKS, and E. J. SMITH, Initial results from the ISEE-1 and 2 plasma wave investigation, *Space Sci. Rev.*, **23**, 103, 1979b.

HALL, L. S. and W. HECKROTTE, Stability of longitudinal oscillations in a uniform magnetized plasma with anisotropic velocity distribution, *Phys. Rev.*, **134**, A1474, 1964.

HARRIS, E. G., Unstable plasma oscillations in a magnetic field, *Phys. Rev. Lett.*, **2**, 34, 1959.

HUBBARD, R. F. and T. J. BIRMINGHAM, Electrostatic emissions between electron gyroharmonics in the outer magnetosphere, *J. Geophys. Res.*, **83**, 4837, 1978.

HUBBARD, R. F., T. J. BIRMINGHAM, and E. W. HONES, Jr., Magnetospheric electrostatic emissions and cold plasma densities, *J. Geophys. Res.*, **84**, 5828, 1979.

HULTQUIST, B., Some experimentally determined characteristics of the turbulence in the magnetosphere, in *Physics of the Hot Plasma in the Magnetosphere*, edited by B. Hultquist and L. Stenfio, Plenum Press, New York, 1975a.

HULTQUIST, B., The aurora, in *The Magnetospheres of Earth and Jupiter*, edited by V. Formisano, p. 77, D. Reidel, Dordrecht, Holland, 1975b.

JAGGI, R. K. and R. A. WOLF, Self-consistent calculation of the motion of a sheet of ions in the magnetosphere, *J. Geophys. Res.*, **78**, 2852, 1973.

KAMIMURA, T., T. WAGNER, and J. M. DAWSON, Simulation study of Bernstein modes, *Phys. Fluids*, **21**, 1978.

KARNEY, C. F., Stochastic heating by a lower hybrid wave, *Phys. Fluids*, **21**, 1534, 1978.

KARPMAN, V. I., JU, K. ALEKHIN, N. B. BARISOV, and N. A. RJABOVA, Electrostatic waves with frequencies above the gyrofrequency in a plasma with a loss cone, *Phys. Lett. Sect. A*, **205**, 1973.

KARPMAN, V. I., JU. K. ALEKHIN, N. D. BARISOV, and N. A. RJABOVA, Electrostatic electron-cyclotron waves in a plasma with a loss-cone distribution, *Plasma Phys.*, **17**, 361, 1975.

KENNEL, C. F., Consequences of a magnetospheric plasma, *Rev. Geophys.*, **7**, 379, 1969.

KENNEL, C. F. and F. ENGELMANN, Velocity space diffusion from plasma turbulence in a magnetic field, *Phys. Fluids*, **9**, 2377, 1966.

KELLEY, M. C. and P. M. KINTNER, Evidence for a two dimensional inertial turbulence in a cosmic-scale low-β plasma, *Astrophys. J.*, **220**, 339, 1978.

KENNEL, C. F. and H. E. PETSCHEK, Limit on stably trapped particle fluxes, *J. Geophys. Res.*, **71**, 1, 1966.

KENNEL, C. F., R. W. FREDRICKS, and F. L. SCARF, High frequency electrostatic waves in the magnetosphere, in *Particles and Fields in the Magnetosphere*, edited by R. M. McCormac, p. 257, D. Reidel, Dordrecht, Holland, 1970a.

KENNEL, C. F., F. L. SCARF, R. W. FREDRICKS, J. H. McGEEHEE, and F. V. CORONITI,

VLF electric field observations in the magnetosphere, *J. Geophys. Res.*, **75**, 6136, 1970b.

KINDEL, J. M. and C. F. KENNEL, Topside current instabilities, *J. Geophys. Res.*, **76**, 3055, 1971.

KURTH, W. S., Ph. D. Thesis, University of Iowa, 1979.

KURTH, W. S., J. D. CRAVEN, L. A. FRANK, and D. A. GURNETT, Intense electrostatic waves near the upper hybrid resonance frequency, *J. Geophys Res.*, **84**, 4145, 1979a.

KURTH, W. S., M. ASHOUR-ABDALLA, L. A. FRANK, C. F. KENNEL, D. A. GURNETT, D. D. SENTMAN, and B. G. BUREK, A comparison of intense electrostatic waves near f_{UHR} with linear instability theory, *Geophys. Res. Lett.*, **6**, 487, 1979b.

KURTH, W. S., L. A. FRANK, M. ASHOUR-ABDALLA, D. A. GURNETT, and B. G. BUREK, Observations of a free energy source for intense electrostatic waves, *Geophys. Res. Lett.*, **7**, 293, 1980.

KURTH, W. S., D. D. BARBOSA, D. A. GURNETT, and F. L. SCARF, Electrostatic waves in the Jovian Magnetosphere, *Geophys. Res. Lett.*, **7**, 57, 1980.

LANDAU, L. D., *J. Phys. USSR*, **10**, 25, 1946.

LUI, A. T. Y. and C. D. ANGER, A uniform belt of diffuse auroral emission seen by the ISIS-2 scanning photometer, *Planet. Space Sci.*, **21**, 799, 1973.

LUI, A. T. Y., P. PERREAULT, S. I. AKASOFU, and C. D. ANGER, The diffuse aurora, *Planet. Space Sci.*, **21**, 857, 1973.

LYONS, L. R., Electron diffusion driven by magnetospheric electrostatic waves, *J. Geophys. Res.*, **79**, 575, 1974.

LYONS, L. R. and R. M. THORNE, Equilibrium structure of radiation belt electrons, *J. Geophys. Res.*, **78**, 2142, 1973.

LYONS, L. R., R. M. THORNE, and C. F. KENNEL, Electron pitch angle diffusion driven by oblique whistler mode turbulence, *J. Plasma Phys.*, **6**, 589, 1971.

LYONS, L. R., R. M. THORNE, and C. F. KENNEL, Pitch angle diffusion of radiation belt electrons within the plasmasphere, *J. Geophys. Res.*, **77**, 3455, 1972.

LYSAK, R. L., M. K. HUDSON, and M. TEMERIN, Ion heating by strong electrostatic ion cyclotron turbulence, *J. Geophys. Res.*, **85**, 678, 1980.

MCILWAIN, C. E., Auroral electron beams near the magnetic equator, in *Physics of the Hot Plasma in the Magnetosphere*, edited by B. Hultquist and L. Stenflo, Plenum, New York, 1975.

MENDE, S. B. and R. H. EATHER, Monochromatic all-sky observations and auroral precipitation patterns, *J. Geophys. Res.*, **81**, 3776, 1976.

MENG, C. I., B. MAUK, and C. E. MCILWAIN, Electron precipitation of evening diffuse aurora and its conjugate electron fluxes near the magnetospheric equator, *J. Geophys. Res.*, **84**, 2545, 1979.

MOSIER, S. R., M. L. KAISER, and L. W. BROWN, Observation of noise bands associated with the upper hybrid resonance by the Imp 6 radio astronomy experiment, *J. Geophys. Res.*, **78**, 1673, 1973.

MOZER, F. S., Power spectra of the magnetospheric electric field, *J. Geophys. Res.*, **76**, 3651, 1971.

NAMBU, M., Negative pressure effect on low frequency waves, *Phys. Lett.*, **39A**, 347, 1972.

NAMBU, M., Nonlinear theory of plasma instability of $(n + \frac{1}{2})\Omega_e$, *J. Geophys. Res.*, **78**, 764, 1973.

NAMBU, M., Electrostatic electron cyclotron waves with an anti-loss cone distribution, *Geophys. Res. Lett.*, **2**, 76, 1975.

NAMBU, M. and T. WATANABE, Electrostatic electron-cyclotron waves with an anti-loss-cone electron distribution function, *Geophys. Res. Lett.*, **2**, 176, 1975.

OYA, H., Turbulence of electrostatic electron cyclotron harmonic waves observed by OGO-5, *J. Geophys. Res.*, **77**, 3483, 1972.

PETSCHEK, H. E. and C. F. KENNEL, Tail flow, auroral precipitation and ring currents (abstract), *Trans. AGU*, **47**, 137, 1966.

ROEDERER, J., in *Earth's Particles and Fields*, edited by B. M. McCormac, p. 193, Reinhold Publishing Corpl, New York, 1968.

RONNMARK, K., Nonlinear generation of ion loss cone waves by electrostatic turbulence in the magnetosphere, *Planet. Space Sci.*, **25**, 149, 1977.

RONNMARK, K., H. BORG, P. J. CHRISTIANSEN, M. P. GOUGH, and D. JONES, Banded electron cyclotron harmonic instability—a first comparison of theory and experiment, *Space Sci. Rev.*, **22**, 401, 1978.

ROSENBLUTH, M. N. and R. F. POST, High-frequency electrostatic plasma instability inherent to "loss-cone" particle distribution, *Phys. Fluids*, **8**, 547, 1965.

SATO, T. and T. E. HOLZER, Quiet auroral arcs and electrodynamic coupling between the ionosphere and magnetosphere, *J. Geophys. Res.*, **78**, 7314, 1973.

SCARF, F. L., R. W. FREDRICKS, C. F. KENNEL, and F. V. CORONITI, Satellite studies of magnetospheric substorms on August 15, 1968, *J. Geophys. Res.*, **78**, 3119, 1973.

SCARF, F. L., D. A. GURNETT, and W. S. KURTH, Jupiter plasma wave observations: an initial Voyager 1 overview, *Science*, **204**, 991, 1979.

SENTMAN, D. D., L. A. FRANK, C. F. KENNEL, D. A. GURNETT, and W. S. KURTH, Electron distribution functions associated with electrostatic emissions in the dayside magnetosphere, *Geophys. Res. Lett.*, **6**, 781, 1979.

SHAW, R. R. and D. A. GURNETT, Electrostatic noise bands associated with the electron gyrofrequency and plasma frequency in the outer magnetosphere, *J. Geophys. Res.*, **80**, 4529, 1975.

SHIMA, Y. and L. S. HALL, Electrostatic instabilities in a plasma with anisotorpic velocity distribution, *Phys. Rev.*, **139A**, 1115, 1965.

SMITH, G. R. and A. N. KAUFMAN, Stochastic acceleration by a single wave in a magnetic field, *Phys. Rev. Lett.*, **34**, 1613, 1975.

TATARONIS, J. A. and F. W. CRAWFORD, Cyclotron harmonic wave propagation and instabilities, *J. Plasma Phys.*, **4**, 231, 1970.

THORNE, R. M. and C. F. KENNEL, Quasi-trapped VLF propagation in the outer magnetosphere, *J. Geophys. Res.*, **72**, 357, 1967.

TOMASSIAN, A. D., T. A. FARLEY, and A. L. VAMPOLA, Inner zone energetic electron repopulation by radial diffusion, *J. Geophys. Res.*, **77**, 3441, 1972.

VASYLIUNAS, V. M., A survey of low energy electrons in the evening sector of the magnetosphere with OGO-1 and OGO-3, *J. Geophys. Res.*, **73**, 2839, 1968.

VASYLIUNAS, V. M., Independent work described in C. F. Kennel, Consequences of a magnetospheric plasma, *Rev. Geophys.*, **7**, 379, 1969.

WARWICK, J. W., J. B. PEARCE, A. C. RIDDLE, J. K. ALEXANDER, M. D. DESCH, M. L. KAISER, J. F. THIEMAN, T. D. CARR, S. GULKIS, A. BOISCHOT, C. C. HARVEY, and B. M. PEDERSON, Voyager 1 planetary radio observations near Jupiter, *Science*, **204**, 995, 1979.

WEST, H. I., Jr., R. M. BUCK, and J. R. WALTON, Electron pitch angle distributions throughout the magnetosphere as observed on OGO-5, *J. Geophys. Res.*, **78** 1064, 1972.

WRENN, G. L., J. F. E. JOHNSON, and J. J. SOJKA, Stable "pancake" distributions of low energy electrons in the plasma trough, *Nature*, **279**, 512, 1979.

YOUNG, T. S. T., Destabilization and wave-induced evolution of the magnetospheric plasma clouds, *J. Geophys. Res.*, **80**, 3995, 1975.

YOUNG, T. S. T., J. D. CALLEN, and J. E. MCCUNE, High frequency electrostatic waves in the magnetosphere, *J. Geophys. Res.*, **78**, 1082, 1973.

YOUNG, T. S. T., Electrostatic waves at half electron gyrofrequency, *J. Geophys. Res.*, **79**, 1985, 1974.

SUBJECT INDEX